Water System Science and Policy Interfacing

Water System Science and Policy Interfacing

Edited by

Philippe Quevauviller
European Commission, Brussels, Belgium
and Vrije Universiteit Brussels, Belgium

RSC Publishing

ISBN: 978-1-84755-861-9

A catalogue record for this book is available from the British Library

Published by The Royal Society of Chemistry,
Thomas Graham House, Science Park, Milton Road,
Cambridge CB4 0WF, UK

Registered Charity Number 207890

For further information see our website at www.rsc.org

Foreword

I am particularly honoured to preface this book because current issues in the management of natural resources, water and aquatic environments raise questions that cannot be solved without innovative and effective partnerships between scientists, policymakers and people in the field. These questions are increasingly complex and their urgency means that we cannot wait to have all the desirable scientific information before taking action. They call for new types of relations between scientists and policymakers.

Today's challenges require the continuous transfer of knowledge as it becomes available and, in return, stakeholder participation is essential in finding answers for up-coming challenges. Managers and policymakers need tools and scientific advice for which the necessary knowledge is complex and not always fully or adequately available. At the same time, it is important to mobilize the scientific community on research challenges often located at the interface between several disciplines or requiring integrated, multi-disciplinary analysis.

To achieve a number of goals at local, national and European levels, for example defining and achieving good ecological status, adapting to climate change, halting the loss of biodiversity, ensuring sustainable and equitable management of water resources, providing good governance frameworks for water and sanitation services, *etc.*, we must be able to understand processes, use diagnostic tools and scenarios to predict and anticipate the future, measure discrepancies between current status and objectives, and finally prioritize and set guidelines for corrective action and measures.

We must learn how to make this interface between science and policy operational and effective so that better understanding can lead to better action.

Water System Science and Policy Interfacing, published by the Royal Society of Chemistry, offers a unique opportunity to take stock of the progress made in recent years to promote more effective use of science to assist policymaking and implementation of water-related regulations. The feedback collected from many initiatives at European, national and regional levels, as discussed in this

Water System Science and Policy Interfacing
Edited by Philippe Quevauviller

Published by the Royal Society of Chemistry, www.rsc.org

book, demonstrates that current practices are not sufficient and that more efforts are needed. The authors were able to identify key factors that have the potential to improve the effectiveness of the interface between research and public water policies, such as identifying and formulating research questions and associated research and development programmes between scientists and the end-users, transparency, access to information and science advice, the role of new players, so-called 'translators', who can translate scientific findings into usable information for decision-makers, communication and educational needs.

The recommendations provided in this book will be of the utmost value for our own strategy to strengthen collaboration between research, business, managers and consumers in the field of water and aquatic environments. These findings are of particular importance for our national strategy in France and at the European level in the light of an initiative launched with the agencies of other Member States and with the European Commission to organise consultations between the main European structures interfacing between research, policymakers and major stakeholders in the water field. Building on synergies spanning Europe, the goal is to make the best use of research results to reach the ambitious objectives of European water policy, itself tailored to the economic, societal and environmental issues facing us.

I warmly thank all the authors of this outstanding volume on the interface between science and policy in the water field. I would also like to extend my thanks to those who took the initiative in bringing together these contributions, notably Philippe Quevauviller, editor of the book, whose unfailing dynamism is a source of progress for the interaction between science and public policy in the environmental field. This synthesis work is the result of many exchanges among the scientific community, policymakers and stakeholders over the last five years, which were strongly facilitated by series of EU-funded research projects that are described in this book and reflected by the authorship.

Whether we are scientists, policymakers, local managers, elected officials, business or association leaders, it is now up to us to build fully operational interfaces to meet the future challenges of the field of water and aquatic environments.

Patrick Lavarde
Director General of the French
National Agency for Water
and Aquatic Environments

Preface

This book would probably never have been produced, at least in its present form, if I had remained solely in the scientific sphere. I had grown up as a researcher in chemical oceanography, then evolving as a scientific officer at the European Commission for many years in the sector of environmental analytical chemistry. A need to change and to 'see something else' made me move to the policy world in 2002 where I have learnt to face different realities, *i.e.* what is behind the design, development, negotiation and implementation of an EU policy.

With a strongly anchored scientific background, my very first days as a policy-maker gave me the impression of landing on another planet with different rules and codes. On my former scientific planet, discussions were often about finding the best methods to measure, evaluate or understand environmental pathways or pollution impacts. We were striving for the most accurate ways to better understand and protect our environment and our communication route was exclusively through the international scientific literature. On the policy planet, the development of legislation often relies on 'best possible political compromises' where scientific progress is only one of the many components of policy design, and later implementation. In this respect, while it is generally well accepted that legislative developments have to fully take into account the existing scientific foundation, I soon realised that the way scientific information is made accessible to policy-makers, and the way policy-makers perceive scientific inputs, turn this 'logical link' into a very complex machinery and sometimes mutual incomprehension.

Discussions with scientists, policy-makers, stakeholders and colleagues of the European Commission confirmed my feeling that science and policy are often living in two separate worlds due to a lack of a clear transfer mechanism. This debate is not new and we can trace discussions back to before the 1990s. I have been, however, in the position of a 'science user' in a very practical case – namely the development of the Groundwater Directive 2006/118/EC – in which

Water System Science and Policy Interfacing
Edited by Philippe Quevauviller

Published by the Royal Society of Chemistry, www.rsc.org

the question of transfer of scientific outputs to various steps of the policy process (design, negotiation, implementation) was very critical. In a way, I started to face the 'two planets' problem when I mixed scientists, stakeholders and policy-makers, asking them naive questions such as 'What do we know, and what scientific information may I use to justify policy orientations?'. This primarily resulted in difficult debates, and it took several meetings and a lot of diplomacy to establish and develop a constructive dialogue.

These interactions gave rise to the consciousness that the two worlds have difficulties in communicating directly and that they need 'intermediaries' to cohabit and evolve together. From 2003 onward, the need to develop an operational interface among the scientific and policy communities has grown up, as reflected by joint works.[i] This latter evolved in practical experiences mixing science and policy in support of legislative developments, namely the EU Groundwater Directive adopted in 2006, which are illustrated by a separate volume published by RSC Publishing.[ii]

In 2008, after several years of discussion, a 'Science–Policy Interfacing' group was proposed in the water sector in the framework of the Common Implementation Strategy of the Water Framework Directive. This was linked to various research and supporting initiatives establishing links between different communities at EU level. This development was initiated while I was still living on my policy planet at DG Environment. Later on, I took an opportunity to move back to my previous (research) planet, where I am now looking at the issue from another angle.[iii]

This book project arose from these many exchanges and projects. It was supported by the positive welcome of a groundwater science–policy book, which encouraged me to embark on this new editorial venture. Behind this project, and behind interfacing in a general sense, there are open-minded people who have agreed to discuss shortcomings and reflect on possible ways to progress together. This book reflects this 'human side' of the issue. I would like to briefly explain why I thought of the authors who I invited to contribute to this volume:

- André Van der Beken has been very active in water engineering-related training, and we have often discussed science–policy issues in the framework of teaching duties (in particular the IUPWARE programme, which is an International Master programme jointly coordinated by the Katholieke Universiteit Leuven and the Vrije Universiteit Brussel, where I am currently

[i] Ph. Quevauviller, P. Balabanis, C. Fragakis, M. Weydert, M. Oliver, A. Kaschl, G. Arnold, A. Kroll, L. Galbiati, J. M. Zaldivar and G. Bidoglio, *Environ. Sci. Policy*, 2005, **8**, 203.

[ii] *Groundwater Science and Policy* – An International Overview, ed. Ph. Quevauviller, RSC Publishing, Cambridge, ISBN: 978-085404-294-4.

[iii] Since October 2008, I have been in post at the 'Climate Change and Natural Hazards' Unit of DG Research, where I am following up projects concerning climate change impacts on water and water-related natural hazards.

associate professor besides my EC position) and EC-funded training programmes (*e.g.* TECHWARE). His life-long experience made him the natural choice for setting the scene of Section 1 of the book (Chapter 1.1).

- In another context, I have been interviewed as a policy-maker in the framework of the 'Science meets Policy' initiative that is described by John Holmes and Alister Scott. Their summary of discussions, in particular the London report written in 2005, has been a source of inspiration that is reflected in many places in this book and is developed in Chapter 1.2.
- Judy Payne arrived later in the book project. I was very interested to hear a presentation she made at a conference of the IWRM.Net project (Chapter 3.2) in Brussels in February 2009, and invited her to write a chapter in a very short time period, which she did (Chapter 1.3).
- It was natural that a chapter on EC undertakings be included in the book (Chapter 1.4). This is written in association with two colleagues, Christos Fragakis and Panagiotis Balabanis, with whom I have been in close contact for many years and with whom I share the same awareness regarding efforts needed for a better transfer of scientific outputs to policy users. Behind this chapter, there are many other EC colleagues who brought ideas and contributions to the discussions, *e.g.* Elisabeth Lipiatou, Cathy Eccles, Wanda Gaj (from the Research Directorate-General), Patrick Murphy, Irja Vounakis (from the Environment Directorate-General), Giovanni Bidoglio and colleagues from the EC Joint Research Centre; the list is not exhaustive!
- I have shared many dynamic discussions with Bob Harris who has, like me, an exotic profile with experiences in both the scientific and policy-making arenas. He is certainly one of the 'translators' and enabling intermediaries that are evocated in this book. Along with Ben Surridge and Alison Holt, he brings us discussions on the evidence base for Integrated Catchment Management (Chapter 1.5).
- One of the elements that popped up from the discussion concerning science–policy interfacing was the need to analyse the policy relevance and possible impacts of research (from the 5th and 6th Framework Programmes) and demonstration (LIFE programme). This has been carried out in the context of the SPI-Water project and is described by Kees Kramer and Xenia Schneider in Chapter 1.6.
- Edi Interwies arrived at the same time as Judy Payne in the book project – as a result of discussions over a coffee break at the IWRM. Net conference. We had in the past many exchanges about the importance of economics in policy-supporting research, and it was natural that he describes experience gained, in association with Ilke Borowski, in the present book (Chapter 1.7).
- The SKEP initiative was the natural choice for introducing Section 2 of the book. This ERANET project looks at the issue of improving the integration of science into the policy process, and, in this respect, I had fruitful discussions with Simon Gardner and the SKEP team from a policy perspective.

- The AQUATERRA project pioneered the idea of including a specific task about science–policy exchanges that had been discussed in the JOINT event in Orléans at the end of 2002. This resulted in many contacts with Tony Chapman, Adriaan Slob, Jos Brils and others. Chapter 2.2 describes the EUPOL subproject, which aims to establish operational links between science and policy.
- Another example of project that considered the need to enhance communication between scientists and policy-makers is the RISKBASE project. It will be no surprise to see that the drivers are people already mentioned, *e.g.* Jos Brils and Bob Harris, along with a range of other dedicated scientists, *e.g.* Winfried Blum, Werner Brack, Dietmar Müller, Philippe Négrel, Vala Ragnarsdottir, Wim Salomons, Thomas Track and Joop Vegter. This project is described in Chapter 2.3.
- The Harmoni-CA initiative has been the source of discussions in the water sector that are now growing into operational developments. From the workshop held in 2005 in Ghent (of which the proceedings were published as a special issue of *Environmental Science & Policy*, vol. 8) until now, Michiel Blind, Ilke Borowski, Jens Christian Refsgaard, Wim de Lange and many others have contributed to develop the science–policy concept linked to the WFD implementation. This is described in Chapter 2.4.
- In the framework of Harmoni-CA, the OpenMI project has also greatly contributed to bringing together scientists and policy-makers. It is one of the few projects that brought a model-interfacing tool to the knowledge of policy-makers and industry stakeholders (through meetings and various contacts) and made efforts to demonstrate its applicability through a dedicated LIFE project. Roger Moore tells us about this experience in Chapter 2.5.
- The involvement of stakeholders in the participatory approach of the WFD and research in support of adaptive integrated water management are core activities of Claudia Pahl-Wostl's group at the University of Osnabrueck. With Ilke Borowski, Britta Kastens and Darya Hirsch, she describes the experience gained in the NeWater project (Chapter 2.6).
- The third section of the book opens with an example of SPI at national level (which is also reflected in the foreword of the General-Director of ONEMA, Dr Patrick Lavarde). This experience is described in Chapter 3.1 by Marie-Perrine Miossec and Patrick Flammarion, who are both very active in the development of the CIS-SPI initiative described in Chapter 5.3.
- I have always thought that an improved transfer of scientific information could be better relayed at national level, and the ERANET funding mechanism is interesting in this respect. The science–policy interfacing issue has been widely discussed within the IWRM.Net network, which is described by Natacha Amorsi and co-authors (Chapter 3.2). Personal exchanges with Peter Allen-Williams, Stephen Migdley, Michiel Blind,

Gilles Neveu and others within this group have been extremely mind-opening.

- Linking water science to policy is not solely a concern for Europeans. Similar debates are taking place on the other side of the Atlantic and this is reflected by the Canadian experience described in Chapter 3.3 by Karl Schaefer, Alex Bielak and Leah Brannen.
- I had been positively impressed by a regional platform mixing representatives of research institutes and universities, industry and local authorities in my home region, *i.e.* the Adour-Garonne area. The benefit of such a regional platform is described by Philippe Vervier and co-authors (Chapter 3.4). It is certainly an excellent example of how transfer of scientific outputs to regional authorities and stakeholders could be ensured.
- Section 4 of the book deals with communication and education needs. Resulting from research projects that initiated discussions in the water sector, *e.g.* Harmoni-CA, the WISE-RTD web-portal brings a gateway for information on water scientific projects, and represents an extremely useful tool (Chapter 4.2). I have enjoyed participating in discussions around this development with Wim de Lange, Patrick Willems, Fred Hatterman, Guido Vaes, Frank Provost, Patrick Swartenbroeckx and many others.
- EUGRIS is another web portal, which follows similar aims to WISE-RTD, with a focus on soil–water interactions. I have also had very constructive discussions in this respect with Jörg Frauenstein, Paul Bardos and Tony Chapman, who describe this tool in Chapter 4.3.
- Raising awareness for improving science–policy links in the water sector is also linked to education and training at international level. In this respect, an example of an outstanding initiative is the EuroAquae network resulting from an Erasmus Mundus project, which Philippe Gourbesville and Jean Cunge describe in Chapter 4.4.
- I met Skye Duncan at a conference in Leuven in September 2008, where she highlighted that SPI also occurs at the level of urban development. I was interested by this different viewpoint and invited her to express her views regarding New York city developments – and she has done so in Chapter 4.5.
- While I was a research correspondent for the Water & Marine Unit at EC-DG Environment, topics could be defined in the framework of 'Scientific Support to Policies' projects (6th Framework Programme), and the unit hence designed terms of references for a support action that is mentioned in several places in this book (the so-called 'SPI-Water' project). Behind the SPI-Water project there is a large network that works hard to ease links among scientists and policy communities. I had many fruitful exchanges with this group, including Patrick Swartenbroeckx, Guido Vaes, Kees Kramer, Xenia Schneider, Katherine Kober, and many others. All these exchanges are reflected in considerations expressed in Section 5 of the book.

My hope is that this book will serve as an inspiring source of information to develop an operational and sustainable science–policy interface in the water sector, and that this will result in building a strong bridge between the two different planets of science and policy. All the pieces of this complex puzzle are on the table. It is now up to us to bring them together and to transform an energy-losing patchwork into an efficient working party mixing scientific disciplines, policy sectors and stakeholder's inputs.

Philippe Quevauviller

Contents

Water System Science and Policy Interfacing
Edited by Philippe Quevauviller

Published by the Royal Society of Chemistry, www.rsc.org

Section 3 Links to Water National or Regional Research, Policies and Management

List of Contributors

Peter Allen-Williams	Environment Agency, Waterside House, Waterside North, Lincoln, LN2 5HA, United Kingdom. E-mail: Peter.Allen-williams@environment-agency.gov.uk
Natacha Amorsi	Office International de l'Eau (OIEAU), 15 rue Edouard Chamberland, 87065 Limoges Cedex, France. E-mail: n.amorsi@oieau.fr
Hugues Ayphassorho	CEMAGREF, 50 avenue de Verdun, 33612 Cestas, France. E-mail: Hugues.Ayphassorho@bordeaux.cemagref.fr
Panagiotis Balabanis	European Commission, DG Research, rue de la Loi, 200, 1049 Brussels, Belgium. E-mail: panagiotos.balabanis@ec.europa.eu
Paul Bardos	r3 Environmental Technology Ltd, Room 120, Department of Soil Science, The University of Reading, Whiteknights, PO Box 233, Reading RG6 6DW, United Kingdom. E-mail: paul@r3environmental.co.uk
Alex T. Bielak	Environment Canada, Science and Technology Branch, 867 Lakeshore Road, P.O. Box 5050, Burlington, Ontario L7R 4A6 Canada. E-mail: alex.bielak@ec.gc.ca
Michiel Blind	Deltares, PO Box 85467, 3508 AL Utrecht, The Netherlands. E-mail: michiel.blind@deltares.nl
Ilke Borowski	University of Osnabrueck, Institute of Environmental Systems Research, Barbarastrasse 12, 49076 Osnabrueck, Germany. E-mail: borowski@usf.uos.de
Leah E. Brannen	Environment Canada, Science and Technology Branch, National Hydrology Research Centre, 11

Water System Science and Policy Interfacing
Edited by Philippe Quevauviller

Published by the Royal Society of Chemistry, www.rsc.org

Innovation Blvd. Saskatoon, SK. S7N 3H5 Canada. E-mail: leah.brannen@ec.gc.ca

Jos Brils Deltares, Daltonlaan 400, 3584 BK, Utrecht, The Netherlands. E-mail: jos.brils@tno.nl

Antony Chapman wca environment ltd., Brunel House, Volunteer Way, Faringdon, Oxfordshire SN7 7YR, United Kingdom. E-mail: Tony.Chapman@wca-environment.com

Michel Combarnous ENSAM – LEPT, Esplanade des arts et métiers, 33405 Talence cedex, France. E-mail: michel.combarnous@laposte.net

Jean A. Cunge 31 Rue Doyen Gosse, 38700 La Tronche, France. E-mail: jacunge@orange.fr

Willem J. de Lange Deltares, PO Box 85467, 3508 AL Utrecht, The Netherlands. E-mail: wim.delange@deltares.nl

Jérôme Depasse ECOBAG, 15 rue Michel Labrousse, BP 42353, 31 023 Toulouse Cedex 1. E-mail: jerome.depasse@ecobag.org

Skye Duncan Graduate School of Architecture, Planning and Preservation, Columbia University, 1172 Amsterdam Avenue, New York, NY 10027. E-mail: skyejduncan@gmail.com

Marie-Perrine Durot ONEMA, the French National Agency for Water and Aquatic Environments, Hall C, 5 square Felix Nadar, 94300 Vincennes, France. E-mail: marie-perrine.miossec@onema.fr

Erik Fellenius Swedish Environmental Protection Agency, SE-106 48, Stockholm, Sweden. E-mail: Erik.Fellenius@naturvardsverket.se

Patrick Flammarion ONEMA, the French National Agency for Water and Aquatic Environments, Hall C, 5 square Felix Nadar, 94300 Vincennes, France. E-mail: patrick.flammarion@onema.fr

Christos Fragakis European Commission, DG Research, rue de la Loi, 200, 1049 Brussels, Belgium. E-mail: christos.fragakis@ec.europa.eu

Jörg Frauenstein Federal Environment Agency, II 2.6, P.O. Box 1406, D-06813 Dessau-Rosslau, Germany. E-mail: joerg.frauenstein@uba.de

Eeva Furman Finnish Environment Institute, Mechelininkatu 34a, P.O. Box 140, FI-00251 Helsinki, Finland. E-mail: eeva.furman@ymparisto.fi

Martine Gaeckler Agence de l'Eau Adour-Garonne, 90 rue du Férétra, 31078 Toulouse Cedex 4, France. E-mail: martine.gaeckler@eau-adour-garonne.fr

Simon Gardner Science Strategy, Environmental Protection Directorate, Environment Agency for England and Wales,

Block 1, Government Buildings, Burghill Road, Westbury-on-Trym, Bristol, BS10 6BF, United Kingdom. E-mail: simon.gardner@environment-agency.gov.uk

Philippe Gourbesville Polytech'Nice-Sophia, Université de Nice-Sophia Antipolis, 930, route des Colles, 06903 Sophia Antipolis, France. E-mail: gourbesv@unice.fr

Bob Harris Catchment Science Centre, The University of Sheffield, North Campus, Broad Lane, Sheffield S3 7HQ, United Kingdom. E-mail: r.harris@sheffield.ac.uk

Fred Hattermann Postdam Inst. For Climate Impact Research, Telegrafenberg A51, PO Box 60 12 03, 14412 Potsdam, Germany. E-mail: hattermann@pik-potsdam.de

Darya Hirsch University of Osnabrueck, Institute of Environmental Systems Research, Barbarastrasse 12, 49076 Osnabrueck, Germany. E-mail: darya.hirsch@usf.uni-osnabrueck.de

Daniela Hohenwallner Umweltbundesamt, 5 Spittelauer Lande, A-1090 Vienna, Austria. E-mail: Daniela.Hohenwallner@umweltbundesamt.at

John Holmes University of Oxford, Department of Earth Sciences, Parks Road, Oxford OX1 3PR, United Kingdom. E-mail: John.Holmes@earth.ox.ac.uk

Alison Holt Catchment Science Centre, The University of Sheffield, North Campus, Broad Lane, Sheffield S3 7HQ, United Kingdom. E-mail: a.holt@sheffield.ac.uk

Irene Huber PTKA, Hermann-von-Helmholtz-Platz 1, 76344 Eggenstein-Leopoldshafen, Germany. E-mail: irene.huber@ptka.fzk.de

Eduard Interwies Intersus – Sustainability Services, Chodowieckistr. 2, 10405 Berlin, Germany. E-mail: Interwies@intersus.eu

Marc Jarry Université de Pau et des Pays de l'Adour, IBEAS, Avenue de l'Université, BP 1155, 64013 Pau, France. E-mail: Marc.Jarry@univ-pau.fr

Britta Kastens University of Osnabrueck, Institute of Environmental Systems Research, Barbarastrasse 12, 49076 Osnabrueck, Germany. E-mail: britta.kastens@usf.uni-osnabrueck.de

Paula Kivimaa Finnish Environment Institute, Mechelininkatu 34a, P.O. Box 140, FI-00251 Helsinki, Finland. E-mail: paula.kivimaa@ymparisto.fi

Päivi Korpinen Finnish Environment Institute, Mechelininkatu 34a, P.O. Box 140, FI-00251 Helsinki, Finland. E-mail: paivi.korpinen@ymparisto.fi

Kees J. M. Kramer	Mermayde, P.O. Box 109, NL-1860 AC Bergen, The Netherlands. E-mail: Kees.Kramer@mermayde.nl
Pirjo Kuuppo	Finnish Environment Institute, Mechelininkatu 34a, P.O. Box 140, FI-00251 Helsinki, Finland. E-mail: pirjo.kuuppo@ymparisto.fi
Xavier Lafon	MEEDDAT, 20 avenue de Ségur, 75007 Paris, France. E-mail: Xavier.LAFON@developpement-durable.gouv.fr
Patrick Lavarde	ONEMA-DG, the French National Agency for Water and Aquatic Environments, Hall C, 5 square Felix Nadar, 94300 Vincennes, France. E-mail: Patrick.lavarde@onema.fr
Hanna Mela	Finnish Environment Institute, Mechelininkatu 34a, P.O. Box 140, FI-00251 Helsinki, Finland. E-mail: hanna.mela@ymparisto.fi
Corinne Merly	BRGM, Avenue C. Guillemin, 45060 Orléans, France. E-mail: c.merly@brgm.fr
Stephen Midgley	SNIFFER, 25 Greenside Place, EH1 3AA, Edinburgh, United Kingdom. E-mail: Stephen@sniffer.org.uk
Roger Moore	Centre for Ecology and Hydrology, Crowmarsh Gifford, Wallingford, Oxon, OX10 6HU, United Kingdom. E-mail: rvm@ceh.ac.uk
Claudia Pahl-Wostl	University of Osnabrueck, Institute of Environmental Systems Research, Barbarastrasse 12, 49076 Osnabrueck, Germany. E-mail: pahl@usf.uni-osnabrueck.de
Daniela Past	Umweltbundesamt, 5 Spittelauer Lande, A-1090 Vienna, Austria. E-mail: Daniela.Past@umweltbundesamt.at
Judy Payne	Hemdean Consulting, 308 Kidmore Road, Caversham, Reading RG4 7NG, United Kingdom. E-mail: judy@hemdean.co.uk
Jurgen Plyson	2MPact, Kortrijksesteenweg 1007, 9000 Ghent, Belgium. E-mail: jurgen.plyson@2mpact.be
Frank Provost	ProvConsulting, Zakstraat 111A, 9112 Sinaai, Belgium. E-mail: f.provost@telenet.be
Philippe Quevauviller	[1]European Commission, DG Research, rue de la Loi 200, 1049 Brussels, Belgium; [2]Vrije Universiteit Brussel (VUB), IUWPARE, Building T, Pleinlaan 2, 1180 Brussels, Belgium. E-mail: philippe.quevauviller@ec.europa.eu
Jens Christian Refsgaard	Geological Survey of Denmark and Greenland, Øster Voldgade 10, 1350 Copenhagen K, Denmark. E-mail: jcr@geus.dk

Marc Rijnveld TNO Innovation & Environment, PO Box 49, 2600 AA Delft, The Netherlands. E-mail: marc.rijnveld@tno.nl

Jennie Savgård Swedish Environmental Protection Agency, SE-106 48, Stockholm, Sweden. E-mail: Jennie.Savgard@naturvardsverket.se

Karl A. Schaeffer Environment Canada, Science and Technology Branch, 867 Lakeshore Road, P.O. Box 5050, Burlington, Ontario L7R 4A6 Canada. E-mail: karl.schaefer@ec.gc.ca

Xenia Schneider XPRO-Consulting Ltd, 7 Nelson Street, CY-2021 Strovolos, Lefkosia, Cyprus. E-mail: xenia-schneider@xpro-consulting.com

Alister Scott SPRU, University of Sussex, Freeman Centre, Falmer, Brighton, BN1 9QE, United Kingdom. E-mail: scott.alister@gmail.com

Adriaan Slob TNO Innovation and Environment, PO Box 49, 2600 AA Delft, The Netherlands. E-mail: adriaan.slob@tno.nl

Ben Surridge Catchment Science Centre, The University of Sheffield, North Campus, Broad Lane, Sheffield S3 7HQ, United Kingdom. E-mail: b.surridge@sheffield.ac.uk

Patrick Swartenbroeckx Hydroscan, Tiensevest 26/4, 3000 Leuven, Belgium. E-mail: patrick.swartenbroekx@hydroscan.be

Guido Vaes Hydroscan, Tiensevest 26/4, 3000 Leuven, Belgium. E-mail: guido.vaes@hydroscan.be

André Van der Beken F. Laurentplein 45, 9000 Gent, Belgium. E-mail: Andre.van.der.beken@pandora.be

Thomas Vansteenkiste Katholieke Universiteit Leuven, Department of Civil Engineering-Hydraulics Section, Kasteelpark Arenberg 40, 3001 Leuven, Belgium. E-mail: thomas.vansteenkiste@bwk.kuleuven.ac.be

Philippe Vervier ECOBAG, 15 rue Michel Labrousse, BP 42353, 31 023 Toulouse Cedex, France. E-mail: directeur@ecobag.org

Patrick Willems Katholieke Universiteit Leuven, Department of Civil Engineering-Hydraulics Section, Kasteelpark Arenberg 40, 3001 Leuven, Belgium. E-mail: Patrick.Willems@bwk.kuleuven.ac.be

Section 1: General Introduction

CHAPTER 1.1

Reflections on Fundamental and Policy-oriented Research of Water System Knowledge over the Past 25 Years

ANDRÉ VAN DER BEKEN

F. Laurentplein 45, 9000 Gent, Belgium

1.1.1 Introduction

In philosophy, reflection means thinking in context, by experience and with a view of evaluation and decision. Since thinking is linked to the language and the meaning of its words, a correct understanding of the words is essential, especially in scientific communication. Hence, it is necessary to start with a few definitions, or at least descriptions, of the words and terms used in this reflection on water system science within a perspective of science–policy interfacing.

Scientific research, policy and management are different fields of human endeavor, each with their own objectives, methodologies, dynamics, finality and evaluation. In this chapter we focus on scientific research of the water system. Scientific research – in short "research" – relates to "science," defined as a set of well-established laws or theories and methodologies in a given domain or discipline. The wording "investigation" is more appropriate for examining non-scientific matters. The wording "water system" refers to the interconnected and complex arrangement of all components of the hydrological cycle, including its relation to human activities of all kinds. Both system

Water System Science and Policy Interfacing
Edited by Philippe Quevauviller

Published by the Royal Society of Chemistry, www.rsc.org

approach and operations research or system design are methodologies commonly applied to the water system: the former being a mathematical approach to the study of the components and how the system will behave under various conditions and leading to "integrated water resources management;" the latter being the study of objectives and how to accomplish them most efficiently, leading to so-called "decision support systems."

It must be recognized that water research and water policy and management are not activities with an end in themselves: water research serves scientific progress and thus has an indirect impact on society, while water policy and management serve society directly.

1.1.2 Policy, Management and Knowledge

It is important to distinguish between policy and management:

- *policy* is legislation (*i.e.* elaborated and adopted by institutions), regulations, strategy and choices, selections and decisions taken by policy-makers (*i.e.* government) within the limits of the legislation.
- *management* is the preparation of choices (scenario building) for policy implementation, elaboration of rules, procedures, methods or specifications that will be put into effect by legislation and regulations at policy level, implementation of decisions taken at policy level, making decisions (and their implementation) on the management level, supervision and monitoring, maintenance and renewal of resources, including human resources (*i.e.* continuing education and training, professional development of personnel); management also includes operation and maintenance (O&M) of all infrastructures and the preparatory studies, control and follow-up of new infrastructures.

This distinction between *policy* and *management*, not always well acknowledged or appreciated, is based on the democratic principles of our society. Crucial aspects are:

1. clear differentiation between issues, and related decisions, on a policy level and on a management level;
2. the presence of a management structure (*i.e.* the executive body) with sufficient financial and human resources;
3. the quality, merit and usefulness of the legislation;
4. last, but not least, wise choices and decisions made by the policy-makers.

The wording *knowledge* in the title of this contribution also needs attention together with its corollaries *skill, competency* and *expertise*:

- *knowledge* is an attribute of the individual person, resulting from processes of learning, understanding, reflecting, comparing, selecting the right information and critical and strategic thinking; it is also "scholarship": the command of learning;

- *skill* is the ability to perform mental and/or physical routine tasks, it is space and time dependent and less universal than *knowledge*;
- *competency* is the right balance between *knowledge* and *skills* to perform a given job, but it includes other attributes of the individual, such as *behavior, emotion, intuition, attitude, judgment, ethical perception, imagination . . .* ;
- *expertise* is obtained by a competent person through practice in his/her particular sphere of activity. In a more popular way it has been said: "An expert is somebody who has made all the mistakes which can be made in a narrow field".

It follows that *information*, a fully neutral and impersonal set of facts or data, and its corollaries such as *information management* or *information transfer* are totally insufficient for any kind of water policy or management activity if *dissemination of knowledge* is not guaranteed.[1] All kinds of information on the water system may be available and are often abundant, but if the knowledge to understand, compare and select the right information, and especially the critical and strategic thinking is not at hand, water problems cannot be solved. *Knowledge* acts as a "sluice" to mitigate the "flood" of *information*.

Furthermore, there are many water issues and problems for which research can do *nothing*:

- *on the policy level* that takes off through inadequate legislation and regulations or unwise decisions;
- *on the management level* that arises from misunderstanding or unwise use of existing knowledge, lack of skills, competency and expertise, misuse of resources or wrong management, bad operation or neglected maintenance; in all these cases only better education and training and correct application of appropriate, efficient and effective quality assurance methods can help.

These issues and problems can be alleviated by effective communication between science, policy and management, or an *interface*: the subject of the present book. Therefore, fundamental and policy-oriented research of the water system should be understood as the action to extend our knowledge and/or to apply existing knowledge to new or emerging problems that have not yet got a full scientific understanding or a solution for better policy and management.

1.1.3 Scientific Research

1.1.3.1 Objectives

The general aim of scientific research is moving the borders of our knowledge and thus contributing to the understanding of phenomena of all kind. We try to get insight into relations, their origin and consequences. For the water system this means understanding the interaction between all its components. This general objective can be pursued both from a philosophical point of view of curiosity and from the utility point of view of problem solving. Both points of view have implications for what follows and are also reiterated in a well-known classification of scientific research:

- *fundamental research* pursues no specific application;
- *applied research* aims at applications of existing or new knowledge. For water it means: to develop methodologies and technologies for better protection against excess or shortage of water, for efficient use and re-use of water, for protection and improvement of water quality.

Other much used terms for this classification are: **basic research* and **targeted research*.

Fundamental research is the basis for any *applied research*: how could one expect to find correct, long-lasting or "sustainable" applications if the phenomena, which are at the origin of the problems to be solved, are not well understood or consequences are uncertain? Sometimes applied research may be initiated too early to obtain reasonable or highly-expected results, because fundamental research of the phenomena lags behind: an issue that creates frustration with both end-users and researchers.

Policy-oriented research belongs to the second category where the expected results will be employed in the field of policy. Given the essential distinction between policy and management, one should not call all types of applied water research by this name. *Management-oriented* research is for many research topics a more correct name. Both types of research will have their specific result-expectations.

An example: research towards a new methodology for detecting a given contaminant is – in its development phase – not policy-oriented, but rather management-oriented: a better monitoring may be expected. But once the methodology has been tested, validated and verified – a process that may take years – a procedure with related cost–benefit analysis may be then prepared to incorporate the new methodology into regulations. However, it may be that doubt exists about the relevance of the given contaminant in matters of health: now a policy-oriented research on risk assessment might be needed to prepare for a wise decision.

Many similar examples could be found in, for instance, flood protection research: if the methodologies for computing/predicting/forecasting floods – typically *management-oriented* research – are not first tested, validated and verified, *policy-oriented* research for decision support systems is premature or, in the worst case, dangerous.

Policy-oriented research should also not forget about the management issues: it is not the researcher who will execute the policy-decision but the manager or practitioner. Therefore, scenarios for policy-makers should in the first place be prepared and elaborated by the practitioners and not by the researchers.

Of course, scenarios should be prepared with the latest knowledge about the compounds of the water system and their management issues and if the knowledge is not yet available or insufficient at management level, *management-oriented* research should help. When it comes to finding the best or most efficient or most "plausible" scenario, systems analysis or a system design approach can be used in *policy-oriented* research. In other words: *management-oriented* and *policy-oriented* research should always go hand in hand. Each research project should have its clear specific objectives. It goes without saying that water policy and management should also rely on contributions of disciplines that are counted traditionally as human sciences, also called "soft" sciences: history, sociology and psychology, jurisprudence and economic sciences. Both *policy-oriented* and *management-oriented* research may need this "soft" support, which is often forgotten or underestimated.

1.1.3.2 Methodology

Scientific research applies a proven methodology of *observation, measurement, analysis, hypothesis and synthesis.* Because of this methodology science distinguishes itself from the speculative approach. Galileo Galilei (1564–1642), a pivotal figure in intellectual and scientific history, is quoted: "Measure what is measurable, and make measurable what is not so." Research is also empirical, where the experiment with reproducible measurements forms an essential test for the value of the research. Feynman,[2] winner of the 1958 Nobel Prize in Physics and famous teacher, wrote:

The principle of science, the definition almost, is the following: the test of knowledge is experiment.

If this reproducibility is not possible by the nature of the phenomena (like for example rainfall), then the number of measurements must be such that valuable statistical interpretations become possible.[i] Measurements must also be representative for the phenomena observed in different conditions or locations and should not disturb the phenomena. Uncertainties about the measurements should be recognized and not hidden.

Measurements must be analyzed with the help of known basic laws of physics (in the broadest sense) and theories/hypotheses that must be tested against the results of the measurements. The quality, precision and representative character of the measurements play a primordial role. This is mostly already research in itself: it includes the selection and kind of measurements, locations, duration and frequency of the measurements, the choice of measuring equipment and the treatment and handling of the data. Good scientific research also takes into account the data and results of the past if these are available: hence the utmost

[i] Earth sciences in general (climatology, geology, hydrology, meteorology, vulcanology) share this limitation of experimental methodology. Incidentally, astronomy, also a science with clear limitations for experiments, has avoided any connotation with the speculative approach (astrology) through its word-ending.

importance of a permanent inventory of past measurements and data derived from these measurements.

The analysis leads to hypothesis and synthesis: in the past the hypothesis/synthesis following careful experiments had been necessarily limited either to empirical formulas, graphical or tabular solutions, or to verbal conclusions; now the hypothesis/synthesis will appeal especially to mathematical or other models that, by use of the computer, can generate simulations under different scenarios, or even forecast results in "real-time." However, the hypothetical character of many results obtained in this manner is often concealed.

Validation, verification and interpretation of thus-obtained results are always essential and must lead to a return to observation, monitoring design, analysis and models to introduce improvements in the methodology. A particularly difficult problem today concerns computer models that are not available in the public field and are also not accessible for modifications.

Research methodology also consists of the "management" of resources allocated to the research project, of the milestones and deliverables described in the research contract.

1.1.3.3 Dynamics

The methodology of research requires specific knowledge, skills, competency and expertise of the researcher. Seldom will one researcher combine all the necessary attributes and for this reason team-work is practically always essential and cooperation with teams of other disciplines will often be appropriate. It is not enough to bring together researchers from several disciplines (*pluri-* or *multi-disciplinarity*[ii]) to study the same subject: there should be *inter-disciplinarity*[iii] and possibly also *trans-disciplinarity*,[iv] which could create, under good circumstances, new research methodologies and even a new discipline. New disciplines are very important for the progress of research and, for solving new problems, are sometimes the only possibility. But the creation of *inter-disciplinarity* or *trans-disciplinarity* demands much time; likewise, acquiring the necessary knowledge, skills, competency and expertise of the individual researcher requires time: serious scientific research will be always a long-term activity and never an activity that starts and finishes with a research contract.

In addition, research occurs on an international scale and cooperation with teams outside the local borders is obviously a must: research networks offer many advantages, especially when they operate in the long term with a suitable structure and code of conduct for the members of the network.[3] Training of young researchers, continuing education and professional development are also essential components in the entire dynamics of the scientific research process.[4]

[ii] *multi-disciplinarity* relates to several disciplines that need to help each other and co-operate, *e.g*. in some regions of physics and chemistry, where the same phenomenon is studied but where the disciplines themselves remain as they are.

[iii] *inter-disciplinarity* is present where unifying theories exist of fundamental problems within different disciplines.

[iv] *trans-disciplinarity* is often used when a problem is looked at in such a holistic way that there is no longer room any for specific narrow view of existing disciplines: a new discipline may emerge.

1.1.3.4 Finalization

The obtained results of the research must be reported in an understandable manner – differing according to the target group and following the milestones and deliverables defined in the contract – and presented with a complete description of the entire methodology. Where appropriate, measurements and other data must be reported or, at least, made available to the research community if requested for testing hypotheses, theories or models. Especially, the quality of the data must be critically presented. According to the objective of the research the report will describe and comment on both positive and negative results: negative results can be very important for the continuation of the research so that a recurrence of the same wrong or inappropriate methodology can be avoided. Unfortunately, often successes only are reported or the applied methodology is not entirely described so that the results cannot be fully assessed or compared to the results of other researchers. The worst case is where scientific communication "has become accustomed to an "off-the-peg" writing style, stringing together pre-digested commonplaces and ersatz for real argument, debate and controversy".[5] A correct code of conduct of the researcher must be observed and "responsible scientific behavior" must be requested.

A vital point forms the applicability of the results when the objective of the research is solving problems, *i.e.* in applied or targeted research. The manner of reporting and presentation of the results should be laid down in advance in the research contract with a view to the applicability of the results. It is the responsibility of the researcher to obtain the best possible results, but the commissioner – manager or policy-maker – is responsible for the application/exploitation of the results: *research/management/policy* each have their own characteristics, responsibilities and finalization and these should not be fused. Also, the commissioner must be aware that research will not always produce useful results: the outcome is, by definition of research, not predictable. If one expects prompt solutions for problems, then one should not look for research, but for *consultancy* according to agreed codes or rules of practice. The applicability of the results has to do also with the training of the end-users: efficient dissemination of knowledge and use of results is more than streamlining information or transfer of information from the research community to the policy-maker or to the practitioner. How and by whom dissemination of the results will occur and who ensures the training also belongs to the finalization of the research and should be described in the research contract. Beside the immediate reporting to the commissioner or supporting agency, communication of the results of the research in scientific journals is an obligation to which absolutely no researcher can withdraw him/herself. Whether the results of the research will be public domain or should remain confidential must be defined in the contract, as well as who is the owner of the results and of the measurements/data or models that led to the results. The dissemination of knowledge and of new or renewed understanding on a broader scale – popularization – is as important, but can best be delivered by those who have skills and talent for this activity, not necessarily by the researcher him/herself.

1.1.3.5 Evaluation

Like any activity, research must also be evaluated. Firstly, one may compare the specific objectives of a given research activity with the general objectives of the research mission or programme of the research institute, organization or funding agency. Those general objectives may be derived from a "foresight" study, in itself a difficult activity, which could also be evaluated many years later. At research project level, result-evaluation is a comprehensive term that includes both the evaluation of the research methodology, the practicality of the results and the impact of the research and its results on the socio-economic development ("impact assessment"). Result-evaluation is done both by the researcher and by the end-user. The former will, anyhow, be evaluated by the research community through the scientific publications.[V] The latter will test the results in practice. However, it is often very difficult, sometimes impossible, to evaluate if the application –or non-application – of the results of the research will be positive or negative, now or in the future. Evaluation of results in practice goes paired with the correct "translation" of scientific results into practical use and thus with the training of the end-users. A participatory model and interaction with the researcher must be pursued and with this aim in mind the principle of networking between researchers and end-users is an important tool for evaluation. Cost–benefit analysis of research, taking into account the dynamics of the research, is seldom possible. In contrast, an evaluation of the research policy is very well possible: comparing the objectives, dynamics, finalization and contract conditions of the research projects, implemented in a given period, could offer interesting elements for an improved research policy in the future.

1.1.4 Water Research Progress and Problems

1.1.4.1 Progress in General Research based on Technological Developments

Huge progress has been achieved in the field of monitoring thanks to several technological developments that increasingly pervade research in general but also the whole society: *digitalization, information techniques* and *communication*. Telemetric systems together with new types of sensors, including remote sensing techniques of all types, and tele-transmission through radio and satellite communication allow for wider and more intensive monitoring activities, including data handling. The advent of the PC (personal computer) in the early 1980s has changed profoundly not only the methodology of research but

[V] Good quality scientific publications are more important than the number of publications: unfortunately, the present trend in research-evaluation is to count the numbers. In many cases this has a negative effect on the quality of the research itself, because the researcher has less time to 'think' about new experiments while writing yet another scientific paper about the same results of his former experiments.

also the dynamics of the researcher, his/her relation with both the research team and community and with the end-user of the research results. Ten years later an even bigger change took place with the increasing use of the Internet and the availability of user-friendly software programs for both scientific analysis and reporting of results. Geographical Information Systems (now called Geospatial Information Systems-GIS) were introduced under this name as early as 1962 in Canada, but the first GIS-software in the public domain was released 20 years later, and it was the PC that made GIS widely accessible in the 1990s. Software, both in the public domain and in the commercial domain, and digital elevation models became very popular and are nowadays essential tools for both the researcher and the manager of the water system.

1.1.4.2 Problems in Water Research

With respect to the development of *information technology* two aspects applied to water research and management are considered here.

1. *Quality and uncertainty of data.* All sensors need calibration and, in the case of remote sensing techniques, this means "ground truth:" this is especially crucial for rainfall measurements derived from microwave detection methods, for evaporation, evapotranspiration and soil moisture data, for certain water quality variables and for sediment suspension in rivers and other water bodies. Another typical example is the monitoring of flow rates in rivers and canals, which is essential information for any water management issue: sophisticated measuring equipment based on advanced electronics, combined with tele-transmission techniques, offer unprecedented possibilities for real-time information. However, the uncertainty of the flow data remains and is fully related to the flow rating curve in the control section of the river/canal. The establishment and maintenance of such a rating curve in a well chosen, often constructed, control section is a difficult and costly procedure. Knowledge about this lengthy process and the uncertainty of the data is *not* only a matter for researchers but should be recognized also by managers and policy-makers. Enhancement of skills and competency and an appropriate organizational structure are needed to guarantee the quality of measurements and the monitoring activities as a whole.
2. *Safeguarding of data.* Digitalization means hardware support and software to store, retrieve and handle the data; the speed of development of these tools is such that special attention must be given to the permanent safeguarding of measurement data and related handling/analysis of these data to avoid *digital cemeteries.* Once again, managers must recognize fully this great danger and create depositories with permanent maintenance and easy access for researchers; policy-makers must decree regulations for making depositories compulsory to all actors in the field of water measurements and observations, including the very important geological and geophysical observations derived from drilling or construction activities.

Mathematical models of the water systems, using more and more sophisticated modeling techniques made possible through the computer software developments, found their first applications in *management-oriented* water research since the early 1960s, *e.g.* the Stanford model of 1964 introduced a full hydrological model at catchment scale, which was the example till today for such "lumped" models. But the real breakthrough was in the 1980s when distributed models at catchment scale like the SHE model of 1981 and water quality models for river networks like QUAL2 of 1984 were first introduced. The former was a commercial product with restricted access to the source codes, the latter a typical product in the public domain where users are able to adjust or modify the components of the programme: a distinction that continues to characterize software development in general and software for water systems in particular.[vi] Sophisticated models using "real-time" measurements of rainfall or even forecasting rainfall, for predicting future high river flows and controlling flood reservoirs, were top-notch in the early 1980s. Similar models were developed for urban water systems consisting of storm drains, combined sewer overflow structures and retention basins. Integrating water quality in those models is still a headache for the simple reason that water quantity is well defined, but water quality is undefined unless a very limited number of compounds governing the water quality are selected.[vii] Moreover, water quality interacts with biological processes, which are much more difficult to describe in mathematical terms. Processes such as soil erosion and sediment transport add greatly to the complexity of the water system. Soil- and groundwater, the hidden source, were also the target for impressive software development, even more than was the case for surface water. Since the end of the nineteenth century the mathematical description of soil- and groundwater flow had been a beautiful and successful domain for mathematicians while simplifying the endless diversities of the soil, in which the rain infiltrates, and the largely unknown geological layers, in which the water is stored. GIS finally offered a tool to describe in digital format the spatial variability in three dimensions (3D), provided correct observations and measurements of the characteristics of the soil and the underground are available.

Operations research or system design also found its way into *policy-oriented* water research as early as the late-1950s (the Harvard Water Program). But again, it was in the 1980s that further developments were made when multicriteria methods became fashionable and were applied in water policy. Nowadays they are often called "decision-support" models: they need huge amounts of data whenever several scenarios are involved since water management has an impact on many aspects of society.[viii] Hence the term

[vi] "Open source" describes the practice that gives production and development of computer programs free entrance into the source material (the source) of the end product. Some people see it as a philosophy, others consider it as a pragmatic methodology.

[vii] The French philosopher Michel Tournier expresses the difference between quantity and quality as follows: '*Sur la qualité, on peut discuter à l'infini, tandis que la quantité, elle est indiscutable*' (one can discuss quality endlessly, while quantity is not to be discussed).[6]

[viii] Many so-called "decision support models" are tools that can be used for simulations in scenario development, but do not include multi-criteria algorithms.

"integrated water management" emerged. However, more conspicuous is the fact that multi-criteria methods need a weighing of criteria or values expressing the interest of the different stakeholders of the water system: the selection of these weights is the bottleneck, and hence decision support models can offer any policy proposal until an agreement is reached on the weights of the criteria.

Mathematical models have been called "the language of the gods," but we must recognize that models can never represent the real processes fully in all their details and thus are never flawless. Modeling is always a trade-off between complexity, describing all processes as best we can, and simplicity with unknown processes represented by parameters to be found (or guessed). There is definitely an element of imagination or art in modeling, hence the modeler must be competent and have expertise. The great danger exists that sophisticated models are sometimes used by people who know little about rivers and its hydraulics, about soils and geology, about water quality and biological processes or were never in the field observing nature's mysteries, and nevertheless take the results of the models for granted and present them to policy-makers as reliable "decision support tools."

1.1.5 By Way of Conclusion

The above reflections do not represent an exhaustive review on issues of the water system knowledge, research needs and the much needed science–policy interface. They are certainly biased by the author's personal experience. Notwithstanding this limitation, the following statement may summarize the exercise: "We must reemphasize the value and importance of observational and experimental skills".[7] Or in a more popular wording: "Too many models, not enough experiments", a claim that should be examined very seriously in water research. Fundamental research is needed to understand better the nonlinear behavior of most processes and the scales at which laws or equations can be applied. Turbulence, boundary layer dynamics, diffusion and dispersion are basic phenomena in all water processes and still rely on unknown parameters that can only be found by careful and painstaking research in laboratory experiments or with many field studies. The interaction of water quality with the biological processes is another field where basic research is continuously needed because new compounds are steadily added or detected in the water bodies: progress in ecotoxicology and the search for optimality criteria depend on experiments and the same holds for any improvement of water treatment or purification methods. If our basic knowledge of these processes is not increased, applied research or targeted research of any kind of the water system is doomed to be endlessly repetitive and its results at worst unreliable or, at best, of limited and local value. An efficient and operational "science–policy interface" should not only play its role in dissemination and communication but has also a critical role in the balance between fundamental research and applied research. No doubt water problems and related management issues need targeted research for solutions or decisions in the short term, and for transferring knowledge into innovation on the market. But it is also a "wise" policy to transfer money into new knowledge through fundamental research because without new knowledge there will soon no longer be any innovation.

References

1. A. Van der Beken, *Dissemination of Knowledge and Experience: the Heart of the Matter*, Keynote address at the TRANSCAT Conference "Integrated Water Management of Transboundary Catchments", Venice, 24–26 March 2004.
2. R. P. Feynman, *Six Easy Pieces – Essentials of Physics*, Perseus Books, Cambridge, MA, 1963.
3. J. Bogardi *et al.*, *Networking in the Water Sector: Suggestions and Examples for Good Practice*, UNESCO, International Hydrological Programme IHP-VI, Technical Documents N°63. Paris, 2004.
4. A. Van der Beken, Water-related education, training and technology transfer, in *Knowledge for Sustainable Development, Vol. II*, Encyclopedia of Life Support Systems Publ., Oxford, ISBN:0 9542989 0 X; http://www.eols.net (2002).
5. M.-Cl. Roland, Communication practice and responsible behaviour in scientific research, *Research'EU*, N°53, 2007.
6. M. Tournier, *Le Miroir des idées- Traité*, Mercure de France, Paris, 1995.
7. National Research Council, *Opportunities in the Hydrologic Sciences*, Commission on Geosciences, Environment and Resources, National Academy Press, Washington D.C., 1991.

CHAPTER 1.2

Bridging the Gaps between Science and Policy: A Review of the Evidence and some Principles for Effective Action

JOHN HOLMES[a] AND ALISTER SCOTT[b]

[a] University of Oxford, Department of Earth Sciences, Parks Road, Oxford OX1 3PR, United Kingdom; [b] SPRU, University of Sussex, Freeman Centre, Falmer, Brighton, BN1 9QE, United Kingdom

1.2.1 Introduction

An understanding of the relevant natural and social systems, appropriately reflected in environmental management decisions, is more likely to result in the desired outcomes than decisions made in the absence of such understanding. This proposition lies behind the substantial investment made annually by the European Commission and European Union Member States in research to improve our understanding of the natural and social systems that determine the state of the environment, and to support environmental policy-making and regulation.

However, the two worlds of science and policy/regulation do not always get on as well as they might. Concerns are frequently expressed that there is a gap between them that can lead to several kinds of problems, including:

- policies can be developed that are not sufficiently evidence-based;
- policy processes can fail to address problems being highlighted by science;

Water System Science and Policy Interfacing
Edited by Philippe Quevauviller

Published by the Royal Society of Chemistry, www.rsc.org

- science can fail to take account of, and respond to, urgent policy needs;
- science can be conducted in ways that do not produce policy-relevant results; and
- inadequate communication can fail to bridge these worlds of science and policy.

At the same time, the context for environmental policy-making and regulation is one of increasing challenge from the private sector, citizens' groups and trading partners. Pressures are increasing for more accountability and for the evidence base on which policies are based to be clearer. This is driving a more evidence-based approach and requiring the more sophisticated use of science.

Consequently, a series of official European 'Science Meets Policy' workshops and 'Bridging the Gap' conferences have been held since 1998, which have sought to address the relationship between science and policy in the environmental domain in Europe (http://www.sciencemeetspolicy.eu/site/11.asp). These workshops and conferences have brought together people working at the science–policy/regulatory interface in the European Commission and the environmental ministries and regulators of the European member states, and have been driven by the concern that the relationship between science and policy has not always been as effective as it should.

This chapter draws on the authors' work on the boundaries of science and policy over the last ten years and more. In particular, it summarizes research carried out over a nine month period in 2005, as a continuation of the Science Meets Policy initiative, to take stock of the relationship between science and environmental policy making in Europe and to consider how that relationship could be improved. The work was sponsored by the UK's Department for Environment, Food and Rural Affairs (Defra) as the main input into the London Science Meets Policy workshop that was held as part of the UK's presidency of the European Union.

A first stage of the study was to review recent developments in the use of science in environmental policy making and regulation in Europe. At the same time, the authors reviewed and brought together the outcomes of the previous workshops and conferences in the Science Meets Policy and Bridging the Gap series. A central second stage consisted of interviews with around 100 policy makers, science advisers and scientists in European member states and the European Commission. These interviews focused on two main issues:

1. the 'front end', the planning and management of research programmes; and
2. the 'back end', the dissemination and uptake of science.

In conducting these interviews, the authors were aware that it is hard to make clean distinctions between the front end and back end of both research and policy – both roll forward with their own logic and timescales, and are rarely able to be neatly boxed in cleanly defined stages. Equally, the authors were aware of the work of Gieryn[1] and Wynne[2] amongst others, who challenge the idea that there is a clear boundary between science and policy, but instead contend that the boundaries between the two worlds – for there are many – are constantly created and hotly contested. Players in both worlds want to influence each other while needing to appear to remain separate:

- Policy-makers need researchers to provide evidence, advice and legitimacy for policy decisions. They also often control research funds, and will seek to obtain 'relevant' research by influencing research agendas and even methods. At the same time, the more sophisticated policy-makers involved in science policy realise that they need the science they procure to appear to be independent of such influences, so invoke such terms as 'basic' science to describe the research they are supporting.[3,4]
- For their part, researchers need the resources that policy organisations, such as the European Commission, can provide, while wanting to maintain their appearance of 'objectivity' and independence. They need to do these things partly to enhance their authoritativeness in the world of policy that they are seeking to influence, and partly to establish and expand their credibility among their scientific peers. In this respect, 'basic' science is overwhelmingly seen as higher grade – and more easily publishable in top scientific journals – and such publishing is *the sine qua non* for progression in scientific careers, along with the ability to raise funds.[5]

Nevertheless, the previous discussions in the Science-Meets-Policy and the Bridging-the-Gap series seemed to categorise themselves under the convenient headings of 'front end' and 'back end', and we found these to be points in the research and policy cycles that respondents found useful as handles on what can be a complex and confusing area to discuss.

Taking a report of these interviews as its starting point, the third stage of the study consisted of a 'Science Meets Policy' workshop held over three days in November 2005 and involving 80 policy makers, science advisors and scientists. The workshop considered the most effective ways forward for the science policy interface in respect of these two issues. The workshop developed several recommendations for action by the European Commission and by environmental ministries, regulators and other bodies in European Union member states.

The work on the front-end and back-end issues is summarized in Sections 1.2.3 and 1.2.4, respectively. Several cross-cutting issues also emerged, which are discussed in Section 1.2.5. By way of context setting, developments in Europe over the last ten years in respect of the use of science in environmental policy making and regulation are summarized in the next section. Section 1.2.6 reflects on the findings of the study in relation to the literature and briefly considers developments since 2005. Finally, Section 1.2.7 presents principles of effective science into policy practices.

1.2.2 Developments in Europe

1.2.2.1 European Commission

Over the period since the first Science-Meets-Policy workshop in 1998, a number of significant developments have reflected an increased awareness in Europe of the need to improve the role that science plays in environmental policy-making. The publication by the European Commission in 2001 of a white paper on European Governance[6] led to an Action Plan on Science and Society[7] and the publication of guidelines on the collection and use of expertise by the Commission.[8] Such efforts were pre-dated and possibly encouraged by initiatives within individual countries, such as the guidelines developed by the Chief Scientific Advisor in the UK.[9,10]

The white paper recognised the need to open up the EU policy making process to get more people and organisations involved in shaping and delivering policy, and to boost confidence in the way expert advice influences policy decisions. The Science and Society Action Plan had as one of several aims to 'put responsible science at the heart of policy making'. Specific actions proposed to meet this aim included:

- promote dialogue between the scientific community and other stakeholders on ethical and sustainability issues arising from scientific and technological developments;
- initiate exchange of experience across Europe on risk assessment, management and communication of scientific uncertainty; and
- enhance mechanisms to provide scientific support to policy-makers (leading to the SINAPSE initiative – a web-based communication platform enabling the exchange of information between the scientific community and policy-makers – see europa.eu.int/comm/research/science-society/science-governance/sinapse_en.html).

The guidelines on expertise apply to the collection and use of expertise by Commission departments and set out three core principles of quality, openness and effectiveness.[i] The motivation for the guidelines included the issues of how to deal with conflicting expert opinion and to ensure that the processes for the collection and use of expertise are credible. However, a subsequent review by the European Policy Centre[11] of the use of science in policy

[i] As an aside, it is hard to see how 'effectiveness' can be a principle; it might be an aim, a result or a judgement, but to declare it as a principle seems to beg the question. If we are being really critical, we could say the same about quality – surely the aim is to achieve quality, not just declare it as if we know ex ante what it will be. Both quality and effectiveness have many potential dimensions, some of which are likely to be in tension (*e.g.* effectiveness may be in tension with the quality dimension of efficiency). Such bland assertions have been a common feature of science-policy debate and rarely help participants to think more clearly about either ends or means. See Scott for a discussion of how such factors affect the practice of peer review.

making by the Commission identified several weaknesses in practice, including the need to:

- formalise the requirements of the guidelines;
- revise procedures for the selection of scientific advisors;
- ensure wide consultation when framing policy related questions; and
- strengthen reporting requirements.

The European Union's Environment Action Programmes have a central role in driving environmental improvements in Europe. The Commission's communication on the 6th Environment Action Programme[12] indicates that 'the complexity of our environment is such that this knowledge is critical to the development of effective policies', and that 'underpinning proposals and decisions with scientific advice of the highest quality available is fundamental to winning acceptance by stakeholders and to avoid unnecessary conflicts with our trading partners'.

The communication makes a commitment to broaden dialogue and give interested groups the opportunity to influence decision-making. It indicates that this should be supported by sound scientific and economic assessment that 'will require a significant upgrading of the quality of the environment and economic data, assessments and policy evaluations available to support our decision making'. It points to the need to:

- improve the dissemination of research results so that they are both more useable by policy-makers and also help to increase the public's understanding of environmental issues;
- gain more benefit at Community level from research carried out at member state level through enhanced coordination.

The European Union's Framework Programmes represent a substantial investment in research and have a significant element concerned with environmental issues. The five-year assessment of the European Union Research Programmes 1999–2003[13] concluded that the Commission should address more clearly the contribution of the Framework Programmes to the broader EU policy formulation process and criticised their planning: 'It lacks an explicit logic connecting the highest objectives to the specific research and knowledge goals, to the amount of resources needed…' It also concluded that 'Coordination between the FP [Framework Programme] and national RTD [Research and Technology Development] programmes, however, remains weak' and recommended that 'the coherence between national science and innovation policies and the FPs must increase.

Subsequent Framework Programmes, the sixth and seventh, have put increased emphasis on enhancing the connection between the research funded and European environmental goals and challenges.

1.2.2.2 European Union Member States

The period since 2000 has also seen increased emphasis on the effective management and use of science to support environmental policy-making and regulation in European Union member states.[14,15] For example, in the United Kingdom a commitment in 1999 to enhance the use of evidence in governmental policy-making[16] has been followed up by a series of initiatives including the updating of governmental guidelines on the use of scientific evidence in policy-making,[17] audits of the management and use of science in governmental departments and agencies, and the installation of Chief Scientific Advisers in senior positions in government departments.

Improving structures and processes for the planning, management and uptake of science in policy-making has been a concern of several of the ERA-Nets established under Framework Programmes 6 and 7. The ERA-Nets bring together the funders of research in European Union member states around specific topic areas to enhance the coordination of research across Europe and to enable collaboration on issues of common interest. The SKEP ERA-Net (www.skep-era.net) in particular has focused on these issues of enhancing the effectiveness of science in supporting environmental policy-making and regulation.

1.2.2.3 Science Meets Policy and Bridging the Gap

Prior to the work described in this chapter, six Science Meets Policy workshops (London 1998, Paris 2000, Amsterdam 2001, Vienna 2002, Crete 2003 and Gdansk 2004) and three Bridging the Gap conferences (London 1998, Stockholm 2001 and Dublin 2004) had been held.[ii] A review of the outputs and conclusions of these workshops and conferences was conducted in the first stage of this study. It revealed several persistent issues that had recurred as central and on-going concerns:

- problems of communication between the science and policy communities given that they have different cultures, languages, motivations and constraints;
- the need for improved mechanisms to link the European Environment Action Programmes and the Framework Programmes;
- difficulties for policy makers and regulators arising from the sheer volume of research outputs, the way that research is often published, and the diversity and variable quality of information;
- insufficient embedding of good practice guidelines intended to improve the transparency of science-into-policy processes with the intention of building stakeholders' confidence in government decision making on environmental management;

[ii] Alister Scott was a rapporteur at the Stockholm event and speaker at the Vienna event.

- a desire to further open up the science-into-policy process to enable better engagement with stakeholders, and to access a broader range of views and different kinds of knowledge; and
- the need for more inter-disciplinary working, particularly bringing in more socio-economic perspectives to complement the natural sciences.

1.2.3 The 'Front End': Planning and Managing Research Programmes

The initial steps in planning and managing research projects and programmes are key to setting them on a course that will lead to their eventual uptake in informing environmental policy-making and regulation. The interviews and workshop identified three important issues in respect of these initial steps: the identification of research needs and the reflection of those needs in establishing research agendas and questions, the need for effective engagement between researchers, research users and stakeholders more generally, and the processes of research project selection. The findings and conclusions of the study on these issues are summarized in the following sub-sections.

1.2.3.1 Identifying Research Needs and Setting Research Agendas

It is common to find that multiple research agendas operate simultaneously in countries at any one time. In addition to the research agendas set specifically to support policy development, research not driven by policy concerns is also conducted in organisations such as universities and research institutes, typically funded by research councils and national scientific academies. While the research here may be in line with the overall national research agenda there is greater autonomy in decisions relating to research, which may also be critical of this agenda and of the government policy of the day.

The definition of research agendas within the scientific community can often be primarily within the confines of the traditions, theories and methods of a single research discipline. These often have only a poor match with the shape of the 'real-world' social problems that policy is grappling with. Concerns were expressed, for example, that much of the environmental research in the European Union's Framework Programme 6 was driven by the interests of the science community rather than the policy community.

However, policy-makers often reported difficulties in identifying their research needs over the medium term (*i.e.* 2–5 years) – precisely the timescale over which much academic research is conducted. Short-term pressures, turnover of staff and changing political priorities were suggested as contributing factors within the policy world. In addition, policy-makers can find it hard to make their research needs heard among those designing research agendas and commissioning research. These factors can create problems for building policy concerns into the design of research programmes.

Those involved in policy need to develop processes for identifying their research needs and for feeding these into the development of research programmes run by 'arms-length' research funders. Such programmes need to be developed collaboratively so that they 'dovetail' appropriately with any research portfolio that the policy bodies commission directly themselves. Identifying the policy needs for research can be assisted by knowledge audits in policy-making bodies.

The complex nature of environmental research and policy was highlighted in many interviews; environmental issues are of importance to various parts of government. As a result, inter-Ministerial cooperation over policy and in setting research agendas is required. This is achieved in various ways, the two most commonly reported being (i) specific structures established for this purpose and (ii) the use of *ad hoc* working groups. In addition, some countries reported national plans and strategies to support inter-Ministerial collaboration. Such collaboration can encourage inter-disciplinary research on cross-cutting issues, and this, in turn, may assist environmental ministries to access additional research funding.

Many interviewees mentioned the usefulness of a bottom-up approach in addition to the more traditional top-down approach in identifying policy needs, developing research agendas and designing the actual research. The bottom-up approach is broadly consultative and seeks involvement from a wide network of those with an interest or involvement in a particular area. It recognises that policy processes are increasingly diffuse, involving a broad set of players. Industry and non-governmental organisations were viewed as two of the most important groups.

This approach also draws strength from the idea that researchers need to be interactive in the way that they do their research: policy makers and practitioners have much substantive expertise of their own, so researchers need to consult broadly if the research is to be relevant, timely and thorough. This contrasts strongly with the narrow, disciplinary and closed style of research that is still strong in many parts of the research world. There are fears that an interactive style of research can undermine the independence of the researchers involved. Much of the devil here is in the detail. All too often, the desire to ensure scientists' autonomy can mean that other voices and non-scientific relevance criteria can struggle to be represented in decisions over research priorities and practices.

1.2.3.2 Engagement and Stakeholder Dialogue

Policy-makers and other stakeholders need to be closely engaged throughout research and policy processes. Such engagement should involve joint framing and planning of the research, the presence of policy-makers and stakeholders on research steering boards, policy mentors for research teams to enable linkages directly into policy networks, and regular exposure for scientists to the policy-making process.

While such close engagement can be argued for on instrumental grounds, *i.e.* to make the linkages more effective, it can also be argued for on substantive grounds – because it leads to better research, policy and learning. Those involved in policy – and other stakeholders – often have an enormous wealth of knowledge themselves that researchers can do well to tap into. An example given at the workshop was of research conducted in the Marche hills in Italy, where a process of dialogue with local stakeholders over the design of research and policy for addressing pollution from agriculture led the researchers to completely change their research design, the questions they asked and the policy recommendations they produced. Such close engagement is therefore also vitally useful for 'quality control' – verifying results and testing the credibility of any policy recommendations that the research might be developing.

Engagement also helps researchers to build familiarity with the imperatives involved in policy-making and of the place of science in policy-making. Government scientists could help this, for example by writing about science–policy processes in scientific magazines.

The European Commission's Thematic Strategy on air pollution was considered to be a good example of the way scientific evidence is collected and used for new environment policy initiatives (http://europa.eu.int/comm/environment/air/cafe/index.htm). Behind this strategy, there were four years of intensive work, including over 100 meetings with experts, an internet-consultation that attracted an unprecedented 11 000 responses, and several studies based on best available science and economics. Information was also used for the impact assessment underpinning the thematic strategy and for the definition of existing knowledge gaps and future research needs.

1.2.3.3 Research Project Selection

It was widely agreed that the peer-review process used in selecting research projects in open calls for research programmes often comes to be dominated by criteria relating to scientific excellence (see, for example, the online debate conducted by the journal *Nature* in 2006). But where research is specifically being developed to assist policy, additional criteria relating to policy relevance, timeliness and usefulness need to be developed and put to use if the research is to be 'fit for purpose'. This does not mean sacrificing scientific quality; research selection processes need to incorporate a broader notion of 'quality' by giving appropriate weight to both scientific and relevance criteria.[18]

To achieve this, research funders will need to issue guidelines to peer reviewers and selection committees to encourage them not to fall back on familiar routines. Policy-makers should sit on project selection committees. An example of this in action was cited as the Unit for Research and Foresight Studies in the French Ministry of Environment. Future Framework Programmes in the European Union should develop appropriate proposal evaluation guidelines.

Ironically, some respondents pointed to the Framework Programmes as having tough selection criteria and processes in this respect: applicants have to show how they are going to engage with decision-makers beyond academia during their research (particularly within the private sector, raising concerns about the privatisation of public knowledge) and how they are going to communicate their findings. Unfortunately, respondents also consistently indicated that these requirements were usually regarded as a burden rather than an opportunity, and were the first tasks to be dropped when time or other pressures started to make themselves felt. So this may be another example of what Gieryn[1] terms 'boundary work' – scientists doing what is required of them to gain access to research resources, but not having the time, skills or in some cases the desire to make engagement and communication a significant part of their work.

1.2.4 The 'Back End': Dissemination and Uptake

In respect of 'back end' processes the interviews and workshop focused on three main issues: the role of intermediaries and translators; ways of enhancing the accessibility of knowledge; and ensuring research quality and the transparency of the science-into-policy process.

1.2.4.1 Intermediaries and Translators

A key point made repeatedly in the interviews and workshop was the need for more effort to be devoted to processes aimed at 'translating' research results into inputs that could be useful to policy-makers. This recommendation was based on the observation that the findings of most research are currently published in academic papers, journals and books, few of which get read by policy-makers. Often the language used is technical and the findings are not presented in ways that make clear the policy implications to non-specialists. As a result, a process of 'translation' is required to turn scientific results into knowledge that is accessible to the non-specialist.

However, while scientists themselves should be encouraged to interact more closely with policy-makers and other stakeholders, there are many who lack the skills and the motivation to do so. In any case, without a process of translation, much useful knowledge is likely to remain locked up in inaccessible forms. There is therefore great potential for various types of intermediaries and 'translators' to do this work. Such roles can potentially be found within both policy and research organisations, or in separate agencies and even private sector providers. These are not easy positions to occupy, since both policy-makers and scientists can sometimes feel that such intermediaries are encroaching on 'their' territory, but it was thought that such people can add great value to the science–policy interface and they need to be involved from as early in the process as possible.

They require different skills to those normally held by researchers. Translators are typically more 'horizontal' than 'vertical', with an understanding of both the science and the needs of policy, and of the languages and conventions of both communities. They require analytical, synthesis and communication skills to assimilate large amounts of information, identify the essence and communicate this for non-specialists. The process of assimilating and synthesising relevant information and providing this information in user-friendly forms was widely thought to be a considerable challenge. The act of translation often means that the finer points of science are omitted; the risks of doing so need to be carefully considered.

These roles need to be ongoing rather than only brought into play when an urgent problem emerges; it has to be someone's job to monitor the findings emerging from science, bring these together and translate them into forms that are accessible to policy makers and practitioners. Researchers can and do sometimes undertake these tasks, but others might not be equipped to do so; some thought it more effective to involve bodies such as scientific academies and the European Environment Agency. It was considered that the European Commission might usefully look at ways to introduce activities into Framework Programmes to translate research findings to support policy making.

1.2.4.2 Enhancing the Accessibility of Knowledge

The view was widely expressed that more effort needs to be put into improving the accessibility of research outputs. Several interviewees suggested that it should be obligatory for researchers to include dissemination plans and a set percentage of their budget for communication in project proposals. Some policy making bodies, such as the Department for International Development (DfID) in the UK, have started to require researchers to spend a mandatory percentage of the research project budget (10% in the case of DfID) on communication with non-academic partners. However, more usually, research dissemination is neglected in research evaluation, which tends to focus on the scientific dimensions.

It was also felt that there is a need for better databases to ensure that scientific knowledge is more readily available to policy-makers. These databases need to be easily searchable and written in plain language. There is a fear that there is a lack of memory about what research has been conducted and where the results can be found. Electronic databases of scientific publications only go back 10 or 20 years and mostly cover English-language reports and papers.

Different audiences need different outputs, and traditional forms of communication within the science community do not meet policy-makers' needs. Many interviewees felt that two key stages are currently 'neglected': *dissemination* of scientific knowledge and *synthesis* of findings together with interpretation of their implications for policy. It was considered that the job is usually only half done on completion of a research project.

Research should not just be published in the form of academic papers. The *dissemination* of research needs to include policy briefs from policy-relevant research projects. It should be a requirement that all policy-relevant projects funded by the Framework Programmes produce a policy briefing; a similar requirement could be introduced by funders in Member States. Policy-makers are also interested in work in progress; significant policy-relevant projects should be required to produce an annual summary of their policy implications.

Other modes of communication, for example science cafes, which allow two-way communication and an exploration of the issues with stakeholders are increasingly widespread and attracting good participation in some locations. It was suggested that a high profile, open access journal be established that is dedicated to interdisciplinary policy-relevant research on environmental issues. The usefulness of the broadcast and print media was highlighted for informing the public, raising awareness of the existence of research on a topic and for lobbying purposes, as well as for challenging researchers to identify their key messages and learn how to put these over to non-specialist audiences.

Synthesis is needed across the research portfolio. Scientific findings being published on any particular topic that is currently of policy relevance need to be brought together. Such findings can be widely scattered in different disciplinary journals and using widely different methods and frameworks. One approach is for policy and research organisations to fund intermediaries to synthesise what is known that is relevant to current policy questions. An example of such synthesis work is the dissemination of 'Science for the Environment' reviews of policy-relevant research by DG Environment in the European Commission. Also, researchers should be encouraged and funded to publish more synthesis reporting in peer-reviewed literature. Policy-makers find it hard to make use of research outputs from the Framework Programmes – sources of information on research such as CORDIS, when used alone, are of limited value. Responding to these needs, Framework Programme 7 is providing funding for review and synthesis. Also, the European Commission's Directorate General for Research regularly organises workshops to communicate research results of clusters of projects to policy-makers and other user groups. Making good use of these initiatives will require the European Commission's Directorate General for Environment and environmental ministries in European member states to clearly articulate their needs.

1.2.4.3 Quality and Transparency

As has already been noted, there is a widely held perception that policy-making processes need to build stakeholder confidence and trust. This requires transparency in policy processes but also in the way in which research-based evidence is brought to bear. Guidelines intended to improve transparency have been developed by the European Commission and some member states. However, their implementation is in most cases still at an early stage. Suggestions for enhancing transparency include:

- improving public participation in policy-making;
- clarifying the place of experts and stakeholders in the advisory chain; and
- ensuring that political targets are kept separate from scientific mechanisms.

It was questioned whether research reports can be used directly to inform policy, or whether they need to be tested for quality beforehand (*e.g.* through peer review). This is to some extent to address the question of the 'shades of grey' – the uncertainties that attach to research results. Again, this points to the need for transparency in the process of using research in policy so that all can see the evidence used, the assumptions made and the questions asked.

1.2.5 Cross-cutting Issues

Several issues emerged that cut across the specific front end and back end concerns: the mutual understanding of the two worlds of science and policy; training and development of researchers, policy makers and intermediaries; the development of and support for inter-disciplinarity; and the need for 'strong science'.

1.2.5.1 Two Worlds

Although it was felt that the situation is improving, the view was expressed that there is still a significant level of 'mutual incomprehension' between the policy and science communities. Reasons cited for this included:

- different demands and delivery/ gestation times;
- work overload: engagement is seen as extra work;
- policy-makers regularly move on and hence tend not to build relationships with researchers;
- policy-makers assume that the science community exists to support policy: researchers are upset if their research is labeled as 'applied';
- scientists tend to focus on remaining uncertainties (and the next research grant) rather than on offering solutions or stating what they know with some certainty; and
- the lack of people with a science background in senior policy positions.

There remain significant cultural and institutional factors that can explain many of the difficulties of creating effective science–policy interaction. Any attempt to make the relationship between science and policy more effective needs to bear in mind these differences between the two communities. Table 1.2.1 characterizes some of these barriers.

Table 1.2.1 Cultural and institutional factors affecting science–policy relationships.

Factor/barrier	*Science*	*Policy*
Aims	New and valid knowledge, rigorous research, publication in academic literature, additional research funding	Respond to political priorities, balance interests and values in policy-making, cope with events, manage resources
Timescale	Slow, need to conduct research and publish. Timescale often in years	Fast, need to react to events, difficult to see years ahead
People	Specialized, narrow focus, disciplinary, stay in field	Generalist, broad view, move often. Don't want to get bogged down in detail. May have little understanding of science
Communication	Mainly to other academics via academic publications and conferences, some respond to policy consultations, wary of media	Mainly with other policymakers, don't read research publications, increasingly consult when making policy, strongly affected by media
Success measures	Publication, new funding, discovery of new knowledge	Make problems go away, avoid crisis, marshal evidence to support policies
Evidence	Based in experimentation, data collection, monitoring and disciplinary traditions	Not just science but also relating to economic, political and other interests and priorities. What is feasible
Quality control	Central to science. Conservative; via peer review/publication. Often a narrow focus on the scientific aspects; little role for 'relevance' criteria	Via policy machine; timetables often too fast for peer review; consultants' research often not reviewed externally. Have to make do with available evidence; thinks scientists too cautious
Synthesis	Few incentives – who conducts it? A lot of work; doesn't lead to new insights; can be hard to publish	Want scientific consensus and 'strong science'. Set up mechanisms such as IPCC to achieve synthesis
View of interaction	Wary, jealously guard independence and autonomy, takes time, skills and resources they might not have	Willing to interact with researchers if useful. But information overload and little time, so need to target efforts
View of 'relevance'	Want control of research agendas, suspicion of 'applied' research, lower reputational rewards than 'pure' research	Why else do research? Frustration at the relevance of much research and vagueness of researchers
Research problems	Often set within disciplines, can be narrow and/or methodological	Want a focus on real-world problems – surely these are interesting for research?
Inter-disciplinarity	Institutionally difficult; often organised in disciplinary units, few incentives or resources for interdisciplinary working, often less highly regarded and harder to publish. Barriers from different disciplinary world-views, assumptions, methods	A problem focus requires inter-disciplinarity. Frustration at the lack of interdisciplinary research. Critical of researchers' policy prescriptions for lacking realism. But also face problems in being inter-disciplinary themselves

1.2.5.2 Training and Development

Both scientists and policy-makers were considered generally to lack training in the skills required at the science–policy frontier. This is a long-term issue that needs several types of initiative. Training for researchers on science communication and how to make effective contributions to policy debates needs to be built into scientists' education and career development. Some research funders provide, for example, media training, but this is the sort of activity that is the first to suffer when budgets are tight.

Equally, there is room for policy-makers to draw more on scientific evidence. In some countries, this is becoming one of the skills that forms part of civil servants' annual performance evaluations; an initiative to identify and share good practices and experience in Member States would be useful. This could usefully be coordinated at EC level, through specific coordination activities of the EU-funded research.

There was enthusiasm for developing specialists who can bridge the science–policy gap. One suggestion was the idea of a post-graduate course in 'science translation and synthesis'. This would be distinct from those courses on science journalism that are currently available. This could be addressed separately by individual Member States but would benefit from a more coordinated approach at EU level. A Masters course would require a host institution to pick this up and promote it and would need backing from environmental ministries and regulators in Member States and from the EC. Achieving a critical mass may require it to be an EU-level initiative.

There was support also for the use of secondments to help both researchers to gain experience in policy settings and policy-makers to gain experience working with scientists. Both policy and research organisations need to create favourable conditions to make such secondments work. Some of the specific training needs, including secondments, might be implemented at EU level under the Marie-Curie activities.

1.2.5.3 Developing Inter-disciplinarity

Inter-disciplinarity is becoming increasingly important for addressing current environmental issues due to their complex and inter-connected nature, and the need to investigate the 'natural' components together with the 'social' (including economic factors). Research funders need to design inter-disciplinary programmes and to ensure that inter-disciplinary research is not penalised in project selection, publication and research assessment. Funders can require inter-disciplinarity in project proposals, and should help overcome the barriers to inter-disciplinarity through providing seed funding with which researchers can initiate collaborations and develop proposals.

Specific funding streams might require inter-disciplinary projects; this has been explored in various countries. Whilst one might expect research funded by ministries and regulators automatically to be problem focused and therefore inter-disciplinary, achieving inter-disciplinarity can still be a problem due to the power of disciplinary allegiances and traditions within peer-review processes.

Inter-disciplinary research teams need to share a focus on a particular problem and develop a clear idea of what their goals are. Some organisations are developing management tools to assist in such situations – 'logical frameworks' that link specific parts of the research to the functions and needs of policy organisations. This is a substantive challenge that could be developed further – research funders and policy organisations need not only to support inter-disciplinary research but also to review how effective their means of procuring inter-disciplinary research are.

1.2.5.4 'Strong Science'

The relationship between science and policy is influenced by the nature of the scientific results or the stage that a particular scientific or policy field may have reached. Many issues go through an 'attention cycle' where policy and scientific attention may initially be low, then increases due to some discovery or political event, then receives resources in terms of scientific research and policy development, and then may recede if the problem is seen to be solved, or for other political reasons, perhaps to re-emerge later.

Interviews indicated that strong science (*i.e.* a critical mass of clear evidence and consensual results from many scientists) is more readily taken up and utilised by policy-makers, while emerging science (new problems identified, tentative and/or conflicting results) may need to be strengthened and more widely disseminated before being acted upon.

The relationship between science and policy is also influenced by contextual factors such as the interests of the Minister, political imperatives and the science culture in a country. Few interviewees mentioned national guidelines for the science-into-policy process, although it was thought important to educate policy-makers about how science is conducted. As one interviewee said, 'science at the cutting edge' is not science as found in school textbooks where 'everything is factual'.

A 'policy-into-policy' model was also reported to be used in smaller countries where science systems are not yet well established in certain fields and where existing infrastructure limits the extent to which large research programmes may be undertaken. In these cases, countries adopt or adapt policies from other countries, rather than developing them from scratch.

1.2.6 Discussion

This chapter began with the proposition that environmental management decisions informed by an understanding of the relevant natural and social systems are more likely to achieve their goals than those that are not. While this may be a relatively uncontroversial proposition, the principles and practice of the use of science in environmental policy making and regulation are anything but. For example, Parsons[19] questions whether research can provide objective answers to policy questions and policy making can become a more rational process, and asserts that 'evidence based policy-making (EBPM) is a missed opportunity for improving government and has only served to make the relationship between knowledge and policy-making in a democratic society *more* muddled rather than less confused'.

Rayner[20] points out that science may raise new questions, making policy less, rather than more, tractable, and contradictory views founded on different interpretations of the science may lead to unproductive arguments whereby scientific inputs are cancelled out and political or economic interests prevail. Similarly, Owens *et al.*[21] remind us of the social and political research that has exposed the complexities of knowledge and policy processes, and the shortcomings of "the linear-rational model in which 'sound science' is straightforwardly translated into policy." They suggest that it is more helpful 'to think in terms of a continuum of influence and utility, ranging from clear and immediate impacts to long term, subtle processes in which problem definitions and modes of thinking change'.

Similarly, many of the problems around scientific decision-making remain such as the difficulty of building 'relevance' criteria into peer review, pointing to some deep structural influences in the culture and institutions of science that will take much effort, and probably a long time, to shift. Players on all sides need to engage with these questions of how best to spend the huge annual investments in science; it is not enough to leave it to the experts. One of the authors has commented in detail on these challenges in a paper entitled 'The Relevance Gap: Ten Suggestions for Policy-makers', which has been favourably circulated at a high level within the European Commission's research directorate and elsewhere.[22] We are under no illusion that the existence of some clear analysis of the problems means that we are any closer to solving them – this will take concerted efforts over a long period on the part of many people in many institutions and in many countries.

The study reported in this chapter has served to point up some of the practical difficulties in enabling science to play an effective role in informing environmental policy making and regulation, but also has pointed the way to how some of these issues may be addressed. Since its completion, Framework Programme 7 has been initiated, which puts increased emphasis on securing a better linkage between the research funded and its uptake in policy making, and on supporting the practical activities that will enable this to happen. In addition, significant progress has been made in establishing the European Research Area. A notable development here has been the creation of ERA-Nets connecting research funders across European Union member states. Many of these have undertaken work on science-into-policy processes.

Notwithstanding these various initiatives, a short review carried out in preparation for the most recent Bridging the Gap conference (held in Slovenia in May 2008) revealed some frustrations on the part of some of the participants in the Science Meets Policy workshops and Bridging the Gap conferences of the slow progress that has been made over the ten years science the workshop and conference series began. While recognising that closer links have been developed between the policy and research communities in establishing the needs for, and communicating the results of, research, the rate of progress was contrasted with that made on the collection, analysis and presentation of environmental data to support policy making and regulation. More then remains to be done.

1.2.7 Principles of Effective Science-into-Policy Practices

The study of science-into-policy practices summarized in this chapter has enabled the development of nine principles that underpin good practice. While there is a good degree of consensus around these principles, the reflections of the previous section point to the need for sustained effort in putting them into practice. The principles draw on European and UK guidelines[23] and are as follows:

1. *Opportunities for dialogue*: All of the parties involved in the science–policy interface need to create opportunities for dialogue. This helps to improve not only communication but also mutual understanding between the policy and science communities. Learning is an interactive, two-way process in which both decision-makers and scientists stand to learn from each other. Dialogue helps with aspects of knowledge sharing that are widely under-estimated in their importance: familiarity, the building of trust and informal interaction.
2. *Dissemination is not dead*: Dissemination is sometimes criticised for being based on an unrealistic and ineffective linear notion of knowledge creation, diffusion and uptake. Although a style of research in which researchers and policy-makers interact throughout emerges as a preferred model in this study, there is still an important role for dissemination. As policy processes and networks become more diffuse, open and consultative, there are more opportunities for researchers to make contributions – but this requires dissemination. The accessibility of scientific knowledge can be enhanced through various types of meetings and publications/websites, and the careful use of mass media. These things take time and need to be built into research plans. Translators of science – as well as scientists themselves – can play significant roles.
3. *Transparency and openness*: Problems arise where there is a feeling that the relationship between science and policy has been opaque and secretive. Transparency is where the workings of decision-making groups and discussions are made visible and accessible; openness is where these processes bring in a wider range of interested and affected groups than the traditional categories of 'experts' and 'policy-makers'.

   Science-into-policy can be 'opened up' through the use of working groups and committees that include both scientists and end-users. In addition, the use of the mass media and translators of science can foster a better understanding of why science is important for policy. Such openness and transparency can help to build the legitimacy of research-based interventions.

4. *Strength of evidence*: Policy-makers need to be able to persuade their constituencies that an issue is important. Therefore, they need to be presented with 'strong science' in a meaningful way, the costs of not addressing the problem, likely solutions to the problem and the budget required for implementing these solutions. Not all of these can be supplied by researchers on their own; researchers need to interact to 'road-test' the viability of any policy prescriptions they are suggesting. The need for strong science calls for processes that can bring evidence together (particularly where knowledge comes from different fields of inquiry and using different methods), as well as identifying areas of remaining uncertainty and ignorance.
5. *Relevance of research*: The social or policy relevance of research programmes and proposals needs to become a more central consideration in related decision-making. Currently, scientific criteria tend to dominate decision-making around research. No-one wants to support 'bad' science, so excellence will remain a central consideration, but relevance criteria such as significance, urgency and dissemination effort need to be actively considered alongside scientific criteria.
6. *Problem focus*: If the above principles are implemented, research that is aiming to be policy relevant will inevitably have a focus on a real-world problem. This is a necessary distinction because academic disciplines often define research problems in narrow ways, leading to research that is abstract, narrow and irrelevant to society's pressing challenges. This problem focus relates closely to the question of inter-disciplinarity, discussed next.
7. *Inter-disciplinarity*: The need for greater inter-disciplinarity has consistently been identified as a priority for creating knowledge that is valuable for policy. Yet there remain many institutional barriers to inter-disciplinarity within many research organisations and disciplines – and also policy organisations. Greater inter-disciplinarity can be facilitated through the use of Inter-Ministerial collaboration, international and national networks, shared research budgets between funders with responsibility for different parts of the disciplinary pallet, pump-priming grants and even a requirement for proposals to demonstrate inter-disciplinarity or an attempt to build inter-disciplinary capacity. As a matter of principle, research funders and policy-makers should pursue these methods and develop them further. If they fail to do so, the strength of researchers' commitment to their 'virtual colleges' (disciplines) will tend to undermine the achievement of inter-disciplinarity.
8. *Asking the right questions*: Researchers need to interact with those interested and affected parties that have a stake in the issue they are investigating. This is not merely to enhance communication, but also to identify the right questions to address in research. It is a tragedy when a completed

research project is unable to answer the 'ordinary' questions that those in policy, the public or the media want to ask. Researchers need to ask relevant constituencies 'if we are doing research in this area, what questions would you want answered?' By consulting all the relevant groups over this, researchers will build a more complete picture of the topic under investigation and the various interests involved; they will conduct better research.

9. *Staying independent*: A problem with the interactive model of research is that researchers can be accused of losing their independence. The answer to this is two-fold. First, all researchers are already affected in their research by external factors such as policy discourses and the priorities of funding sources, so it is not a question of whether they are affected but how, by whom and to what extent. It is probably healthier to acknowledge and face up to such influences on research than to try to ignore them. Second, if researchers actively interact with the full range of interested groups, it enhances their ability to 'see the whole picture', identify salient questions and extreme views, and provide independent assessments of policy options. Thus the diversity of interactions is a key to retaining and enhancing the independence so rightly valued by researchers.

References

1. T. Gieryn, Boundaries of Science, in: *Handbook of Science and Technology Studies* ed. S. Jasanoff, G. Murkk, J. Peterson and T. Pinch, Sage, London, 1995.
2. B. Wynne, Risk as globalizing 'democratic' discourse? Framing subjects and citizens, in *Science and Citizens: Globalisation and the Challenge of Engagement*, ed. M. Leach, I. Scoones and B. Wynne, Zed Books, London, 2005, pp. 66–82.
3. J. Calvert, Goodbye Blue Skies? The concept of basic research and its role in a changing funding environment, SPRU (Science and Technology Policy Research: http://europa.eu.int/comm/research/science-society/pdf/ss_ap_en.pdf), Brighton, University of Sussex: also at www.centres.ex.ac.uk/egenis/staff/calvert/publications.php (2001).
4. J. Calvert, The idea of 'basic research' in language and practice, *Minerva*, 2004, **42**, 251–268.
5. A. Scott, Relevant' social science? The Case of Global Environmental Change Research in UK Universities, Doctoral Thesis, SPRU (Science and Technology Policy Research: http://europa.eu.int/comm/research/science-society/pdf/ss_ap_en.pdf). Brighton, University of Sussex, 2004.
6. European Commission, European Governance: a white paper. COM (2001) 428 final (2001).
7. European Commission, Science and Society Action Plan (2002) http://europa.eu.int/comm/research/science-society/pdf/ss_ap_en.pdf.
8. European Commission, Communication from the Commission on the collection and use of expertise by the Commission: principles and guidelines. COM (2002) 713 final (2002).

9. R. May, The use of scientific advice in policy making: UK Chief Scientific Advisor's advice to government, Office of Science and Technology, London, 1997.
10. R. May, The use of scientific advice in policy making: Implementation of the guidelines. Second annual report by the Chief Scientific Adviser, Office of Science and Technology, London, 1999.
11. European Policy Centre, Enhancing the role of science in the decision-making of the European Union, EPC working paper no 17 (2005).
12. European Commission, Communication from the Commission on the sixth environment action programme of the European Community, Environment 2010: our future, our choice, COM(2001)0029 (2001).
13. European Commission, Five-year assessment of the European Union Research Framework Programmes 1999-2003. (http://www.euractiv.com/29/images/five_year_assessment_tcm29-135412.pdf) (2004).
14. E. Furman, P. Kivimaa, P. Kuuppo, M. Nykanen, P. Vaananen, H. Mela and P. Korpinen, Experiences in the management of research funding programmes for environmental protection, SYKE: http://www.skep-era.net/site/files/WP3_best_practice_guidelines_final.pdf (2006).
15. J. Holmes and J. Savgård, Dissemination and implementation of environmental research, Swedish Environmental Protection Agency Report 5681, February 2008; available at http://www.skep-era.net/site/files/WP4_final%20report.pdf (2008).
16. HM Government, Modernising Government, Presented to Parliament March 1999, Cm 4310 (1999).
17. Office of Science and Technology, Guidelines on Scientific Analysis in Policy Making, October 2005: http://www.dius.gov.uk/partner_organisations/office_for_science/science_in_government/strategy_and_guidance/~/media/publications/F/file 9767.
18. A. Scott, Peer review and the relevance of science, *Futures*, 2007, 39(7) 827–845, also published in the SPRU Electronic Working Paper series at www.sussex.ac.uk/spru/documents/sewp_145.pdf. September (2007).
19. W. Parsons, From muddling through to muddling up – evidence based policy making and the modernisation of British Government, Public Policy and Administration **17**(3), (2002) 43–60.
20. S. Rayner, What drives environmental policy? *Global Environ. Change*, 2006, 16, pp 4–6.
21. S. Owens, J. Petts and H. Bulkeley, Boundary work: knowledge, policy and the urban environment, *Environ. Planning C: Government Policy*, 2006, **24**(5), 633–643.
22. P. Nightingale and A. Scott, 'The Relevance Gap: Ten Suggestions for Policy-makers. Lead paper in a special edition on peer review.' *Sci. Public Policy*, 2007 (October).
23. A. Scott, Communicating sustainability research – theoretical and practical challenges, in *Innovation, Education and Communication for Sustainable Development*, ed. W. L. Filho, Peter Lang, Frankfurt, 2006, vol. 24, pp. 535–557.

CHAPTER 1.3

Research into Practice – An Organisational Learning Perspective

JUDY PAYNE

Hemdean Consulting, 308 Kidmore Road, Caversham, Reading RG4 7NG, United Kingdom

1.3.1 Introduction

There is a gap between research and practice in almost all disciplines where there are researchers and practitioners.[1] Whole books are devoted to the subject, practitioner–academic networks have been set up to try and bridge the gap, and whole organisations operate at the boundaries between academia, research, policy-making and practice. Yet the gap persists.[2]

Various authors (most of them academics working in management research) have suggested explanations for the existence of the gap. Common themes include the incentives, cultures and values of researchers – which do not encourage relevant research – and the argument that it is difficult to be good at being both rigorous and relevant. 'Rigorous' research usually means positivist science – which means researchers focus on narrow questions they can answer through careful quantitative data collection and analysis, often over periods of many years. In contrast, practitioners want tools that will help them manage their daily tasks – and they want them now. Practitioners' research questions are complex (with political, social and emotional dimensions) and are often better addressed through case studies than through quantitative research.

Water System Science and Policy Interfacing
Edited by Philippe Quevauviller

Published by the Royal Society of Chemistry, www.rsc.org

The study of knowledge and of learning has the potential to help both researchers and practitioners understand the interface between research and practice. A key theme in knowledge management is the interplay and tension between exploration (broadly defined as the creation of new knowledge) and exploitation (broadly defined as the application of existing knowledge).[3]

If we conceptualise research as exploration and practice as exploitation, an understanding of these issues and how they can be managed might help us understand how research led by one individual, group or organisation can be put into practice by another individual, group or organisation.

The chapter is organised as follows. Section 1.3.2 describes three different types of gap that exist between researchers and practitioners. Section 1.3.3 describes the tension between exploration and exploitation and relates this to the research–practitioner gaps identified in Section 1.3.2. Section 1.3.4 describes an organisational learning framework that includes the tension between exploration and exploitation, uses the framework to explain the three types of research–practitioner gap, and describes different approaches to closing them. Section 1.3.5 gives a summary and the implications for the design of research programmes.

The chapter draws on the author's experience, which has included water quality research and modelling, collaborative research management, business consulting and knowledge management research.

1.3.2 How Many Gaps?

Research has identified two gaps between research and practice: 'lost in translation' and 'lost before translation'.[4] In this section I suggest that a third gap, 'lost after translation', also exists.

1.3.2.1 Lost in Translation

'Lost in translation' frames the gap as a knowledge transfer problem that might be solved by improving the way research is translated into guidance for practitioners. In this type of gap, researchers are producing material that is relevant to the needs of practitioners – but the practitioners are either unaware of the research, don't find it useful or don't understand it. Research published only in scholarly journals, for example, is unlikely to be read by practitioners. Researchers who publish in practitioner journals are often anxious not to over-simplify their findings, which makes it difficult to present their results in ways that are accessible to practitioners – leading to a stand-off in which the researchers argue that practitioners don't deserve to understand the work without some effort on their part, and the practitioners argue that the researchers don't live in the real world and are making things too complicated.

1.3.2.2 Lost before Translation

'Lost before translation' frames the gap as a knowledge production problem that might be solved by closer collaboration between researchers and practitioners, especially in the definition and scoping of research. In this type of gap, researchers are producing accessible material, but the questions being answered are not the ones practitioners are interested in. This leads to accusations from practitioners that researchers inhabit ivory towers, don't understand real-world problems and are spending (often) public money on their own esoteric research interests. Researchers typically counter such accusations with the argument that significant scientific breakthrough is usually the result of blue skies research that doesn't pander to the short-term demands of practitioners.

1.3.2.3 Closing the 'Lost in Translation' and 'Lost before Translation' Gaps through Collaboration

Researcher and practitioner responses to the different types of gap described above can polarise the two communities, making the problems worse.

Because the two types of gap have different causes, different solutions are needed to close them – and for researchers to produce accessible answers to relevant questions, both gaps have to be closed.

Various mechanisms exist for closing the two gaps, often centred on some kind of collaboration between researchers and practitioners. Examples include academic–practitioner networks and (in the UK) the industry research associations created in the 1960s to deliver relevant research to groups of practitioners. Collaboration models vary, but typically involve a group of researchers and a group of practitioners from different organisations that co-fund the work. Some collaborations focus on making research relevant to practitioners, some on making research results accessible. Some do both. Some openly acknowledge the existence of tension between researcher and practitioner needs and seek to find a balance that is acceptable to all the participants, for example by producing both academic and practitioner outputs.

1.3.2.4 Lost after Translation

Even in the most successful collaborations, where the 'lost in translation' and 'lost after translation' gaps are closed, there is a further difficulty. Once a group of practitioners and researchers have completed some work together, the practitioners take the results back into their own organisations – where they are faced with the challenge of explaining the work to busy colleagues (who have had little involvement in the work) and persuading them to change the way they do things.

My experience of working in research associations and academic–practitioner networks suggests that at this point a third gap appears, which I will call 'lost after translation'. In this type of gap, accessible and *generally* relevant information is available. Although its relevance is understood by the individuals who were involved in the collaborative research, the research findings only become part of organisational practice if:

- the individuals involved in the collaborative research are highly skilled communicators;
- the organisations have well-developed mechanisms for spreading understanding;
- the organisations have effective mechanisms for changing working practices.

Spreading the understanding requires both resources and motivation for learning. Because the available information is of *general* relevance (*e.g.* to all organisations operating in a particular industry sector or to all people working in a particular organisational function) it is often not sufficiently contextualised to resonate with the individuals and groups that have the potential to learn from it, develop their own knowledge and put it into practice. This makes it difficult even to attract interest in the research results. It is far easier to dismiss the research ('it won't work here') and accuse the individual involved in the collaboration of wasting the organisation's time by attending unnecessary conferences and meetings.

1.3.2.5 Summary

In summary, I suggest there are three gaps that separate research from practice:

- Lost in translation – research is relevant to practitioner needs, but the results fail to reach practitioners in an accessible form.
- Lost before translation – research is not relevant (or perceived as not relevant) to practitioners.
- Lost after translation – relevant, accessible research findings fail to become embedded in organisational practice.

For relevant research to be put into practice, all three gaps need to be closed.

1.3.3 The Tension between Exploring New Knowledge and Exploiting Existing Knowledge

1.3.3.1 Exploration and Exploitation

Organisations need to explore and learn new ways of working whilst also exploiting what they have already learned.[5] In practice this is difficult, because the capabilities (skills, mindsets, environment, processes and other structures) needed for exploration are different from those needed for exploitation and the two compete for scarce resources. As a result, many organisations focus on being good at one or the other – but this too has its problems. Too much emphasis on exploration results in an organisation following every interesting idea without ever pausing to work out how to use it; too much emphasis on exploitation makes an organisation inflexible and closed to new ideas.[6]

1.3.3.2 The Tension

Organisations that try to be good at both exploration and exploitation often experience tension. There are two sources of tension.[6]

First, processes for developing the different capabilities can send conflicting messages to individuals. A common strategy for dealing with this is to separate activities so that exploration (*e.g.* research and development) happens in one place and exploitation (typically day to day operations) happens in another. This can be achieved by creating an ambidextrous organisation with highly differentiated sub-units,[7] or through partnerships between organisations so that one can focus on operations and another on developing new ideas.

A second source of tension is the need to 'unlearn' before new knowledge is accepted.[6] Knowledge becomes embedded in processes, habits and behaviours – which can be beneficial in the short term, but makes it difficult to change in response to new knowledge. Separating exploration and exploitation can solve the problem of developing and nurturing different capabilities, but can exacerbate the problem of needing to unlearn before new knowledge is accepted – because there are two sets of processes, habits and behaviours to unlearn. To resolve the 'unlearning' problem, the differentiated units or partnering organisations need somehow to be strategically integrated in a way that encourages them to collaborate with each other.

1.3.3.3 Exploration, Exploitation and the Gap between Research and Practice

The tension between exploration and exploitation relates to getting research results into practice in two ways. First, European society tends to think of research and knowledge creation as something that happens in universities and research institutes. This is an example of the separation of exploration from exploitation. This model of knowledge creation has been challenged (see, for example, ref. 8) and is the subject of much debate, revived by the recent explosion in the use of social software[9] and the advent of phenomena such as Wikipedia.

Second, the tension exists within research organisations and within organisations that can put research results into action. In practitioner organisations the need to 'unlearn' structures, processes and habits based on existing knowledge simply gets in the way of accepting research results. In research organisations, existing knowledge that needs to be 'unlearned' is embedded in the processes for selecting, managing, conducting and publishing research.

The more embedded and institutionalised the existing knowledge, the more difficult it is for an organisation to change the way it works.

1.3.4 Organisational Knowledge Creation and Learning

This chapter is concerned with getting research results into practice, and therefore with the interplay between the way in which knowledge is created and the way in which organisations learn new ways of working.

Understanding the tension between exploration and exploitation gives some conceptual insights into how we might close the gaps between research and practice. Being aware of the tension and managing it actively is a good starting point. To understand more, we need to consider how organisational knowledge creation and organisational learning happen.

There are many different theories and models of individual, collective and organisational knowledge creation and learning. The one we use has to include the tension between exploration and exploitation; help us understand the roles of different individuals, groups and organisations; provide a link between understanding and action; and include some learning processes that suggest ways in which we can close the gaps. Crossan, Lane and White's organisational learning framework[10] includes all these features.

1.3.4.1 Crossan, Lane and White's Organisational Learning Framework

Crossan, Lane and White's framework is based on five premises:[10,11]

1. Organisational learning involves a tension between assimilating new learning (exploration) and using what has already been learned (exploitation).
2. Organisational learning is multi-level – it happens in and between individuals, groups and organisations.
3. The levels of organisational learning are linked by social and psychological processes: intuiting, interpreting, integrating and institutionalising (the '4Is').
4. The processes are linked to the tension between exploration and exploitation.
5. Organisational learning involves interaction between cognition (understanding) and action.

The framework is shown in Figure 1.3.1. The 4I processes in the feed forward direction (solid lines) describe how learning happens at the individual, group and organisation levels. The dashed lines in the feedback direction describe the effect of existing learning on the learning processes.

Intuiting is the beginnings of an idea: the pre-conscious recognition of a pattern or possibilities based on personal experience. It is an individual process that can be derived from past experiences or from novel connections between the past and the future.

Interpreting is the way an individual explains the idea to themselves and to others. It is a conscious, social process that creates shared understanding using

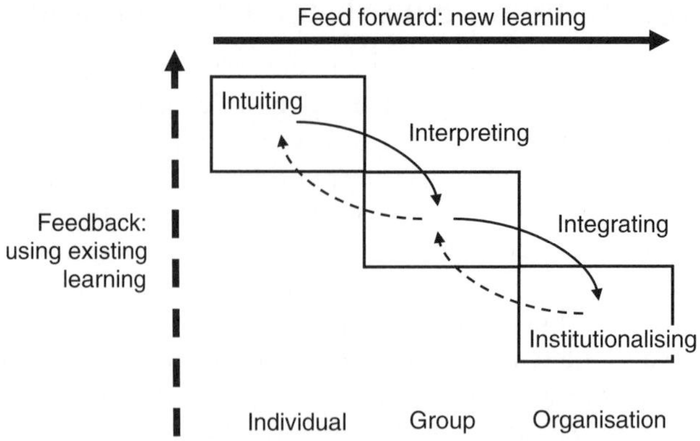

Figure 1.3.1 Crossan, Lane and White's organisational learning framework.[10]

language and images. The process both influences and is influenced by the knowledge and dynamics of the group in which it takes place. A high-trust group of specialists in one discipline, for example, will interpret an idea differently from a group of specialists in another discipline who are meeting for the first time. As the idea takes shape and is understood and accepted by the group, the interpreting process merges into the integrating process.

In the process of *integrating*, the focus shifts from individual and small group understanding to wider understanding and collective, co-ordinated action. Through dialogue and joint action, understanding is shared, developed and adjusted. Like the interpreting process, the integrating process is highly context-specific: what works in one organisation will not necessarily work in another. Once co-ordinated action starts to recur, shared understanding exists and can be institutionalised.

Institutionalising is the process of embedding the learning of individuals and groups in organisational systems, processes and practices. Whereas intuiting, interpreting and integrating are fairly fluid, informal, continuous processes, institutionalising is usually formal and punctuated. Most organisations have a process for assessing a new practice before formalising it in systems and structures.

The whole organisational learning process is a dynamic one that creates tensions between new learning and using what has already been learned. Two particularly difficult parts of the overall process are the move from interpreting to integrating (in the feed forward direction) and the tension between institutionalising and intuiting (in the feedback direction). The move from interpreting to integrating is difficult because it involves changing minds. Institutionalising can prevent intuiting because new ideas challenge existing organisational beliefs and behaviours: individuals have to think beyond accepted ways of working and experiment with new ideas.

1.3.4.2 Using the Organisational Learning Framework to Explain the Research–Practice Gaps

The organisational learning framework was originally developed to describe learning in a single organisation. In this sense, institutionalising is the key organisational learning process: it can only take place within a single organisation. But ideas and learning might begin outside the organisation, or in a differentiated sub-unit that is culturally separate from the practitioner organisation (as described in Section 1.3.3.2). In these cases, the language and mindset barriers that need to be overcome for interpreting and integrating to be effective are likely to be significant.

Figure 1.3.2 describes a simple case in which researchers, policy makers and practitioners are all members of the same organisation (and therefore have some shared language and knowledge to start with). Researchers in Figure 1.3.2 are individuals who develop their ideas through action and dialogue with a group of policy makers. The policy makers develop the ideas further through action and dialogue with the wider organisation, until learning that is considered valuable becomes institutionalised and part of routine practice.

Figure 1.3.3 is an example of a 'lost in translation' gap. The learning process of integrating is missing. Learning is effective at the policy maker level (although learning processes could break down before this), but without integration it isn't reaching the wider organisation – so there is no institutionalising and no change to practices.

Figure 1.3.4 is an example of a 'lost before translation' gap. In this example, both feedback processes (from practitioners to policy makers and from policy makers to researchers) are missing. Learning is effective at all levels, but without feedback processes the learning isn't sufficiently relevant and valuable to be selected for institutionalising. Again there is no change to practices.

Figure 1.3.5 is an example of a 'lost after translation' gap. Relevant, valuable learning is effective at all levels, but it is not institutionalised, so its effect on

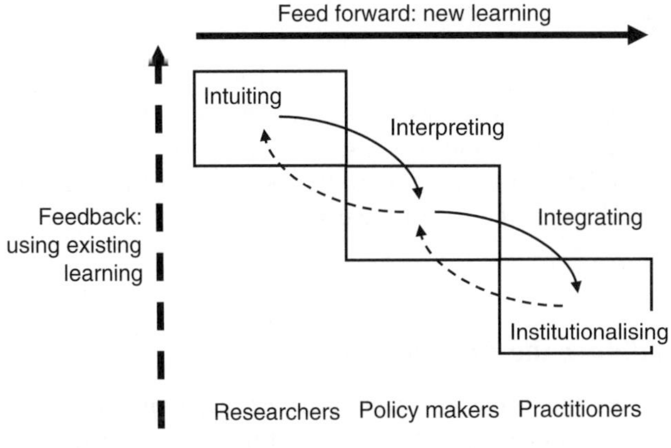

Figure 1.3.2 From research to practice: an organisational learning perspective.

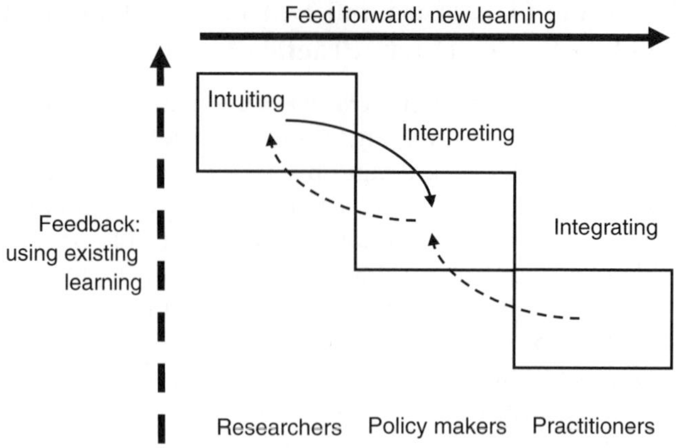

Figure 1.3.3 Lost in translation.

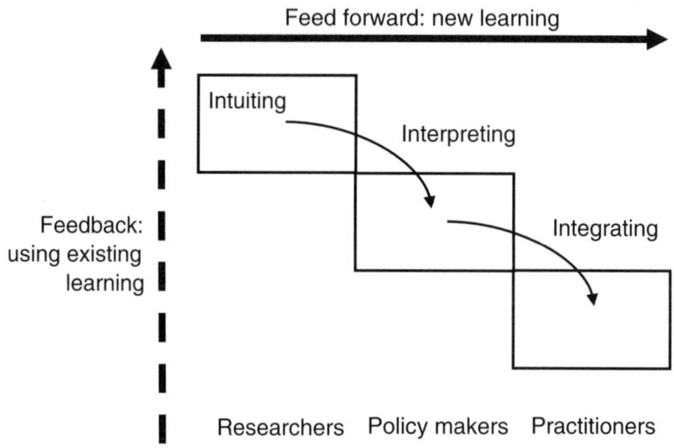

Figure 1.3.4 Lost before translation.

practice is likely to be limited. This could be because the organisation doesn't have effective mechanisms for assessing new ways of working, or for incorporating them into systems and processes. It could also be because practitioners choose not to follow new systems and processes – although this points to weaknesses elsewhere in the overall learning process.

1.3.4.3 Enabling the 4I Processes

Having explained the three research–practice gaps in terms of missing or ineffective learning processes, what can be done to enable the processes and make them more effective?

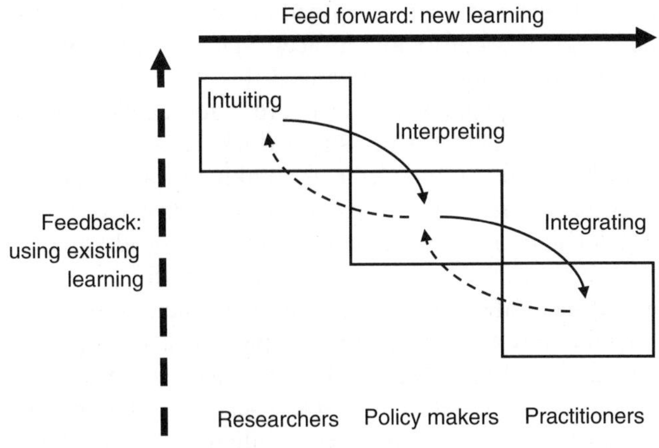

Figure 1.3.5 Lost after translation.

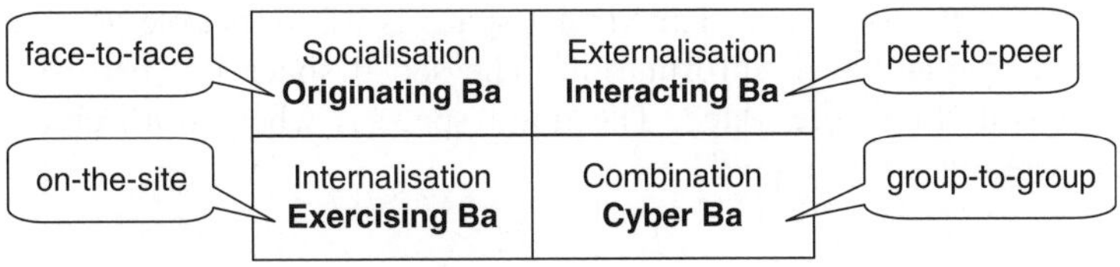

Figure 1.3.6 Four types of ba.[15]

Organisations cannot manage knowledge and learning directly, but can provide environments, conditions and tools to encourage and support learning and knowledge sharing.[12] This section explains the principles of providing 'spaces', conditions and tools for learning, then considers the difficult transition between the 4I processes of interpreting and integrating.

1.3.4.3.1 Enabling by Creating 'Spaces'

Figure 1.3.6 summarises the Japanese concept of 'ba' in relation to knowledge creation. Ba is a shared space that can be physical (such as an office), virtual (such as email or a shared team website) or mental (such as shared experiences and values). The core concept is that different types of ba encourage different knowledge creation processes. The underlying knowledge creation model in Figure 1.3.6 includes four processes: socialisation, externalisation, internalisation and combination (SECI).[13] The four types of ba support the SECI processes.[14]

The 4I processes correspond roughly to the SECI processes. Interpreting includes elements of socialisation and externalisation; integrating includes elements of externalisation and combination; and institutionalising includes elements of combination and internalisation.

Originating ba is the world where individuals share feelings, experiences and mental models. This is traditionally a face-to-face process, but the advent of social networking tools might have changed this.

Interacting ba is the place where ideas are made explicit through dialogue and reflection, typically between peers in a work team.

Cyber ba is where explicit knowledge is combined, catalogued and sorted. It is called cyber ba because it is typically a virtual space, including shared databases and documents. The term might now be misleading, as virtual spaces are no longer used exclusively for managing information.

Exercising ba facilitates the conversion of explicit into tacit knowledge. Training, coaching and mentoring are examples of exercising ba.

Inspired by the concept of ba, Raj Datta[15] has described the environment created by Mindtree, a global IT and R&D services company. The Mindtree environment has four spaces – designed to align people with the company's core values and to enhance sense-making. The spaces are part of an overall organisational 'ecosystem' that supports the whole innovation process, from ideas to practical implementation. The physical space is an inspiring workplace in close proximity to nature. The virtual space is where people go to connect with each other and seek information. The social space is where trust- and community-building take place. The mind space is where both creative and rational thinking are nurtured.

1.3.4.3.2 Enabling by Creating Supporting Conditions

If designing 'spaces' doesn't resonate with an organisation, the work environment can be described in terms of conditions that support learning and knowledge sharing.

In the current era of innovation, the key organisational capability is *collaboration* – needed to drive the knowledge creation and transfer that are at the heart of the innovation process. Developing the ability to collaborate effectively is difficult, because collaboration is voluntary. Collaboration and innovation can not be managed hierarchically, so the whole process is 'organisationally counter-intuitive'. Innovative, collaborative organisations are designed on very different principles from traditional organisations. People need time to explore ideas, make new connections and develop the trusting relationships in which the interpreting and integrating processes are most effective. The new principles of organisational design include self-management rather than hierarchical management; behavioural etiquette rather than pre-set goals and methods; and shared responsibility for strategy. Organisations need to invest in creating the conditions for collaboration to take place.[16]

Further guidance is given by Nonaka and Takeuchi, who developed the SECI model. They describe five enabling conditions for effective knowledge creation:[13]

1. Intention – an organisation's aspiration to its goals, usually in the form of a strategy for managing knowledge. This includes a vision about the

kind of knowledge that needs to be developed and a strategy for implementation.

2. Autonomy – appropriate freedom for individuals and groups to define their own work to pursue organisational goals. Autonomy also includes the concept of multi-disciplinary teams to integrate the knowledge and experience of 'ordinary people'.
3. Fluctuation and creative chaos – to stimulate interaction between an organisation and its external environment. Fluctuation involves breaking down habitual, comfortable thinking by individuals continuously questioning and reconsidering their beliefs about how and why things work. Creative chaos is triggered by a crisis or by fluctuation itself. Creative chaos induces and strengthens the commitment of individuals.
4. Redundancy – the existence of information that goes beyond immediate operational needs. This encourages knowledge creation 'at the edge' of individuals' job roles, and creates links between different parts of an organisation that give people access to new and different perspectives.
5. Requisite variety – internal diversity to match the variety and complexity of the organisation's environment. This can be enhanced by making sure everyone has fast and equal access to a wide variety of information.

1.3.4.3.3 Enabling by Giving People Access to Tools and Techniques

There are many tools and techniques that can enhance learning processes – including facilitation, concept mapping, creativity techniques and storytelling. These are covered in great detail in the learning, knowledge management and innovation literature. Perhaps the most powerful set of tools to emerge in the last few years is social software. Social networking websites, wikis, blogs and social networking services allow people to connect and collaborate from the bottom up, without imposed top-down control – and can therefore provide the spaces and conditions needed for learning and knowledge sharing in one hit.[12]

1.3.4.3.4 Enabling the Interpreting and Integrating Processes

Interpreting is a social, often informal and *ad hoc* process, strongly influenced by the individuals involved. It follows that interpretations can be changed by bringing new individuals into workgroups – and there is empirical evidence to support this.[11] New people bring new perspectives. The downside of this is the need to build new trusting relationships with new individuals. Without such trust, individuals are less likely to share their emerging ideas.

For existing employees, organisations can provide opportunities that encourage individuals to change or broaden their perspectives. Examples include employee rotation programmes, membership of external groups and other experiential learning activities.

The move from interpreting to integrating is challenging. It is difficult to explain a new idea at the best of times, more so to people with existing knowledge and habits. Because integration is focused on co-ordinated joint action Crossan, Lane and White[10] suggest that the key to success at this stage is probably experiential learning, which in turn suggests consideration of different individuals' learning styles (see, for example, refs 17 and 18).

1.3.4.4 Learning from Research across Organisational Boundaries

Learning processes can be problematic even in the simple case where researchers, policy makers and practitioners are all members of the same organisation – and therefore have some shared context, language and knowledge.

In practice researchers, policy makers and practitioners are often in different organisations. In addition, there might also be other groups, such as research managers, who act as co-ordinators in collaborative research programmes. Each group has its own knowledge, language[i] and dynamics and therefore its own unique interpreting and integrating abilities. What is learned through the interpreting and integrating processes in one group cannot be magically transported to another: the second group has to have its own learning experience and make the knowledge its own.

Documents such as stories, pictures and manuals can all help knowledge to flow from one group to another, but deep 're-experience' often happens only through dialogue between groups.[13]

This has several implications for the way collaborative research programmes are designed. For practitioners to re-experience researchers' learning, practitioners and researchers need to talk with each other directly. Designing research programmes so that research managers and policy makers come between researchers and practitioners multiplies the number of potential gaps. Instead, research managers and policy makers should become facilitators of the dialogue between researchers and practitioners. Their role is to create the right spaces and conditions for effective interpreting and integrating (Figure 1.3.7). This can include face-to-face meetings – whose purpose should be social, for trust- and community-building to build interpreting and integrating capabilities. Without the right spaces and conditions, researchers and practitioners are likely to adopt the stand-off positions described in Sections 1.3.2.1, 1.3.2.2 and 1.3.2.4. As explained in Section 1.3.4.3.3, Web 2.0 applications can provide suitable spaces and conditions for ongoing dialogue between researchers and practitioners from different organisations.

At the Environment Agency in England and Wales, Stuart Kirk advocates the use of 'science translators' to help bridge the gaps between researchers and practitioners. These people help the Environment Agency to identify and

[i] 'Language' here refers mainly to the degree of specialisation in the words and other symbols understood within a group. Different national and regional languages add a further hurdle for learning processes.

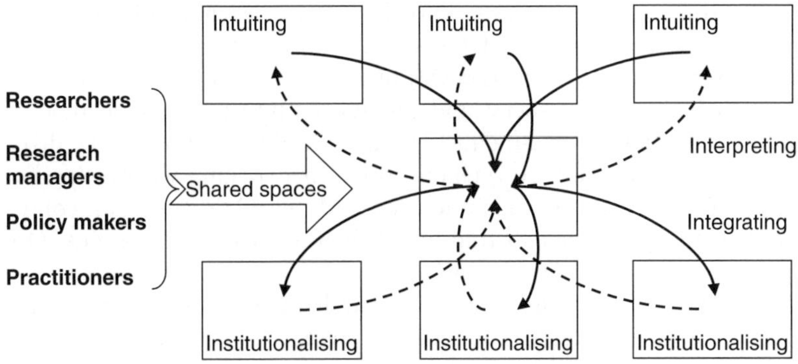

Figure 1.3.7 Shared spaces for dialogue between researchers and practitioners.

express its needs, to seek solutions from the external research community, and to tailor research results to meet the needs of practitioners. Science translators must be knowledgeable in their scientific discipline, but beyond this Stuart has identified communication (particularly listening) skills as their key capability. Many effective science translators have experience of working in policy, operational and consultancy roles – which gives them direct experience of these environments or a strong focus on customer needs and general problem-solving. This is a good example of skilled communicators acting as facilitators of the dialogue between researchers and practitioners. Stuart has also recognised that effective science translators need time to focus on their role, which means organisational investment and commitment that is often hard to secure when resources are stretched. This illustrates some of the difficulties organisations encounter in 'unlearning' existing habits so they can provide conditions that support collaboration and learning (Section 1.3.4.3.2).

1.3.5 Summary and Implications

This chapter began with the proposition that there are three types of gap between research and practice: lost in translation, lost before translation and lost after translation. All three types of gap have to be closed if relevant research is to be put into effective practice.

Using knowledge management and organisational learning principles helps us understand how to close the different types of gap. People and organisations need to 'unlearn' existing processes, habits and behaviours before they can accept new knowledge. This applies equally to researchers and practitioners. Practitioners' existing processes and behaviours get in the way of understanding and accepting research; researchers' existing processes and behaviours get in the way of understanding and accepting operational practices; existing collaboration models get in the way of new methods for bringing researchers and practitioners closer together.

Organisations cannot manage knowledge and learning directly, but can provide spaces, conditions and tools that support learning and knowledge sharing – particularly the 4I processes of interpreting and integrating. Spaces can be physical, virtual, social and mental. Conditions include time, trust, autonomy, fluctuation, redundancy in information and requisite diversity. Tools include facilitation, storytelling, social networking websites and collaborative software such as wikis and blogs. Many of these spaces, conditions and tools seem quite alien to traditional, hierarchical organisations. Organisations need to invest in developing them.

All of this applies to universities, research institutes, researchers, policy-makers, research managers, collaborative research programmes, operations managers and practitioners. Learning across organisational boundaries is difficult. Documents can help knowledge to flow from one group of people to another, but deep 're-experience' often happens only through dialogue. If cross-organisational learning processes are to be effective, researchers, practitioners and people in policy and management roles need the right spaces, conditions and tools for talking to each other.

Spaces, conditions and tools enable learning processes; effective learning processes close the gaps between research and practice. Making sure this happens is everyone's responsibility.

References

1. S. L. Rynes, J. M. Bartunek and R. L. Daft, *Acad. Manage. J.*, 2001, **44**, 2.
2. C. Markides, *Acad. Manage. J.*, 2007, **50**, 4.
3. K. Gupta, K. G. Smith and C. E. Shalley, *Acad. Manage. J.*, 2006, **49**, 4.
4. D. L. Shapiro, B. L. Kirkman and H. G. Courtney, *Acad. Manage. J.*, 2007, **50**, 2.
5. J. G. March, *Organization Sci.*, 1991, **2**, 1.
6. J. McKenzie and C. van Winkelen, *Understanding the Knowledgeable Organisation*, Thomson, London, 2004.
7. M. J. Benner and M. L. Tushman, *Acad. Manage. Rev.*, 2003, **28**, 2.
8. M. Gibbons, C. Limoges, H. Nowotny, S. Schwartzman, P. Scott and M. Trow, *The New Production of Knowledge – The Dynamics of Science and Research in Contemporary Societies*, Sage Publications, London, 1994.
9. N. Cook, *Enterprise 2.0 – How Social Software will Change the Future of Work*, Gower, Aldershot, UK, 2008.
10. M. M. Crossan, H. W. Lane and R. E. White, *Acad. Manage. Rev.*, 1999, **24**, 3.
11. M. M. Crossan and I. Berdrow, *Strategic Manage. J.*, 2003, **24**, 11.
12. J. Payne, *KM Rev.*, 2007, **10**, 5.
13. I. Nonaka and H. Takeuchi, *The Knowledge Creating Company*, Oxford University Press, New York, 1995.
14. I. Nonaka and N. Konno, *California Manage. Rev.*, 1998, **40**, 3.
15. R. Datta, *KM Rev.*, 2008, **11**, 2.

16. E. Miles, C. C. Snow and G Miles, *Long Range Planning*, 2000, **33**(3), 30.
17. P. Honey and A. Mumford, *Using your Learning Styles*, Peter Honey Publications, Berkshire, UK, 1983.
18. D. A. Kolb, *Experiential Learning: Experience as the Source of Learning and Development*, Prentice Hall, New Jersey, 1984.

CHAPTER 1.4

*General Features of the EU Water Policy and Related Scientific Framework**

PHILIPPE QUEVAUVILLER, CHRISTOS FRAGAKIS AND PANAGIOTIS BALABANIS

European Commission, DG Research, rue de la Loi 200, 1049 Brussels, Belgium

1.4.1 Introduction

Policy orientations rely on scientific evidence and the efficient use of science represents an increasing challenge for the scientific and policy-making community, the private sector, NGOs, citizen's associations and professional organisations. The need to improve the role that science plays in environmental policy-making has been highlighted in a number of 'science-meets-policy' events since 1998[1] (Chapter 1.2). In particular, the need to ensure better linkages between policy needs and research programmes has been underlined, with enhanced coordination regarding programme planning, project selection and management, and mechanisms for knowledge transfer to ensure that outputs from research programmes really contribute to policy development, implementation and review. This issue has been discussed in depth in the water sector at European Union level for more than five years,[2] underlining the need to develop a conceptual framework for a science–policy interface related to water, which would gather various initiatives and knowledge.

*The views expressed in this chapter are purely those of the authors and may not in any circumstances be regarded as stating a formal position of the European Commission.

Water System Science and Policy Interfacing
Edited by Philippe Quevauviller

Published by the Royal Society of Chemistry, www.rsc.org

Possible scenarios regarding integrated environmental science & policy interface linked to the implementation of the EU Water Framework Directive[3] have been recently proposed[4,5] and are discussed in depth in the present book. This chapter provides insight into the EU water policy and scientific framework, on the basis of which current initiatives are being developed.

1.4.2 General Features of the EU Water Policy Framework

According to the 6th EU Environment Action Programme (6th EAP),[6] environment policies should be based on the best scientific evidence and its priorities should figure prominently in the Community RTD (Research and Technological Development) programmes (Section 1.4.3). The complexity of environmental problems and their related uncertainties that we are currently facing makes the science approach even more necessary, in particular when designing new policies with strong socio-economic consequences. In this respect, research efforts have been devoted at EU level since 2002 in support of the Water Framework Directive (WFD) and its related milestones.[4,5] This Directive establishes environmental objectives of 'good status' for all waters to be reached by the end of 2015, which are based on the effective implementation of clear milestones (*e.g.* characterisation, monitoring, river basin management plan, programme of measures), each representing technical challenges requiring scientific knowledge and exchange of expertise and best practices. Owing to the complexity of the technical challenges, a forum has been established since 2001 called the 'Common Implementation Strategy (CIS)' which aims to develop a common understanding and approaches, elaborate informal technical guidance, including best practice examples, share experiences and resources, avoid duplication of efforts, and limit the risk of bad implementation of the directive.[7] The WFD provides a well-established policy basis and a stable platform that enables building up communication and best practice exchanges among different players (policy implementers, technology providers, scientific community, industrial stakeholders, NGOs, *etc.*). As described in other chapters of this book, this has resulted in clear improvements during the period 2003–2008, with plans for developing joint initiatives (involving EC Research and Environment General Directorates, scientific actors and Member States through the WFD-CIS) aimed at a better integration of scientific inputs and progress into the water policy implementation process. An in-depth description of the WFD and its related challenges is available elsewhere.[8]

1.4.3 Water in the EU R&D Framework

1.4.3.1 EU RTD Framework Programme

Research on water has been supported by the European Commission in the early days of the Framework Programmes (FP) for Research and

Technological Development (RTD).[9] The basic aim of the FP is to foster scientific excellence, competitiveness and innovation through the promotion of better co-operation and coordination within the European Union, and to produce advances in knowledge and understanding while supporting the implementation of related European policies. Different types of research and development (R&D) activities have been supported over the years:

- Research contributing to medium to long-term policy objectives, *e.g.* developing scientific knowledge on hydrology and climate processes, ecological impact of global change, soil functioning and water quality, integrated management strategies and mitigation technologies, scenarios of water demand and availability, *etc.*
- Research with a clear policy focus funded under the 6th Framework Programme (Scientific Support to Policy) to underpin the formulation and implementation of Community Policies, accommodating specific research needs identified by the policy sector, including water policies.
- Research coordinated by national research programmes (through the so-called ERA-NET scheme), in which funding organisations like ministries and research councils develop networks of national or regional research programmes or innovation programmes in sectors of their choice. This mechanism represents a very valuable mechanism to regroup national funds and to increase the access to scientific expertise available at regional or national level as well as cross border cooperation at the levels of programmes and of projects.

The Seventh Framework Programme (FP7) (2007–2013) covers priority areas reflecting EU research needs in sectors such as health, food and agriculture, information and communication technologies, nanosciences, energy, transport, socio-economic sciences, space and security. Environment and climate change is one of these ten priorities. It focuses on knowledge on the interactions between the biosphere, ecosystems and human activities, and the development of new technologies, tools and services, with emphasis on:

- improved understanding and prediction of climate, earth and ocean systems changes;
- tools for monitoring, prevention and mitigation of environmental pressures and risks;
- management and conservation of natural resources.

Within FP7, the research areas address pressures on environment and climate, impacts and feedback, environment and health, conservation and sustainable management of natural resources (including groundwater), evolution of marine environments, environmental technologies, understanding and prevention of natural hazards, forecasting methods and assessment tools and earth observation. The overall Environment (including climate change) theme has a budget of €1890 million for the period 2007–2013 (out of a total budget of €50 521 million).

1.4.3.2 Research carried out by the Joint Research Centre (JRC)

EC funded research has also been developed through the Joint Research Centre' Multi-annual Work Programme (MAWP) embedded into the EU RTD Framework Programme. Several key actions of the MAWP are directly relevant to water policies, *e.g.* actions related to chemicals in the aquatic environment, ecological water quality, marine ecosystems, water quality information system, European Measurement Infrastructure, and policies and human resources for research.

1.4.3.3 Demonstration Projects

The European Commission also funds demonstration projects in the framework of the LIFE programme – the Financial Instrument for the Environment – which is an EC DG Environment's financial mechanism specifically aimed at assisting the development of environmental policy through its co-financing of demonstration projects. More than 1200 projects related to water policies have been co-funded by LIFE since 1992. This funding instrument represents in principle a natural continuation of research projects, aiming at demonstrating the applicability of innovative methods, solutions and techniques in real environmental cases.[4,5]

1.4.4 Identifying Research Needs in the Water Sector

The identification of research needs in the water sector and its related timing (short-, medium- and long-term) are intimately linked to the way research instruments are being operated. In principle, technical policy milestones (*e.g.* monitoring, programmes of measures, environmental impact assessment) are fed by advances in scientific knowledge while being of course driven by the Directive's agenda. In this respect, research needs in support of water policy may be defined according to timing considerations:

- *On a short-term basis* (~1–2 years), the policy timing is often not adapted to develop new types of research activities (unless very specific needs are identified, which may be sorted out in a 6–12 month period). The key issue, rather, is how to get efficient and user-friendly access to background scientific information required, for example, at the stage of design of a new policy. In other words, specific 'urgent' research needs can hardly be tackled through a FP call for proposals, considering that any project – if selected – would only start the year after. In this case, the only possible way to develop *ad hoc* research on a short term would be either through JRC action lines (Section 1.4.3.2) or national research programmes (*e.g.* through the ERA-NET scheme – see Section 1.4.3.1).

- *On a medium-term basis* (~2–5 years) research may contribute to policies with well-defined milestones. In this context it is indeed possible to anticipate research needs and deliver scientific outputs in time for use by policy-makes. It should be stressed, however, that policy-makers often report difficulties in identifying their research needs over the medium-term,[1] which is precisely the timescale over which much academic research is conducted. This is often due to short-term pressures, turnover of staff and changing political priorities.
- *On the long-term basis* (~5–10 years) scientific progress may either support policy milestones that are clearly identified at the ten-year horizon or the review process of given policies. In the case of the WFD, long-term research needs are typically linked to the research and technological challenges of the development of the successive river basin management plans and to the review process of the technical annexes of the directive and its 'daughter' directives (*e.g.* Groundwater Directive 2006/118/EC).

1.4.5 Examples of Water-related Research Projects and Initiatives

The sub-sections below provide some examples of current RTD projects that are directly or indirectly supporting water policies. These are only selected examples aiming to illustrate the type of FP contributions. The list is obviously not exhaustive, and the reader is invited to consult the literature to get an insight into other types of projects.[8,10]

1.4.5.1 Research on the Knowledge of Groundwater and Dependent Ecosystems

The GENESIS project (Groundwater and Dependent Ecosystems: An Integrated Research Project to Support GW Systems Management) project has been selected for funding under the 7th Framework Programme to establish the scientific basis needed to address the challenges of the new Groundwater Directive and 'prepare the ground' for the review of the latter that is foreseen for 2013. This large Integrated Research Project, which consists of 25 partners from 13 EU states and 4 Associated states, aims to integrate pre-existing and new scientific knowledge into new methods, concepts and tools for better management and monitoring of groundwater (GW), to improve the understanding of how different drivers affect GW, to incorporate new process understanding, including the interactions between GW, surface water and related dependent ecosystems into mathematical models and assessment tools, to simulate scenarios to assess impacts of external pressures in an integrated way taking account of uncertainties, to address also socio-economic issues and legal aspects, and promote effective dissemination and knowledge uptake to relevant stakeholders. The research outcomes from this project are to be tested on carefully selected case study aquifers covering important geological formations and aquifers, different climatic regions, different ecosystems (wetlands, springs, streams, lagoons) and different levels of economical development and drivers for environmental change and they should assist Member States in their formulation, implementation and assessment of cost-efficient packages of measures whilst guaranteeing legitimate uses of groundwater and protection of connected surface water and groundwater dependent terrestrial ecosystems.

1.4.5.2 Research on Catchment Modelling

The development of integrated water resources management policies at river scale requires the development of appropriate modelling tools that are able not only to integrate the different physical, chemical, biological and ecological domains but can also integrate the various socio-economic drivers that affect the water systems and the way people are involved or interested in the management of a water problem. In addition, these tools have also to be able to analyse, compare and assesses the impact of various management plans and policies in an integrated way.

To respond to this challenge, the European Commission has initiated, in the context of the 5th EU Framework Programme for Research and Technological Development (FP5), the CatchMod initiative, with a view of developing and harmonizing integrated modelling tools that should be used by water authorities for planning and managing water resources in an integrated way at the scale of river basin. This very ambitious and long-term objective initiative has been implemented through a dedicated cluster constituted by a core group of several large research projects and coordinated by a concerted action, the Harmoni-CA concerted action, which also has the task of coordinating other related EU-funded projects and national research activities.

Besides major outcomes identified in terms of scientific and technological impacts, this initiative has also proven to be successful in providing some important policy relevant results, *e.g.*:

- advice and criteria (socio-economic, bio-geo-chemical and systems analytical criteria) on the selection, use and evaluation of models to aid the implementation of the Water Framework Directive (WFD) has been provided;
- a European Open Modelling Interface and Environment (OpenMI) that can simplify and formalize the linking of models and hence allow catchment managers to explore the likely outcomes of different policies in a more time and cost effective manner (Chapter 2.5);
- generic, scientifically based quality assurance guidelines for modelling studies in an ontological knowledge base with a view to improving the

quality of model-based river basin management and enhancing the confidence of all stakeholders in the use of models;
- harmonised techniques and representative river basin data for assessment and use of uncertainty information in integrated water management;
- guidance for an appropriate choice of quantification tools to satisfy existing European requirements on harmonisation, reliability and transparency for quantifying diffuse nutrient losses, in particular from agricultural land, to surface freshwater systems and coastal waters, which can facilitate the implementation of the Water Framework and the Nitrates Directives;
- water quality models that can be used to simulate the responses of lakes to future as well as past changes in the weather and develop methods to quantify sensitivity of the lakes to local, regional and global-scale changes in the weather;
- incorporation of the special dynamics of ephemeral and temporary waters in existing in-stream water quality models.

Science–policy integration is another important challenge that has been addressed by the CatchMod initiative and more particularly by the Harmoni-CA concerted action. By initiating and facilitating the discussion between the Common Implementation Strategy Coordination Group of the WFD and the various researchers, Harmoni-CA helped to foster synergies between researchers, policy makers and water mangers on different levels to achieve a common understanding of the various WFD implementation issues.

1.4.5.3 Research on Climate Change Impacts on Water

One of the flagship EU-funded projects in this domain is the Euro-limpacs project (Integrated Project to Evaluate the Impacts of Global Change on European Freshwater Ecosystems) project. It is an Integrated Project, funded under FP6, that aims to evaluate the Impacts of Global Change on European Freshwater Ecosystems. The Euro-limpacs consortium, consisting of 37 partners from 18 countries, adopted an holistic approach, bringing together and for the first time the three major ecosystem components of surface freshwater systems (rivers, lakes and wetlands) within an integrated catchment context covering a full range of spatial and temporal scales and combining observational, experimental, palaeoecological and modelling methodologies. The project considered the key changes to the climate system that are expected to occur in Europe over the next 100 years in addressing:

- the impact of climate change on the physical characteristics of freshwater ecosystems, distinguishing between change that is already being observed and changes that are anticipated in future;
- the ecological consequences of climate change with respect to lakes, rivers and wetlands across Europe, dividing the continent between cold ecoregions (high latitude and altitude), temperate and warm-humid ecoregions and warm-arid regions (principally the Mediterranean region);

- adaptation measures that could be adopted to mitigate the adverse consequences of climate change projected as a result of increasing water temperature, changes in hydrology and hydromorphology, including interactions with problems of eutrophication, acidification and toxic substance contamination;
- implications for policy, especially the implementation of the EU Water Framework and Habitat Directives and the development of cross-sectoral policies for water resources, agriculture, energy, economic development and atmospheric emissions where water quality issues are affected.

Due consideration and emphasis was given to the uncertainties involved in the projections, including those that are not only due to limitations in our understanding of freshwater ecosystems but also due to the inherent difficulty in predicting how the key drivers affecting water quality and freshwater biodiversity, especially those related to agricultural policy, will change in the future.

Another key project concerning climate change impacts on water is the WATCH project ('Water and Global Change'). This 6th Framework Programme Integrated project aims to unite researchers to evaluate the global water cycle's response to current and future drivers of climate change (see www.eu-watch.org). The project brings together 25 leading research centres in climate change throughout Europe to forge new research collaboration projects linking climate and hydrological science, clarify the vulnerability of global water resources, evaluate the response of extreme environments, enhance modelling frameworks to assess water resources and establish links to the international community outside Europe. Besides the on-going research developments, a closer collaboration with policy networks is being sought through contacts taken with the Steering Group on Climate Change Adaptation of the WFD Common Implementation Strategy. In particular, recommendations from the project will be formulated in a section of research perspectives of a policy guidance document on climate change adaptation addressed to the EU Member States.

Finally, the FLOODsite project (also funded under FP6) has developed an integrated European methodology for flood risk analysis and management for rivers, estuaries and coastal environments, as well as techniques and knowledge to support integrated flood risk management (*e.g.* sustainable 'preflood' measures, flood event management, post-event activities) and frameworks for decision support for long-term and flood event risk management. The project was completed in spring 2009. It included a clear science–policy component with close links being established with environmental agencies/ministries (at national level) and with the CIS Working Group on Floods (linked to the implementation of the Flood Directive 2007/60/EC). In this respect, the project made all the necessary efforts to translate research outputs into usable information by policy-makers (*see* www.floodsite.net for details).

1.4.5.4 Technological Platforms

The Commission's Seventh Framework Programme for Research (2007–2013) is designed to play a key role in achieving the renewed Lisbon objectives. European Technology Platforms (ETPs) are playing a major part in this context. ETPs provide a framework for stakeholders, led by industry, to contribute to the identification of research and development priorities, timeframes and action plans related to what are perceived as strategically important issues for the EU. They are also contributing to the development of European research policy, in particular in advising the FP7 to better meet the needs of industry and contribute to the restructuring of traditional industrial sectors.

The Water Supply and Sanitation Technology Platform (WSSTP) addresses the main challenges that European water resources are facing, *i.e.* increasing severity of water stress and shortage of water, urbanisation, lack of significant infrastructure, especially water and wastewater services in rural and underdeveloped countries, with a view to increasing the competitiveness and sustainability perspectives of the European water sector and contributing to the relevant Millennium Development Goals.

The platform was launched in June 2004, bringing together all stakeholders involved in water supply and sanitation in Europe, and has produced already its Vision Document, Strategic Research Agenda (SRA) and Implementation Strategy Plan. The SRA addresses a broad range of research areas, from monitoring, sensor and communication technology, advanced treatment technology, technologies to produce energy and products from wastewater, water saving technology and pollution control at source, asset management to risk assessment and management tools and integrated management solutions.

WSSTP proposes to implement the SRA trough six pilot themes, each addressing a major water challenge in Europe. These pilot themes are:

- mitigation of water stress in coastal zones;
- sustainable water management inside and around urban areas;
- sustainable water management for agriculture;
- sustainable water management for industry;
- restoration of degraded water zones (groundwater and surface water);
- adaptation and mitigation of extreme hydroclimatic events.

The Strategic Research Agenda of the WSSTP is relevant to the Environment (including climate change) Theme of the FP7 Cooperation Programme. In fact the above-mentioned challenges identified in the vision document of the platform are fundamental drivers for action identified in that theme. Therefore, several water technology topics identified in the first annual work programmes (2007, 2008 and 2009) of the Environmental Technology activity of the Environment (including climate change) theme are linked to the SRA of the WSSTP and are expected to contribute in its implementation of intent to support the generic research and the enabling technologies development of the six pilot themes of the Strategic Research Agenda, leaving the full-scale demonstration activities of the pilots to the various members of the WSSTP. Our work programme for 2007 and 2008 then identified specific topics relevant to implementation of the Strategic Research Agenda.

1.4.5.5 Science–Policy Interfacing

As already mentioned, the dissemination and use of research results to various policy-makers is a challenging issue in water management. The Harmoni-CA concerted action addressed this challenge and laid the foundations for building a research–policy interface in the water field. It also helped – through the development of an interactive and powerful web portal, *i.e.* the WISE-RTD portal (Chapter 2.4) – to consolidate knowledge generated by several EU funded research projects and facilitated the take-up of research results in the practice of river basin management.

This SPI-Water Project[11] continued this effort further by evaluating the relevance and usefulness of the results of various EU research projects, as well as projects funded under other EU funding instruments, like LIFE and INTERREG and by selecting information with respect to European water policies, the policy implementation tasks and the related implementation milestones of the WFD. It also strengthens the operational capability of the WISE-RTD web portal, which was fed with WFD-related information. In addition, the SPI-Water project reviewed water policy experiences in non-EU countries and compared them with current WFD practices with a view to facilitating the implementation of integrated water resources management principles in non-EU countries.

References

1. A. Scott, J. Holmes, G. Steyn, S. Wickham and J. Murlis, Report of the 'Science-Meets-Policy' Conference, London, November 2005, www.theknowledgebridge.com (2005).
2. Environmental Science & Policy, Proceedings of the workshop on Research and Technology Integration in support of the Water Framework Directive, *Environ. Sci. Policy*, 2005, **8**, 3.
3. European Commission, Directive 2000/60/EC of the European Parliament and of the Council of 23 October 2000 establishing a framework for Community action in the field of water policy, *Official J. Eur. Communities*, L 327, 22.12.2000, p. 1 (2000).
4. Ph. Quevauviller, P. Balabanis, C. Fragakis, M. Weydert, M. Oliver, A. Kaschl, G. Arnold, A. Kroll, L. Galbiati, J. M. Zaldivar and G. Bidoglio, *Environ. Sci. Policy*, 2005, **8**, 203.
5. Ph. Quevauviller, B. Harris and P. Vervier, in: *The Water Framework Directive – Ecological and Chemical Status Monitoring*, ed. Ph. Quevauviller, U. Borchers, K. C. Thompson and T. Simonart, John Wiley & Sons Ltd., Chichester, 2008, p. 411–444.

6. European Commission, 6th Environment Action Programme, European Commission, 2002–2012, http://ec.europa.eu/environment/newprg/index.htm (2002).
7. European Commission, Common Implementation Strategy for the Water Framework Directive, European Communities, ISBN 92-894-2040-5, 2003. Final CIS document available at http://europa.eu.int/comm/environment/water/water-framework/implementation.html (2001).
8. Ph. Quevauviller, U. Borchers, K. C. Thompson and T. Simonart (eds), *The Water Framework Directive – Ecological and Chemical Status Monitoring*, John Wiley and Sons Ltd., Chichester, 2008.
9. B. Schmitz, P. Reiniger, H. Pero, Ph. Quevauviller and M. Warras, Europe and scientific and technological cooperation on water, European Commission, Report EUR 15645 EN, ISBN 92-826-6464-3, 1994.
10. Ph. Quevauviller (ed.), *Groundwater Science and Policy – An International Overview*, The Royal Society of Chemistry, Cambridge, ISBN: 978-0-85404-294-4, 2008, p. 754.
11. G. Vaes, P. Willems, P. Swartenbroeckx, K. Kramer, W. de Lange and K. Kober, *Science-Policy Interfacing in support of the Water Framework Directive implementation*, in press.

CHAPTER 1.5

Developing the Evidence Base for Integrated Catchment Management: Challenges and Opportunities

BEN SURRIDGE, ALISON HOLT AND BOB HARRIS

Catchment Science Centre, The University of Sheffield, North Campus, Broad Lane, Sheffield S3 7HQ, United Kingdom

1.5.1 Introduction

Water is a fundamental natural resource for humans and for all other organisms, but is under immense pressure from climate change, population growth, changing living standards and rising environmental expectations. Globally, water underpins many of the Millennium Development Goals set by the UN General Assembly in 2000, particularly those associated with reductions in poverty and child mortality. The World Summit on Sustainable Development in Johannesburg in 2002 identified better management of water as a key element in the move towards more sustainable patterns of development. Governance of water has the potential to spark tensions within and between countries, with the prospect of so-called water wars emerging across many continents in the near future.

However, it is increasingly evident that the management of water cannot proceed as an isolated activity. Water is by its very nature an agent of connectivity, moving between physical locations and between human populations.

Water System Science and Policy Interfacing
Edited by Philippe Quevauviller

Published by the Royal Society of Chemistry, www.rsc.org

Connectivity between land and water is now widely recognised. For example, the way in which we farm has important consequences for rivers that receive water after it has run off agricultural land. Our land and water environments are also intimately connected to social, economic and political factors, at scales ranging from the choices made by individuals to the implementation of international legislation. Clearly, the management of water is one element of an interdependent web of environmental, social and economic components. This web may be perceived as a highly complex and interconnected system, where changes in one component can cascade through the system to result in a series of synergistic and conflicting changes in other components. We are currently faced with very significant challenges surrounding the design and implementation of appropriate frameworks for managing these integrated systems, particularly if we are to configure them to cope with an uncertain future.

This chapter is specifically concerned with the management of rivers and the associated catchments that rivers drain. We take a catchment to mean the topographically defined area of land drained by a river and its tributaries, also known as a watershed, river basin or drainage basin. Although based on physical boundaries rather than social, political or administrative divides, a river catchment provides a tangible and meaningful spatial unit within which to address the interplay between environmental, social and economic processes. In this chapter we discuss a new framework for the integrated management of river catchments, and examine the ways in which scientific knowledge can contribute to this framework. We end by exploring what this framework might mean for the way in which scientific research is conducted, and the ways in which science, and scientists, interact with wider society.

1.5.2 Towards an Integrated Catchment Management Approach

Prior to the early 1900s there was little evidence that integrated approaches to catchment management were being developed, either conceptually or practically.[1] Early stages of river management focussed largely on the direct supply of individual services to society, *e.g.* the use of rivers as a source of power for industrial purposes or the abstraction of water for human consumption. These stages were predominantly water-centric and lacked any real appreciation of the links between land and water. The actions of private individuals and organisations, alongside the initial emergence of public bodies charged with ensuring a fairer distribution of water, dominated during these periods.[2] Subsequent stages of management increasingly focussed on the need to improve and protect both the chemical and biological state of rivers, alongside continued control over the volumes of water abstracted from rivers and the risks to human populations due to flooding. These stages were characterised by the use of a range of approaches to secure services from rivers. Both structural approaches, *e.g.* waste water treatment plants, and non-structural approaches, *e.g.* incentive schemes to promote changes in farmer behaviour, were applied. There was also a greater appreciation of the links between land and water. However, the responsibility for management remained largely with public, government-led bodies, and the active engagement of a wider range of actors in the process of catchment management was lacking. Whilst connectivity between land and water emerged as an important concept, wider links between environmental, social and economic components were not effectively incorporated.

Woods[3] describes these two separate stages in the UK where, as in other developed countries, the focus has shifted from 'Sanitisation Provision', with an emphasis on clean water provision and safe sewage disposal, to 'Pollution Control' with an emphasis on water quality improvement by the control of polluting discharges. Brown *et al.*[4] have identified a similar evolutionary pathway, specifically in the context of urban water management. However, each of these authors also identifies a subsequent stage, dominated by the realisation that a holistic approach to management is necessary to meet sustainable development aspirations. Whilst the concept of more integrated catchment management has been developed since the early 1900s, it is only in the last 20 or 30 years that integrated catchment management (ICM) has gained prominence as a potentially valid management approach.[1,5] Whilst we adopt the term ICM in this chapter, others have adopted different terminologies to describe a more integrated approach to management, including Integrated Water Management, Integrated River Basin Management, and Integrated Water Resource Management. Consensus over a definition of what ICM and related terms actually mean is lacking, and this has led to criticism of the concept (see for example, ref. 6). However, the most often used definition was developed by the Global Water Partnership (GWP) in 2000, defining the concept as 'a process which promotes the coordinated development and management of water, land and related resources, in order to maximise the resultant economic and social welfare in an equitable manner without compromising the sustainability of vital ecosystems'. The emergence of such approaches to the management of catchments was initially stimulated by the United Nations Conference on Environment and Development in Rio de Janeiro in 1992, with the publication of Agenda 21 and the Dublin Principles. The continued support for an integrated approach was shown through the subsequent development of several organisations such as the World Water Council and the GWP, and the call for integrated river basin management at the World Summit on Sustainable Development in 2002. During the same period, the principle of taking a whole ecosystem approach to the management of our environment has been developed, recognising that humans are an integral part of ecosystems,[7] and that humans receive a wide range of benefits from ecosystems. These twin drivers of integrated management of catchments and the re-integration of humans within ecosystems provide new challenges and opportunities for future catchment management.

Whilst the concept of more integrated management of river catchments has become increasingly accepted by policy makers, environmental managers and scientists, the translation of the concept into operational terms has proved a significant challenge. Historical and present-day institutional structures, and the way in which national and international legislation has been implemented within these structures, are major factors contributing to the lack of implementation of ICM in practice. We now examine a number of these factors, drawing on our specific experience of the UK context.

1.5.3 Institutional and Legislative Contexts for ICM – Examples from the UK

The creation of Regional Water Authorities in England and Wales in 1974 was an early attempt to deliver ICM. Combining the delivery of water supply, sewage disposal and land drainage services with fisheries, water quality and resources regulation using (non-political) catchment boundaries was a radical step. However, the separation of these services from the local political arena, and most especially from land use planning, began the divorce of environmental concerns from the local populace. This continued with the privatisation of the water companies in 1989, curtailing any further development of an integrated approach to catchment management. The splitting apart of the provision of services (including drinking water and sewage disposal) from the protection of the water environment and associated ecosystems, and the management of flood risk, that was caused by privatisation left an institutional legacy that has made the task of developing ICM all the more difficult in the UK. The National Rivers Authority and the Environment Agency, the government bodies responsible for protection of the water environment and management of flood risk after 1989, both attempted to develop catchment management plans. However, these plans were highly water-centric and were driven by a top-down process that allowed for little engagement of stakeholders within catchments. More recent catchment management plans have been created by the Environment Agency in England and Wales, specifically in the form of Catchment Flood Management Plans and Catchment Abstraction Management Strategies. However, these plans focus specifically on flood management or on water abstraction, and do not integrate broader uses and pressures within catchments. They also continue to be driven by top-down concerns without engaging stakeholders fully in the management process.

A further challenge for the UK is that the centralised, top-down institutional framework that pervades much environmental governance does not create a sufficiently flexible setting for the development of approaches such as ICM. In particular, there is generally poor connectivity between the aspirations emerging from higher political and organisational levels and the implementation activities being carried out at a more local level. National government ministry policies, implemented by national agencies such as Natural England and the

Environment Agency (with their equivalents for the devolved administrations in Scotland and Northern Ireland), dominate. There is a pressing need for better policy integration across sectors, *e.g.* conservation, water quality and flood risk. Currently, policies are largely directed at single components of the environment, and there is little consideration of the possible synergies and conflicts for other components that might be brought about by individual policies. Finally, although local authorities administer spatial planning controls, these are largely disconnected from environmental policy, and local authority involvement in a political, financial or regulatory sense is largely confined to urban issues such as waste, litter, traffic and noise, with little impact upon wider environmental objectives. There are concerns that the potential for the spatial planning system in England to support ICM is not being fully realised, particularly with respect to the water environment, and that greater integration between spatial planning and ICM could offer an effective route forward in the future.[8,9]

For the past 20 years European Directives have set the ideologies, priorities and timescale requirements for the management of the European environment. The Water Framework Directive (WFD)[10] is the primary legislative driver offering opportunities for the development of ICM in Europe. The river basin planning process embedded within the WFD provides a strategic driver for land and water management and, in some areas, land use change of a significant scale. However, the style of implementation of the WFD in the UK is currently directing efforts away from true integration of land and water management. The approach is commensurate with a top-down, government driven process to address major issues in a compartmentalised, structured and controlled way. Competent authorities, policy-makers and river basin managers working with the WFD are driven by timescales and the process of implementation itself. They are currently highly focused on the first round of River Basin Plans (RBPs) to be published by the end of 2009. Often, so-called 'best endeavours' approaches have been adopted, based on existing data, knowledge, systems and structures. Research, new methodologies, models, concepts and approaches can therefore really only influence the second and third rounds of RBPs from 2009 onwards. Combined with concerns over the costs of implementation, a lack of vision and integration amongst policy-makers, and the lack of a significant engagement of science with the implementation process,[11] the WFD may be reduced for the first 10 years to a mere reporting exercise at the catchment scale. This has resulted in significant disappointment at the operational and grass-roots level, where the WFD was eagerly anticipated as a driver to support local initiatives as part of more integrated approaches.

One of the key challenges facing the implementation of the WFD, and more broadly the development of ICM, is to connect top-down, largely government-led processes operating at regional, national and even international scales, with local energy and activity operating at smaller scales. In some administrations in the UK there is little that connects the policy-making level with the operational level, resulting in a two way lack of communication. This lack of vertical integration leads to policy failures due to disconnects between the requisite

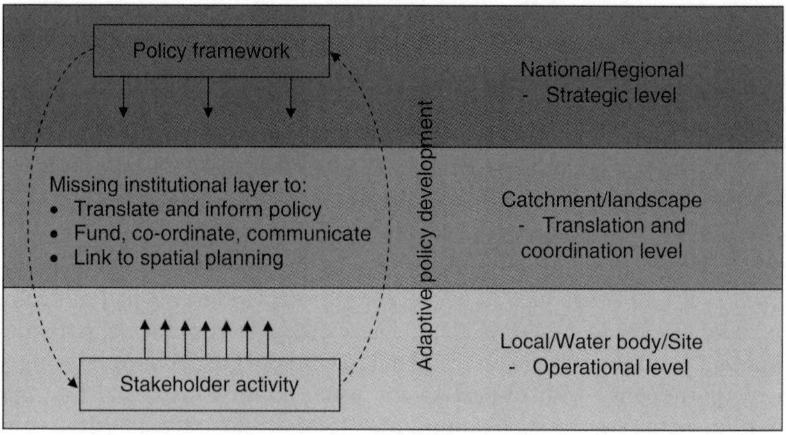

Figure 1.5.1 Levels of governance, and associated roles, relevant to ICM.

levels, and also to poor coordination of local-scale activities. There is a key intermediate layer of governance that is not being sufficiently addressed at present, which we term the catchment or landscape level (Figure 1.5.1). This level has several key roles to play in the future development of ICM. Firstly, in translating national or regional policy and guidance to the people who are implementing activities at the operational level, thereby supporting a traditional top-down flow of information. This may be focused towards a water body, river reach, canal stretch, lake, or a farm or industrial site, but the commonality is that the scale will be local to the participants who relate to it. It will be part of their landscape. Secondly, from a bottom-up perspective the catchment level co-ordinates local activity to help achieve the goals established at the higher catchment, regional or national scale. It is also at the catchment level that coordination can help to fund local activity, *e.g.* by matching proposals with other initiatives that might provide synergy, in other words developing so-called 'win-wins'. The catchment level also has a key role in transferring information back to regional and national scales, ensuring that local-scale activities inform the development of new policy and practice at these broader scales. More essentially, the catchment level provides a link to spatial planning, which operates well in an urban and developmental context, but less well in relation to 'softer' land uses, such as agriculture and forestry. Finally, the catchment is a tangible and relevant scale for the coordination of knowledge related to the integrated environmental, social and economic system operating within its boundaries. This use of the catchment level as a focus for knowledge collation and exchange is a crucial element underpinning the ICM framework described later in this chapter.

Whilst conceptually attractive, the translation of ICM into operational terms is currently constrained by several factors. Fundamentally, the lack of organisational structures and support at the catchment scale severely inhibits ICM in a UK context. However, even if such structures existed, a framework for

ICM would need to be available to inform the processes overseen by such a catchment-scale body. These frameworks are not widely available, and an important challenge for researchers and managers is to engage in their design and implementation.[2] In the following section we describe a new framework for ICM that we believe offers potential for future application in river catchments.

1.5.4 A New Framework for ICM

In Figure 1.5.2 we propose a framework to support the implementation of ICM at the river catchment scale. The framework is built on the recognition that land and water systems are interconnected, and therefore that the management of land systems has consequences for water systems and vice versa. In addition, it recognises that catchments are linked social–environmental systems where traditional divides between humans and the environment are no longer valid.[12] This is analogous to an 'ecosystem approach', a strategy for the integrated management of land, water and living resources that promotes conservation, equitability and sustainable use (Convention on Biological Diversity

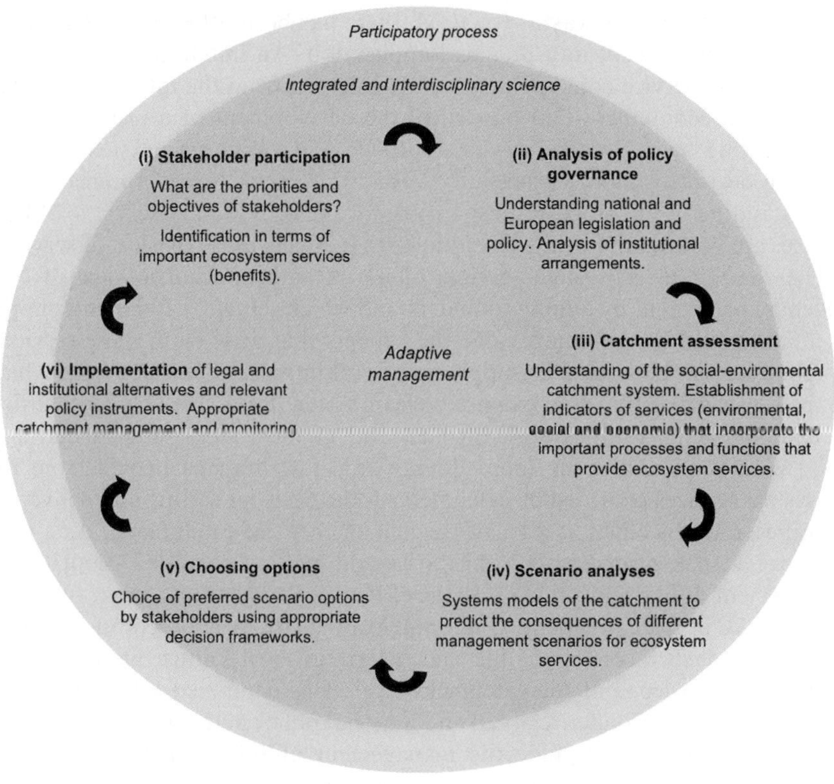

Figure 1.5.2 A framework for Integrated Catchment Management.

www.cbd.int/ecosystem/). The ecosystem approach has recently been adopted by the Department for the Environment, Food and Rural Affairs (DEFRA) in the UK, and importantly it recognises humans as an integral component of the ecosystems found within catchments.

1.5.4.1 Ecosystem Services and the ICM Framework

The ICM framework is also based on the concept of ecosystem services, which is now thought to be central to a successful ecosystem approach. The concept of ecosystem services has grown in popularity since the Millennium Ecosystem Assessment (MA).[13]Although there is no one standardised definition of ecosystem services, they are broadly defined as the benefits that humans derive from the conditions and processes that occur within ecosystems.[13–15] The MA categorised the services provided by ecosystems into: *provisioning services* (*e.g.* drinking water, food), *supporting and regulating services* (*e.g.* nutrient retention, water purification, flood regulation) and *cultural services* (*e.g.* recreation and leisure). However, several other typologies for categorising, describing and valuing ecosystem services have been developed since, and prior to, the MA.[16,17] Although the concept of ecosystem services has received much discussion, in a similar way as for ICM, there has been little progress towards translating the concept into practical application. An initial attempt to apply the ecosystem service concept has recently emerged from the interface between ecology and economics. This has led to the development of a framework for implementing an ecosystem service approach and valuing ecosystem services within a decision-support process.[18,19] This framework is underpinned by an understanding of the processes responsible for the provision of services, of how services are valued, and of the governance that affects the provision of services.

The concept of ecosystem services offers an anthropocentric view of ecosystem management by aiming to understand which services catchments supply to humans, and how humans value the benefits that arise from these services. Catchments can be viewed as supplying several important services from which people derive benefits. For example, clean water, food through agriculture, recreational opportunities both on land and in the water, cultural heritage, flood mitigation and many more. Identifying how humans benefit from the ecosystem services provided by catchments is the basis for setting and delivering priorities and objectives as part of the management of catchments. As a consequence, setting priorities and objectives, and making decisions about future management policy and practice, requires the involvement and input of actors with a stake in the catchment. Our framework for ICM is based on a participatory approach throughout, one that integrates local knowledge with other sources of knowledge at the catchment level. This combined knowledge is in turn embedded in broader regional and national policy and legislative contexts. Stakeholders ultimately guide the management of the catchment, and help to develop integrated policy and practice to achieve the desired objectives that they themselves have shaped. This is a significant departure from the idea that catchment management can be guided by 'natural' reference states, for example as promoted by the WFD. Instead our framework recognises and emphasises that, in a world dominated by humans, catchments should be managed in a way that clearly accounts for the needs of those living within them.

The framework also encompasses the need to use the best possible integrated and interdisciplinary evidence to support understanding and decisions. The natural and social sciences have several roles within the framework, *e.g.* helping to structure the participatory processes underpinning the framework, analysing formal and informal governance structures, providing an evidence base from which future management options can be developed, and reducing the uncertainty of decisions by helping to construct decision support systems. Because scientific knowledge is being used within a fully participatory process as part of the ICM framework, this framework can be seen as a vehicle for facilitating knowledge exchange between the realms of science, policy and wider society. We return to discuss in more detail the challenges for science seeking to support the ICM framework at later stages in this chapter.

The ICM framework has clear parallels with recent approaches that have emerged in water resource management, *e.g.* Participatory Integrated Planning (PIP),[20] decision support procedures (*e.g.* ref. 21), and frameworks for the implementation of an ecosystem services approach to the conservation of natural capital.[18,22] However, to date there has been no development of such frameworks in the context of, or application of these types of approaches specifically to, the spatial scales and diversity of issues inherent to river catchments.

1.5.4.2 Implementing a Framework for ICM

Figure 1.5.2 shows each stage of the framework to support ICM, and we describe each stage below, including a brief introduction to the types of approach that may be used to deliver them.

Stakeholder Participation

Stakeholder participation from the beginning and throughout the process is central to ICM. We define stakeholders very broadly as individuals, groups or organisations that share an interest or stake in a particular issue or system,[23] and participation as where these stakeholders choose to take an active role in making decisions that affect them.[24] Stakeholders may be a combination of, for example, farmers, landowners, water companies, government bodies, conservation NGOs and the general public. Choosing who is relevant and who should be involved in the participatory process is an important step, not least because the composition of the stakeholder groups can have significant consequences for the legitimacy and success of subsequent stages of the framework. There are now several stakeholder analysis techniques that can help to support this stage of the ICM framework, including social network and stakeholder analysis.[23,25]

By bringing together stakeholders we seek to understand which ecosystem services are of importance to them, what their perspectives are on the significant management issues in the catchment, their objectives and aspirations for the catchment. This stage of the framework focuses on the knowledge held by stakeholders regarding their catchment. This knowledge is crucial for helping to create a focus in subsequent stages of the framework, in particular the development of alternative scenarios for future management and the modelling of likely outcomes of these scenarios. Stakeholders can be brought together in focus groups or meetings using novel techniques such as participatory mapping[26] to illustrate how stakeholders use and value ecosystem services, and participatory modelling,[27] for example through the construction of cognitive maps of issues and/or objectives within the catchment. These approaches could be used to create hierarchies of weightings for objectives, as described in Reichert *et al.*[21]

It is important that participation is a two-way process between those who instigate the ICM process and the stakeholders. As a consequence, the ICM process is designed to foster social learning.[28] Through participation in the ICM process, stakeholders learn about complex interlinked problems within their catchments, and have the opportunity to gain knowledge regarding the likely causes of and possible solutions to these problems. This social learning process also supports understanding of the multiple standpoints of stakeholders and their perceptions of catchment issues, which in turn can help to build more sustainable relationships between stakeholders, may reduce conflict and can improve management capacity.[24,29]

Analysis of Policy and Governance

Analysis of existing policy and governance is an essential part of the ICM framework because these factors are often the drivers of current catchment issues and management approaches. A multitude of international and national legislation operates in many catchments to keep environmental systems within certain quality standards. For example, regulations exist in the UK under the Common Agricultural Policy, national Biodiversity Action Plans, a wide range of European Directives, and national water resources strategies. However, a key barrier to more integrated management is that individual regulations are often directed towards single ecosystem services that fall under the control of specific sectors. Overcoming this barrier is dependent on understanding how the existing regulatory frameworks currently interact, and then identifying how more integrated policy and regulation could be developed. The WFD has the potential to stimulate a more integrated legislative approach in catchments. However, the current style of WFD implementation, certainly in the UK, is failing to exploit this opportunity. In particular, there appears to be little desire to consider the trade-offs, conflicts and synergies that exist when multiple policies and regulations affect multiple ecosystem services within catchments. This integrated analysis of policy is a key element of our ICM framework.

Current governance of catchments in the UK is dominated by top-down, government-led structures. At the local-scale, there is energy and activity associated with a range of groups, *e.g.* river trusts, angling clubs and local wildlife trusts. However, this local activity is poorly funded, not co-ordinated and not supported or engaged by top-down governance structures. Understanding the web of governance within catchments is a significant challenge for the ICM framework. This understanding is crucial if revised, multi-scale and polycentric governance structures are to be developed as part of later stages in the ICM process. Polycentric structures mean that stakeholders in different institutional settings, such as private companies, interest groups, individuals and government agencies, cooperate and collaborate in the formulation and implementation of public policy.[30] This is an attempt to move away from the traditional command-and-control style of management, and to promote collective decisions supported by top-down *and* bottom-up institutional arrangements. Ultimately, this may lead to greater acceptance of and commitment to delivery of management solutions by all stakeholders involved.[30]

Catchment Assessment

The properties, or services, that emerge from coupled social–environmental systems, such as catchments, are governed by sets of social rules and sets of environmental rules, and by the interaction of these rules.[31] Understanding these rules is crucial if the current state of the catchment, and the causes of this state, are to be accurately assessed. This understanding is also required in other stages of the ICM framework, for example to set reasonable objectives for the catchment and to develop effective management scenarios to deliver these objectives. However, one of the most significant challenges to the ICM framework is that our understanding of these rules is currently limited. Addressing these limitations is one of the key ways in which scientific knowledge can help to support the ICM framework. There are challenges for fundamental research here, including developing basic understanding of how environmental processes and functions underpin ecosystem services, and how services behave over different spatial and temporal scales. There are also challenges for system-orientated research that seeks to understand the factors governing social–environmental systems, and we return to these challenges in Section 1.5.5.

To enable informed decisions to be made regarding the sustainable management of ecosystem services it is necessary to find indicators of the status and trends in these services.[32] It is important that a robust approach is developed to define the status of the catchment, one that accommodates the various demands of society. If these indicators are to best reflect the needs of society and work for stakeholders, attempts should be made to integrate indicators developed in collaboration with the stakeholders alongside those that are scientifically derived.[33] An assessment of the social and environmental aspects of the catchment will be based on sociological, economic and environmental indicators, or proxies, of ecosystem services. For example, visitor numbers may be used as an indicator of the recreational capacity of a landscape, the value of housing as an indicator of the aesthetic quality of a river, and reservoir level or groundwater level as biophysical indicators of water availability. Novel approaches exist that can be used to assess the health or quality of catchments using multiple indicators of a suite of services, *e.g.* the Holistic Ecosystem Health Indicator (HEHI), and the related Monitoring Ecosystem Health through Trends Analysis (MEHTA),[32] and the more conceptual Ecological Quality Index (EQI).[34]

Development of indicators will necessitate a combination of existing data with the collection of new indicator data. These indicators will be important for assessing the current status of ecosystem services within the catchment, but also for tracking changes in these services through time. Change in the status of services, *e.g.* due to the implementation of new management actions, will provide important information to feed back into decision-making as part of the adaptive management principle underpinning the ICM framework.

Scenario Analysis

Once a baseline assessment of the catchment and its ecosystem services has been established, several different management scenarios can be developed, drawing in part on the objectives and knowledge of the stakeholders established in step (i) of the framework. The impacts of these scenarios on the multiple ecosystem services within the catchment are predicted through the use of integrated modelling systems. An important reason for undertaking this modelling stage is to highlight the trade-offs that emerge as a result of any integrated management scenario. Ecosystem services are not independent of one another, and management that affects one particular service will almost certainly impact on the viability of another in a synergistic or competitive way. These interdependencies must be communicated to stakeholders participating within the ICM framework, and integrated models provide one means of supporting such communication. In addition to using models in this predictive mode, the co-development of models through the participatory processes in the ICM framework is an important way to support knowledge exchange and learning. For example, developing cognitive and conceptual models, an important early stage in any integrated modelling exercise, is undertaken as a participatory exercise within the ICM framework. This approach exposes stakeholders to new knowledge and different perspectives regarding the catchment, and may lead to new learning.

There are very significant challenges for science to address in this stage of the ICM framework, primarily because integrated modelling systems for catchments are not widely available. New modelling technologies are under development that enable a combination of the different sources of knowledge that will exist within a catchment (data, existing models and opinions), and that enable a more realistic incorporation of the human dimension of social–environmental systems within integrated models. The importance of participatory conceptual model development, the formulation of scenarios and the use of integrated models in a predictive mode and to support learning are discussed further in Section 1.5.5.4.

Choosing Options

There will be a need to prioritise between the alternative management scenarios, the outcomes of which are predicted during the preceding modelling stage of the ICM framework. As in previous stages, the selection of preferred options is conducted through a participatory process. The decision as to which option is chosen needs to be made within a framework that will enable a range of scenarios, including business as usual, to be compared against each other. Choices can be made based on a range of criteria that describe each scenario, including traditional economic metrics such as cost effectiveness, but also other criteria such as the degree to which benefits are equitably distributed across different groups of stakeholders. Multi-criteria analysis provides one framework within which decisions about the alternative scenarios can be made. This approach would enable the individual scenarios to be presented in several ways, including through the use of monetary and qualitative values.[35]

There are a number of tools that could be used within the ICM framework to understand stakeholder preferences for alternative scenarios; for example, economic valuation methods including contingent valuation (see ref. 36) and choice modelling (see ref. 37) that may encompasses techniques such as assessing stakeholders' willingness to pay or willingness to accept certain trade-offs. Cost–benefit analysis could also be applied to provide a comparative evaluation of the different scenarios (*e.g.* ref. 38). Non-monetary valuation may also be desirable, for example constructing value functions for scenarios that are derived from the preferences of stakeholders.[21]

Implementation and Monitoring

Once a preferred management scenario has been identified, its compatibility with existing institutional, policy and regulatory structures must be assessed. In many cases these structures are unlikely by themselves to be able to deliver the desired management scenario. The participatory nature of the ICM framework also extends to developing new collaborations between stakeholders to support implementation of the chosen scenario. Such collaborations may include voluntary agreements to change land management in specific areas of the catchment, or more general commitments, for example to further development of networks and partnerships that support the management scenario. These collaborations are underpinned by the concept of polycentric governance,[39] where power is diffused away from centralised, bureaucratic management organisations in an attempt to develop more sustainable and effective governance structures. However, this is only possible if the concept of polycentric governance is given political, legislative and financial support as part of a broader commitment to the ICM framework.

The ICM framework is also underpinned by the concept of adaptive management. Adaptive management stresses that each management action should be seen as an opportunity to learn about the system under consideration, and that the information derived from these management actions should be used to continuously update understanding of the system.[40] Updated understanding can subsequently be used to adjust integrated models of the catchment, and reduce uncertainties when these models are used in a predictive mode to support decision-making. Monitoring of the catchment throughout the ICM process is therefore crucial to adaptive management. An important part of monitoring relates to the collection of data from monitoring networks across a catchment, both in space and time. These networks should focus on data related to the indicators of ecosystem services that are identified in stage (iii) of the ICM framework.

1.5.5 Using Science to Support the ICM Framework

The framework for ICM that we have presented should be based upon the best available knowledge at all stages. Relevant knowledge could be derived from a range of sources, including historical data, individual and collective experiences and perceptions, and the outcomes of future scenarios as predicted through integrated models. Not all of these sources necessarily require the formalised, institutionalised production of new data, information or knowledge. There are significant challenges to address in the mobilisation and integration of the vast knowledge reserves society already possesses. However, the development of frameworks such as ICM is often accompanied by calls for a new, integrated and interdisciplinary knowledge base, providing robust evidence to underpin the implementation of these frameworks. The source of this new knowledge is often seen to be science, or more specifically scientists working within academic institutions. These calls mirror broader perspectives in which science is seen as playing an increasingly important role in providing data, information and knowledge to support environmental decision-making.[41,42] Some would go further and argue that the nature of science itself has changed in recent decades, and that these calls reflect a broader shift towards the production of knowledge in the context of its application, sometimes termed Mode 2 science.[43]

Whilst calls for such knowledge abound, identifying the specifics of the new knowledge that is required remains a very significant challenge. In part this is an inevitable consequence of the participatory nature of the ICM framework. Participation is a key element in defining the services required of the catchment, in identifying and agreeing the current issues to be addressed, and in specifying the range of possible options for future management within a catchment. The knowledge required to underpin the ICM framework within any given catchment is therefore partly dependent on the outcomes of these participatory processes, and partly context-dependent. Whilst there are undoubtedly knowledge requirements that are likely to underpin the ICM framework in many catchments, creating a highly detailed research agenda based solely on scientific perspectives prior to participation would run contrary to the participatory nature of the framework. Instead we believe that the participatory nature of the ICM framework extends to framing the knowledge requirements, and therefore the knowledge gaps, that must be addressed. Note that we do not argue that from the perspective of individual scientific disciplines there are no intellectually valid or stimulating research questions left to address, we do not subscribe to the view that we are at the 'end of science' and that 'pure' science has run its course (see ref. 44). Further, the ICM framework is not designed such that science simply reacts to the outcomes of participation. Instead a more effective integration of scientific knowledge with other forms of knowledge is sought through the participatory process. For example, where scientific knowledge is agreed as valid it can be used to adjust the conceptual models held by stakeholders participating within the ICM framework, and can be used to guide the choice of options to be included within the scenario analyses.[28]

In the following sections we seek to address two generic challenges that we believe face science seeking to support the implementation of the ICM framework. First, we describe the need to understand the environmental rules, the social rules, and their interrelationships that govern the social–environmental systems at the heart of the ICM framework. Second, we examine how modelling tools can be used to support both prediction and learning within the ICM framework. We finish by examining what these challenges mean for the process of conducting scientific research, and for the nature of interactions between scientists and the broader community of stakeholders engaged in the ICM framework.

1.5.5.1 Social–Environmental Systems and ICM

The ICM framework explicitly recognises that managing catchments involves a coupling between social and environmental systems. This reflects the fact that many of the most significant challenges facing society are concerned with coupled social–environmental systems. Adapting to global climate change, generating 'green' energy demand and supply, conserving biodiversity, and delivering social and economic development whilst minimising environmental degradation are all examples of issues that span traditional social–environmental divides. It is highly unlikely that effective and efficient solutions to these challenges will be found solely through the techno-rational solutions that traditionally have emerged from scientific research, or that these challenges will be successfully addressed without technological and scientific advances. In many cases substantial changes in individual, collective and corporate perceptions, choices and commitments, alongside technological and scientific innovation, will be necessary to deliver solutions. Scientific methods and knowledge have a range of important roles to play in supporting any management framework dealing with such systems.

1.5.5.2 Developing Participatory Processes

From the outset, the ICM framework has been designed to be based upon a fully participatory process, reflecting the increased demand for stakeholder participation in environmental decision-making.[24] Knowledge and methodologies from the social sciences are required to identify the participatory processes and methods best suited to certain social and cultural contexts, and to offer insights into best practice for stakeholder participation (see ref. 24). They enable the appropriate typology for the degree to which participation should be sought within the ICM framework to be identified, depending on the decision-making context; an understanding of which stakeholders should be represented in the process, of how to facilitate such a process in practice, of how to link stakeholder and scientific knowledge and of how to institutionalise stakeholder participation (*e.g.* refs. 24,28 and 45). The importance of transparent, inclusive and legitimate processes, and how to build such processes into the ICM framework in practice, is also strongly dependent on perspectives derived from the political sciences (*e.g.* ref. 46). Several techniques are currently being developed that enable an analysis of stakeholders to achieve best practice for participatory decision-making, *e.g.* social network and stakeholder analysis.[47]

1.5.5.3 Understanding Social–Environmental Systems

Subsequent stages of the ICM framework, in particular catchment assessment and scenario analyses, are based upon understanding of social–environmental *systems*, not just of their component parts. These systems are simply too complex to unravel through analysis of individual components, and at a system level our interest lies not only in the functioning of individual components but also in their interaction.[12,31,48,49] To decompose social–environmental systems into their component parts necessarily leads to naive or non-existent treatment of crucial elements of the system, risks changes to unexpected and undesirable states if narrow, discipline-specific knowledge is applied uncritically to a system, and is unlikely to support the development of sustainable solutions. The properties of coupled social–environmental systems emerge as a result of the interaction between sets of biophysical and sets of social rules.[12] Key challenges for research seeking to support the ICM framework are to characterise the nature of these rules, to understand how environmental and social rules interact to determine the characteristics of social–environmental systems, and to incorporate scientific knowledge alongside other sources of knowledge to guide the implementation of the ICM framework.

Approaches to Developing System Understanding

Calls to better understand social–environmental systems present renewed challenges to scientific research that has historically progressed through a reliance on reductionism and ever increasing specialisation and separation of academic disciplines. Perhaps the most daunting challenge is the complexity involved in such systems. As a result of the sheer number of variables, possible states of those variables, and interrelationships with other variables, any attempt to understand 'the system' may seem like an intractable problem. However, one can build (at least) two different conceptual approaches to the challenge of system understanding, approaches that are analogous to the way in which integrated water management has been addressed in the past (see ref. 5). The 'comprehensive' approach emphasises the need to understand *all* variables and *all* relationships between variables within any given complex system. Being comprehensive minimises the chance of 'missing' any key variable or relationship, and stresses the importance of all links between different elements of the system, however large or small the effects of these links may be.

In contrast, the 'integrative' approach focuses on the key variables and relationships that are responsible for most of the variability within any given system. Clear parallels can be drawn between the integrative approach and ideas about complexity and the emergence of system characteristics that have been developed in the ecological literature (see, for example, ref. 48). Whilst the ecosystems at the heart of ICM are complex, Harris argues that behind this complexity lies a set of relatively simple, fundamental rules that govern the emergent properties of these systems. In the context of the ICM framework, Harris' focus on ecosystems as predominantly biophysical systems requires extension to capture broader environmental and, particularly, social components. Therefore, the challenge to science becomes one of identifying and understanding the simple, fundamental rules, the so-called 'key' variables and relationships, that govern the emergent properties of social–environmental systems. Here we are dealing not only with complex systems but with complex adaptive systems in which features such as evolution and feedback are inherent. This means that the relationships between key variables within the system are likely to be described by nonlinear, contingent functions and by feedback loops, rather than by relatively simple linear functions. It also means that the emergent properties that are seen as desirable, the services within the ICM framework, may warrant extension to incorporate services related to the properties of the system itself, rather than focussing only on components 'within' the system. For example, there is growing interest in the capacity of social–environmental systems to absorb change and reorganise, also defined as the resilience of the system, as a desirable system property.[12,50,51] In this case, our understanding must also be extended to capture those factors that govern the ability of social–environmental systems to be resilient, and how these factors can be managed to increase resilience. However, there remain significant conceptual and practical challenges associated with resilience of social–environmental systems that must be overcome before the concept can be operationalised.

The specific gaps in our understanding of social–environmental systems are numerous, have been the subject of extensive reviews (see, for example, ref. 31 and references therein), and extend across both environmental and social sub-systems. One of the broad questions for science to address is what type of research approach is needed to identify the governing rules for these systems? If we accept the premise that the complexity of social–environmental systems emerges from relatively simple origins, the question becomes whether these simple origins can be identified through traditional scientific research, building on fundamental research within a systems framework? In the past, and largely within environmental systems, fundamental research has provided the base for subsequent integrative analysis, and it is this integrative analysis that has yielded the governing rules.[48] Whilst integrative, systems-orientated approaches have been developed quite extensively in the natural sciences, the lack of such approaches and underpinning frameworks within the social sciences limits our current ability to understand the rules that govern social–environmental systems using integrative approaches.[51]

An alternative approach, one based upon adaptive management, uses each management intervention within a system as an opportunity to test hypotheses about how that system behaves, and to incorporate the resulting knowledge into an on-going cycle of learning that includes revision of management approaches as new knowledge becomes available.[12,28,52] This approach to generating understanding of a system can also combine scientific knowledge alongside knowledge held by the stakeholders who design, implement and experience management interventions within the social–environmental system. The ICM framework is built upon the adaptive management approach to learning about social–environmental systems, and also seeks to combine scientific knowledge alongside the knowledge of stakeholders involved in the participatory process.

Linking Knowledge about Social–Environmental Systems to Integrated Modelling

Knowledge about the social–environmental system is required if a shared conceptual model of the catchment is to be developed between participants within the ICM framework. This conceptual model should capture the processes operating within the catchment, how these processes are related to the provision of desirable services, and how current management interventions impact upon the processes. The building of such conceptual models relies on the integration of narrow, discipline-specific scientific knowledge and broader understanding of the relationships between that knowledge and other components of the system. Whilst scientific knowledge is a key component of these conceptual models, their construction also offers the opportunity to integrate scientific knowledge with the subjective mental models held by stakeholders participating within the ICM framework.[51] A conceptual model that integrates both scientific understanding and the knowledge of stakeholders represents a robust, transparent and complete basis to subsequent stages of the ICM framework.

Clearly, holistic frameworks, such as that for ICM, are unlikely ever to be informed purely by data collection and process studies.[42] Integrated modelling, both within and more recently between environmental and human systems, has emerged as a strategy to tackle the complexity and uncertainty involved.

The use of integrated models, both as a predictive tool to enable stakeholders to evaluate possible future management scenarios and as a means of supporting communication and learning, is a crucial element of the ICM framework. Conceptual models, integrating both scientific knowledge and the demands and mental models of stakeholders, are a crucial part of the problem formulation stage of any subsequent integrated modelling exercise.[42,53] These conceptual models, and their use within integrated modelling, represent an important bridge between understanding the social–environmental system within a catchment and the application of that understanding in support of environmental decision-making. We now move on to examine the use of integrated models within environmental decision-making, and how these models contribute to the ICM framework.

1.5.5.4 Scenario Analyses and the Evolution of Integrated Models

Scenario Development

A crucial stage of the ICM framework is the evaluation of alternative scenarios for future management of a catchment, primarily in terms of their impact upon the objectives and priorities of stakeholders participating in the framework. These scenarios should capture a representative spectrum of plausible alternatives,[42] and their formulation is an important means of ensuring assumptions about future developments are made transparent and are documented.[53] Despite the conceptual attractiveness and simplicity of a 'representative set of scenarios', given the size and complexity of any given catchment the formulation of these scenarios is a challenging task. Different scenario typologies can be imagined, for example separated into baseline versus policy, exploratory versus anticipatory, or qualitative versus quantitative.[54] Regardless of the exact typology used, the possibilities considered during the scenario analyses must be kept broad and not constrained. Constraints are often applied during scenario analyses due to preconceptions about future possibilities based on maintaining the status-quo and minimising change and cost to management organisations, or due to a lack of vision or 'evidence' to support innovative approaches. The conceptual models developed in earlier stages of the ICM framework are an important tool to inform the debate between stakeholders when formulating future scenarios. Therefore, scientific knowledge, as part of these conceptual models, has an important role in guiding the choice processes during the development of scenarios, but it should not overly constrain the possibilities considered.[51]

Integrated Modelling

Given the complexity and uncertainty surrounding the social–environmental system within any catchment, of any possible future management scenario for that system, and the need to identify synergies and conflicts between management actions and multiple stakeholder objectives, no single decision maker will be able to conceptualise or solve the necessary components to predict change. These challenges have led to increased development and application of modelling tools to support environmental decision-making.[55,56] Through the process of building an integrated model, knowledge can be organised and communicated and existing knowledge can be mobilised and integrated with new knowledge in a single coherent modelling framework. Developing integrated models as part of the ICM framework also serves a number of purposes, including providing a predictive capability, supporting communication and engagement with stakeholders and stimulating participation in the ICM framework.

Modelling of environmental processes has been conducted for several decades (*e.g.* ref. 57). Whilst clearly of interest as an approach for conducting scientific research, models have also been used frequently to support environmental decision-making. In the past these models were predominantly developed to address single issue decisions within planning and resource management agencies, by predicting the effect of specific management interventions on the single issue of interest.[58] In the context of catchment management, models have focussed largely on biophysical issues, including water resource management, licensing of effluent discharge and estimating the pollutant load derived from different land uses. With the growing recognition of the complexity of environmental processes and their mutual dependence, the complexity of models has increased. They have moved from single to multiple biophysical issues, and more recently to attempts to integrate biophysical, social and economic processes and issues. The use of models within the context of environmental decision-making has also changed, moving from a focus on supporting the decisions of individuals within management agencies, through increased use of models in public arenas of communication and consultation, to current approaches that seek the active participation of stakeholders throughout the model building process.

Changes to the way in which models have been used within environmental decision-making have also been accompanied by changes in modelling technologies. Early issue-specific models were largely deterministic and mathematical, and interaction with the model builder was principally through the modification of computer code. Later integrated models have often drawn on these earlier models, for example by developing coupling approaches to link individual models together. However, the complex environmental, social and economic processes and objectives involved within frameworks such as ICM present two significant problems to model coupling approaches. Firstly, models may not currently exist for many parts of the social–environmental catchment system. Secondly, although models may not exist, other types of knowledge may be available for these parts of the system, such as empirical data and the knowledge of stakeholders. Combining these alternative sources of knowledge with more traditional mathematical models in a coherent modelling framework has led to the development of novel modelling technologies, including neural networks and Bayesian belief networks. The evolving, adaptive nature of social–environmental systems has also resulted in the development of new technologies that can include realistic representations of the human dimension within integrated models, including agent-based techniques.[51,59,60]

Developing Decision Support Systems for ICM

Whilst such technical developments have generated much progress towards greater integration in modelling, support for the complex, poorly structured and uncertain decision contexts that are inherent to the ICM framework involves integrated models as one element within broader decision support systems (DSS). Whilst the specific architecture of individual DSS can be highly variable, common components include an integrated modelling platform and related databases, which are often accessed through a user-friendly graphical user interface (GUI), and a data management structure that usually draws upon the spatial functionalities within geographic information systems.[55] The GUI often represents a key communication tool, simplifying the complex models within the DSS and providing graphical or numerical representations of a large amount of complex information in a way that can be understood by the relevant audience. However, providing a transparent and coherent GUI remains a significant challenge to current efforts to develop DSS. In a similar way as for integrated models, the users of DSS have evolved from individual technical specialists through individual decision makers to broader groups of stakeholders.

The process of building a DSS to support the ICM framework is envisaged as participatory from the outset. There is clearly a role in constructing an integrated model within the DSS for both new scientific knowledge and the mobilisation of existing scientific knowledge, including the use of scientific models, data and expert opinion. However, we believe there are also opportunities to capture other sources of knowledge within the modelling framework. The participants within the ICM framework will hold their own subjective mental models of a catchment, formed as a result of factors such as previous experience and cognitive bias.[28] These mental models play an important role in shaping the decisions that affect the social–environmental system at the heart of the ICM framework, and it is important for an integrated model of the system that these human dimensions are explicitly captured.[51] The elicitation of these subjective mental models, and their integration with knowledge from scientific research, is a key element of the conceptual model building and problem formulation stages that are crucial to the early development of DSS.[42]

Construction of an integrated model within any DSS is faced with similar challenges to those described previously in the context of developing understanding of complex social–environmental systems. It is unlikely to be computationally feasible, conceptually desirable or attractive to stakeholders to build a comprehensive model capturing knowledge about all variables and relationships within a system. Proper investment in problem formulation and conceptual model building at the outset of the integrated modelling exercise is one means of identifying the crucial elements of the system to include within subsequent model building stages.[42] If we accept the existence of governing factors that control the emergent properties of complex social–environmental systems, the identification of these factors is a key step in determining which elements to include within an integrated model. Even if critical system components are identified, it is likely that different spatial and temporal scales will be relevant to different processes, and to different objectives within any modelling framework. These scales will be determined in part by the scales of operation for the factors governing the emergent properties of the system, and abstraction to these scales is likely to give predictive power to the model.[49] An inverse relationship appears to exist between model complexity and probability of use.[49] However, complexity is often directly related to perceived credibility. Therefore, the challenge is to find an appropriate balance between complexity and simplicity at each scale and for each of the elements of the system included within the model. The issue of appropriate scale and modelling approach for different scales is common to both the environmental and social components of a system. The development of multi-scale, hierarchical, modular modelling systems – able to deal with the need for different levels of complexity across individual modules, and to couple these modules using a combination of hard or soft approaches – may offer a promising way to address these issues.[42,56]

Evaluation and Uncertainty in Decision Support Systems

Evaluating an integrated model and/or a DSS that has been developed to support a framework such as ICM is a significant challenge, and is often only considered in an informal way.[55] Without entering the vexed question of the correct terminology, and instead adopting the definitions of Sojda,[61] evaluation is taken to be composed of both verification and validation. Verification largely involves checks of the DSS from a modelling and programming perspective, and relates primarily to the adequacy of the software and computer code used in building the DSS. Validation can relate either to the DSS achieving a stated purpose, which can be extremely varied but is often related to knowledge organisation, communication and support of decision-making, or to whether the DSS provides an adequate representation of the system it is to be used to analyse. The former is largely a subjective judgement taken by experts and/or users of the DSS, as any objective measure to isolate the specific contribution of a DSS to a particular decision-making problem has proved elusive.[56] The latter is more closely aligned with traditional empirical validation of model outputs.

Sojda[61] provides an extensive discussion of the options available for validation of DSS. These include the use of test cases with known prior outcomes against which the DSS may be tested, comparison of DSS outputs with empirical datasets, and the use of expert panels to generate predicted outcomes against which the DSS can be compared. Given the complexity and uncertainty inherent to any social–environmental system, each of these empirical validation approaches is limited when it comes to evaluation of a complete DSS. Test cases are unlikely to exist for the multi-objective social–environmental problems that are at the heart of the ICM framework. The use of expert panels is unlikely to be possible because one of the major justifications for the development of DSS is that conceptualisation and solving of problems involved in the ICM framework are beyond the grasp of any single expert. Long-term empirical datasets are unlikely to be available for the range of issues included within a DSS. Certain components of the DSS will not be the subject of any routine monitoring programme, whilst others may be described by data that are not of sufficient accuracy or precision to enable comparison with the outputs of the DSS. Given the difficulties associated with validating a complete DSS, traditional empirical validation techniques, such as comparison with datasets and the use of sensitivity analysis, can be applied to individual components of the DSS, yet only with the clear recognition that such approaches cannot validate the entire integrated model/DSS.[61]

Assessing the uncertainty associated with model predictions is an important element of any modelling exercise seeking to support environmental decision-making.[42,53,57] Dealing with uncertainty in models has historically been conducted as a specific activity after a model has been constructed and validated.[53] This approach should be modified so that the treatment of uncertainty is an element of the modelling exercise throughout.[62a] Multiple sources of uncertainty may arise at various stages throughout the modelling exercise, from initial problem definition and conceptual model building, through model structure uncertainty, to parameter uncertainty and technical uncertainties associated with computer implementation of the model, such as bugs and numerical approximations.[62b] A broad range of methodologies and tools for assessing uncertainty have been developed, and the reader is directed to Refsgaard *et al.*[53] for a thorough review. Within the ICM framework the development of scenarios to be tested through an integrated model offers a specific opportunity for assessing the uncertainty in predictions of the future state of social–environmental systems. Given the inherent uncertainty involved in any prediction of the future, the selection of a representative spectrum of plausible alternatives during scenario analyses is one means of explicitly and transparently capturing the assumptions about future changes, these assumptions being the ultimate source of uncertainty regarding the future.[42,53]

Models as Predictive Tools and Models to Support Learning within the ICM Framework

The development of integrated models within the ICM framework is conducted for two reasons. Firstly, there is a need for modelling tools that provide predictive power, enabling the impacts of future management scenarios on multiple objectives to be evaluated. Predictive modelling of coupled social–environmental systems is a major challenge, both conceptually and technically in terms of how the human dimension should be incorporated within integrated models,[42,51] and because such models are currently beyond the experience of many modellers, particularly environmental modellers. The principal challenges are to couple the interrelationships between human society, their decisions and the environment in a dynamic sense, allowing for evolution and adaptation to alter these relationships, and for dynamic models of the social–environmental system to be incorporated within DSS. Whilst such challenges are the focus of current advances in modelling technology and application, there are always likely to be substantial uncertainties associated with models applied to such unpredictable and chaotic systems,[63] which may ultimately limit the application of models as predictive tools.

Leading in part from an acceptance of limitations in predictive power, the second reason for developing such models is to structure discussion and support learning as part of the participatory processes within the ICM framework. By developing integrated models based on both scientific knowledge and the subjective mental models of stakeholders, opportunities arise to adjust and develop existing mental models. It is also possible to create new mental models through engaging with the subjective knowledge of other participants and with the outcomes of scenario analyses. Using models to support continual learning is aligned with the adaptive management concept that underpins ICM. It may also bring different perspectives to bear on model evaluation as part of the development of DSS. For example, fully participatory model building may result in the explicit inclusion of stakeholders within the model validation stage of a project,[53,64] and validation is then increasingly concerned with the *process* of model development, and the communication and learning surrounding that process, as much as the predictions made by any particular integrated model. The inclusion of multiple stakeholders, and their subjective knowledge, within a learning process based on participatory model building also brings into focus the exact nature of uncertainty. Whilst uncertainty in modelling is traditionally associated with a lack of knowledge regarding an agreed problem, or with the inherently unpredictable, stochastic nature of a system, using models to structure learning is more likely to be concerned with issues of ambiguity. Ambiguity can be thought of not as a lack of knowledge surrounding an agreed problem, but as multiple, equally valid perspectives held by stakeholders regarding the definition of a problem, its core elements and the relationship between a problem and the wider system.[63] The role of models within a highly ambiguous situation, as is likely to emerge within ICM, can be thought of as one of promoting learning with the aim of moving towards a clearer, more unanimous problem definition and articulation. This use of models as a means of promoting learning and developing agreement is an essential base for subsequent stages of model development that seek to provide predictive power.

1.5.6 Interdisciplinary Science and ICM

1.5.6.1 Calls for Interdisciplinary Research

Developing the knowledge base to support frameworks such as ICM is often seen to rely on new forms of integrated and interdisciplinary science. Here we avoid the distinctions made between multi-, inter- and trans-disciplinary science (see, for example, ref. 65 and references therein) and adopt the term interdisciplinary as a general description for multiple forms of working across disciplines.

Calls for integrated and interdisciplinary research, although not new, are growing in number, in the diversity of sources from which they emanate, and in the strength with which they are delivered. The drive towards greater interdisciplinarity features prominently in science policy. For example, at a UK level the government's recent Science and Innovation Framework identifies the interfaces between separate research disciplines as being the location of the greatest challenges to science in the next decade.[66] National and international research funding programmes are also increasingly structured to promote interdisciplinary research. In the UK, the Lowland Catchment Research Programme and the Rural Economy and Land Use Programme (RELU) are two clear examples of research funders adopting interdisciplinary agendas related to land and water. Similar priorities emerge at the European level, for example in the importance placed on interdisciplinary integration within the European Commission's Fifth Framework Programme[67] and subsequent framework programmes. Interdisciplinarity has become not only a focus of individual research projects and programmes but is also increasingly seen as a means of co-ordinating and re-structuring science at a strategic level.[68] In the UK this is reflected by the call from the Office of Science and Technology for greater collective strategic leadership and cross working amongst research funders, if the UK science base is to deliver its full potential.[69]

These calls for interdisciplinarity do not emerge independently of science itself. Indeed they are often articulated by and are the direct result of existing research activity. The cyclical relationship between research that demonstrates a need for, and promotes, integrated and interdisciplinary approaches and the demands for such research that emerge from science policy makers and funders should be clearly recognised. Understanding such inter-relationships between the supply and demand for science is crucial if more robust decisions about science policy and funding are to be made.[70] For example, it may be that such cyclical relationships promote stability and a lack of innovation in science that seeks to address key social–environmental issues, and this danger should be guarded against by scientists and by funders alike.

1.5.6.2 Changes in the Practices of Science

The development and implementation of frameworks such as ICM is indicative of the challenges stimulating wider debates within science, explicitly about the relationships between humans and ecosystems, and implicitly about the relationships between science and society.[71] The conceptual separation of humans from ecosystems, a historical feature of much biophysical science, is being challenged by a concerted effort to recognise whole ecosystems with humans as integral parts.[72] By bringing humans back into ecosystems, science is also being challenged to address the myriad of social structures, norms and judgements which, by their exclusion, have largely defined science in the past.

The re-orientation of science to consider coupled social–environmental problems and to view humans as integral to ecosystems is accompanied by a breakdown of the traditional divide between science and society. New forms of interaction between scientists and wider society are developing, driven by concepts such as adaptive management and adaptive governance,[12,72] and involving components of civic science, including participation, engagement, and knowledge exchange.[73] These interactions are challenging the historical role of the scientist as a neutral, objective, detached specialist. Scientists are becoming part of a broader network of actors involved in the production, exchange and application of knowledge for the management of social–environmental systems. Further, as science moves into the realm of social–environmental systems, not only do social, economic and political perspectives become important for defining research agendas and for giving context to research findings, but science and scientific institutions become objects of increasing interest to society. There are growing demands for civil society to be more closely involved in setting the strategic direction for science, and for science to become more transparent, 'democratised' and responsive to the concerns and priorities of society.[68]

However, science and the scientific method have historically progressed through increased disciplinary specialisation, rather than through embracing interdisciplinary approaches. Particularly in the last two centuries, challenges to science largely revolved around problems that appeared definable and consensual,[51,74] and were addressed through the development of predominantly technical solutions. The scientific method and resulting solutions were well aligned with the 'command and control' management paradigm, where problems were agreed and solutions developed and implemented.[72] Taking water as a specific case serves to illustrate these points. Many of the advances in water sanitisation and pollution control that were delivered across Europe between the mid-1800s and the late-1900s were underpinned by scientific, technical solutions that addressed the effects and not the cause of the problems. These advances include public water supply, sewerage collection and treatment, and the control of water-borne diseases. Whilst delivering undeniable benefits to society, this historical context has led science towards increased fragmentation of disciplines and of knowledge, a reliance on reductionism and the disassociation of humans from the environment.[71] Such fragmentation of disciplines and knowledge, and the lack of system-orientated frameworks, is a problem common to both natural and social sciences.[51]

The problems that science is now being asked to address, such as supporting the design and implementation of the ICM framework, are complex. They deal with undefined or poorly defined problems, involve a plurality of legitimate perspectives over problems and desirable solutions rather than a consensus, and require flexible, adaptable, system-orientated approaches. Such problem and decision-making contexts are not well addressed by traditional reductionist, experimental science alone. Science and many scientists are, therefore, increasingly torn between a reductionist, disciplinary model of working that seems increasingly detached from pressing social–environmental problems and an alternative model that stresses the importance and relevance of integrative and interdisciplinary approaches, but that seems incompatible with the methodological foundations of science. Such tensions have been clearly articulated in the ecological sciences (see, for example, ref. 7), but are also true of several other disciplines.

1.5.6.3 Supporting Interdisciplinary Research

Whilst calls to embrace interdisciplinarity when developing frameworks such as ICM might appear to present new challenges to science, they are in fact renewed challenges. Interdisciplinary research has been a feature of science since the early twentieth century, and research into the practice of interdisciplinary science since the 1950s.[75] Bibliometric studies based on analyses of academic journals, papers, authors and citations provide evidence of interdisciplinary outputs, but there remain substantial variations between disciplines, and working is often across closely related disciplines rather than between those that are initially more disparate.[65] Bibliometric studies also focus on academic papers as a traditional metric to describe interdisciplinary working. As such they are inherently limited by their inability to capture other forms of outcome that are relevant to the co-production of knowledge through interdisciplinary research. However, identifying metrics other than paper outputs that are able to describe these broader aspects of collaboration remains a significant challenge (see, for example, ref. 76).

Despite the current and likely future demand for interdisciplinary research, and some historical precedent for such research, it appears that both the biophysical and social sciences remain relatively poorly structured and equipped for meeting the challenges of helping to develop and implement frameworks such as ICM. Several key elements must be addressed if the integrated, interdisciplinary science that is required to support ICM is to be produced and incorporated effectively within such frameworks.

Building Interdisciplinary Research Teams

Addressing the challenges within the ICM framework will require the establishment and development of integrated teams of scientists. Such teams should be united by the desire to address a common issue or question – one of sufficient complexity and intellectual rigour to stimulate their multiple inputs. However, this requires more than bringing multiple, separate perspectives to bear on a common issue, problem or scale. It involves a much greater degree of integration during the framing, design and implementation of research, and as a result close collaboration between individual members of the research team. Developing such teams requires effective leadership of the group, and disciplinary equity commensurate with the nature of the issue or question to be addressed. It also requires personal and professional qualities of individual members of these teams, including commitment, energy, receptiveness, trust and confidence.[77] The time and opportunity to develop these qualities, and to develop working relationships between members of the team, is crucial and can be provided through a range of activities, including seminars, workshops, field visits and social gatherings. Relatively small amounts of funding to support these team preparation and development activities can be crucial to subsequent success. Some would also argue for a clearer articulation of the need for, and benefits of, integrated and interdisciplinary approaches. Such articulation should emphasise both the moral imperative associated with the societal challenges that integrated research seeks to address and the intellectual stimulation that can be found through such research.

Managing Boundaries

Research suggests that a decade or more may be required before the impact of science, technology and ideas on the evolution of issues may be reliably identified.[78,79] One reason for such ineffective percolation of the outcomes of research into policy- and broader decision-making is often seen to be the poor organisation and operation of the boundary between science and wider society. The different knowledge frameworks, methodological frameworks, priorities and timescales create both real and perceived gaps between communities who produce scientific knowledge and those who use and apply it. Achieving more effective and efficient use of science in decision-making is dependent on better work at the boundary between these communities, so-called boundary-management, which often relies on the use of boundary-objects, such as models, to facilitate such management.[80] The management of these boundaries can be a role for formalised institutions and procedures that are created specifically to act between separate communities, or such roles can exist within individual communities.[80] However, all too often such boundary management activities are taken on by individuals within separate communities without support from formalised institutions or procedures.[78] The challenge and opportunity is to harness the activities and networks of individuals within a supportive institutional and procedural context, to promote more successful boundary management.

Boundary management is an important element of the ICM framework, and such management must be informed by revised perspectives on 'what' is to be managed and 'when' this management should occur. The traditional model describing the transfer of information from science-producing to science-adopting communities is essentially uni-directional and focuses on the transfer of tangible data and information, often occurring towards the end of research projects.[81] In contrast, the exchange of knowledge involves tangible material but also tacit, non-formalised elements associated with intuition, contextual information and experiences, which are essential to the successful implementation of frameworks such as ICM. The exchange of such knowledge cannot be achieved through uni-directional transfer of data and information, however well a boundary is organised. Instead, new approaches that move towards true co-production of knowledge and bi-directional knowledge exchange are required to support frameworks such as ICM. Greater collaboration between scientists and wider society during the co-production of knowledge highlights one future role for scientists within the broader network of actors that must be brought together to manage social–environmental systems.[28] Co-production of knowledge may also represent a means of managing boundaries and breaking down barriers, such as the often quoted science–policy divide.[82] However, it must be recognised that for organisations in both science-producing and science-adopting communities there are significant structural and institutional barriers to co-production that must first be overcome.[78]

Funding, Institutional Support and Training for Interdisciplinary Research

Whilst the demand for more integrated and interdisciplinary research may appear clear, funding to support such research is often more difficult to access. Developing proposals for truly integrated and interdisciplinary research that compete successfully with more discipline-specific proposals requires skills and experience that are not common within academic institutions. In some cases there is at least an implicit perception amongst both scientists and the funders of science that interdisciplinary projects, in particular those addressing so-called 'applied' questions through the use of collaborative and participatory approaches, are less prestigious and less scientifically rigorous than those focused on more intellectual and theoretical issues.[83] Such perceptions represent problems not only for the engagement of scientists during the writing of proposals, but also during subsequent peer review when interdisciplinary proposals are often seen to be both less scientifically rigorous and more risky than their discipline-specific competitors.[68] Whilst individual funding programmes may seek to specifically support interdisciplinary projects, for example the RELU programme in the UK, they are themselves dependent on arguments for the need and value of interdisciplinarity being successfully made by individuals within the relevant funding organisations.[68] Such programmes are also unable to address the barriers to funding interdisciplinary science that persist within mainstream funding routes.

Negotiating such funding hurdles exposes a number of subsequent difficulties when trying to implement an interdisciplinary research project within existing structures of academic institutions. Support, reward and attractive career trajectories for academics who engage in the time-consuming activities that are fundamental to interdisciplinary teams, and the working relationships that underpin such teams, are often lacking, particularly for academics at an early stage in their career. Performance indicators based solely on the outputs of research, for example in the form of peer-reviewed academic papers that form the core of the UK's Research Assessment Exercise and forthcoming Research Excellence Framework, fail to recognise the broader range of outcomes that are crucial to successful interdisciplinary working. These outcomes include building of capacity, trust and credibility with partners, and knowledge exchange that moves beyond only formalised, tangible papers. Legitimising the existence of interdisciplinary researchers, whose expertise and role spans across disciplines, is sometimes a difficult task within institutions that remain divided along departmental or faculty boundaries.

Finally, training the next generation of interdisciplinary researchers requires new forms of postgraduate research training, supervision and project development, recognising the particular practices that underpin interdisciplinary research. For example, collaborative framing of research projects at the outset of interdisciplinary research programmes is crucial if the programme as a whole is to remain integrated and interdisciplinary throughout. This requires new practice in research, expanding existing foci on synthesis and collation towards the end of research and incorporating true collaboration throughout the research process. Such practice must be supported during the supervision and assessment of new researchers.

1.5.7 Future Challenges Facing Implementation of ICM

The concept of more integrated management of land and water within river catchments is not new, and has a history of at least several decades.[1–2,5] However, the ICM framework we present differs from such management in one particularly important respect. We envisage the framework to be participatory from the outset, incorporating the perspectives, values and knowledge of multiple participants throughout. There are very few examples in the world where such participatory approaches to the management of river catchments have been put fully and successfully into practice. Recognising this, we have sought to map out a possible trajectory towards future management of river catchments based on the ICM framework, and to highlight important roles for science within such a framework. Putting such a framework into practice will be dependent on progress within three main areas:

1.5.7.1 Developing a Revised Conceptual Understanding of ICM

The management of river catchments has traditionally been perceived as the role of government-led planning and regulatory bodies, with wider stakeholder interaction being achieved largely through consultation with interested parties. Implementing a framework such as ICM requires a fundamental shift in this perception, towards one that recognises the importance of participatory, collaborative approaches that actively engage a far broader group of actors throughout the process of catchment management. A framework such as ICM cannot be implemented solely by better co-ordination of public bodies within existing institutional and management frameworks, a 'myth' that has pervaded integrated approaches to catchment management in the past.[2] Whilst better co-ordination of such bodies is undoubtedly required, the participatory approach that underpins the ICM framework significantly extends both the number of actors involved and the depth to which they are engaged.

The water-centric focus of past river basin management must be expanded to recognise the inter-dependencies between land and water within river catchments. Whilst such inter-dependencies are increasingly accepted, frameworks such as ICM also rely on a less widely adopted conceptualisation of river catchments as social–environmental systems. This conceptualisation requires us to consider more than just environmental processes and objectives, and to fully incorporate social rules and the values, interests and preferences of human actors within the management process. The ICM framework is based on the understanding that within social–environmental systems a plurality of understanding, objectives and approaches to the future will exist. Given such complex, 'messy' problems, the power of traditional techno-rational solutions is often reduced, and an increasing emphasis must be placed on alternative solutions, *e.g.* joint commitments to changed behaviour or revised systems of governance.

1.5.7.2 Designing Suitable Frameworks for Implementing ICM

Although the concept of more holistic, integrated and co-ordinated management of river catchments has become increasingly accepted by policy makers, environmental managers and scientists, the translation of the concept into operational terms has proved to be a significant barrier. Whilst conceptually attractive, without being able to implement collaborative and participatory approaches for the management of river catchments they will rapidly lose validity.[6] One cause of this implementation gap is the inertia associated with current understanding of ICM. Successful implementation of ICM will not be achieved when the process continues to be perceived as one concerned with environmental systems, one focussed on techno-rational solutions and one achieved through better co-ordination of government-led bodies. However, researchers and managers must also engage in the development of frameworks that enable the concept of ICM to be operationalised.[2] The framework that we present here is an initial attempt to engage in such a task. We fully envisage an iterative development process before such a framework is ready to be tested through application to real-world situations.

1.5.7.3 Supporting the Implementation of Frameworks for ICM

Whilst revised conceptualisations of ICM and the development of frameworks that enable these conceptualisations to be operationalised are significant challenges, addressing these challenges will not by itself ensure that ICM is effectively implemented. There are several other key steps that we believe must be addressed in the future:

- The current roles and responsibilities for catchment management, certainly in a UK context, result in a centralised and top-down structure to decision-making, focused within a relatively small number of government-led agencies and organisations. Successful implementation of a framework such as ICM requires greater release and decentralisation of this power, promoting new forms of polycentric governance that are aligned with the participatory and collaborative basis to ICM. Whether the relevant agencies and organisations are willing or able to release these powers, certainly under current institutional structures and constraints, is doubtful. There is a requirement for political, financial and legislative commitment to the concept of ICM at the catchment-scale.
- Despite the need for decentralisation of power and a greater commitment to participation, a framework such as ICM will still require substantial management and facilitation. This immediately raises the question of 'who' should be responsible for such activities. It is our opinion that the management of such a framework must be the responsibility of an independent and impartial body, agreed by the participants in the process to be valid. Such a body must be supported with the necessary political, financial and legislative power to ensure that the framework provides sufficient teeth with which to take decisions and ensure their subsequent delivery. In the context of the UK such bodies are almost entirely lacking, particularly at the catchment scale, a scale that is intermediate between the larger river basin districts defined under the WFD and the multitude of smaller-scale, local activities occurring within many catchments. This intermediate scale remains a crucial organisational gap that will constrain the implementation of a framework such as ICM within river catchments. Enlightened exceptions in the UK include the Mersey Basin Campaign (http://www.merseybasin.org.uk/) and the Tweed Forum (http://www.tweedforum.com/).
- Frameworks such as ICM rely on the commitment of participants to actively engage in the process. Obtaining this commitment is a significant challenge, one that is often underestimated because the assumptions are made that participants will obviously be willing to engage, and will be able to resource the time and financial commitments necessary to do so. These should be reconsidered and it should be assumed that the commitment of participants to a potentially slow, emergent and iterative process, such as ICM, is far from certain and will require sustained negotiation and facilitation.[2]
- There is a need to be able to demonstrate the advantage of implementing a framework such as ICM. These advantages must be highlighted to sustain broad commitment to the concept of ICM as compared to other potential management paradigms, and more practically to ensure continued commitment from participants to the process. Particularly during the early

stages of the ICM framework, the emphasis will be on developing networks of actors and on sharing knowledge of the catchment. Such activities are less likely to yield tangible products such as reports, strategies and management decisions that have traditionally been key metrics of the progress of management frameworks. There is a requirement to recognise and value other outcomes that emerge as a result of the ICM framework, *e.g.* the organisation of joint workshops, meetings and conferences that are crucial for stimulating knowledge exchange and capacity building. These are also important benefits of the ICM framework, and they should be included within judgements as to the value of such frameworks alongside more tangible outputs, and ultimately the quality of decisions made as a result of the framework.

References

1. N. Watson, G. Walker and W. Medd, Critical perspectives on integrated water management, *Geograph. J.*, 2007, **173**, 297–299.
2. N. Watson, Integrated river basin management A case for collaboration. International, *J. River Basin Manage*. 2004, **2**, 243–257.
3. D. Woods, *The Evolution of River Basin Management in England and Wales*, Foundation for Water Research, London, 2006.
4. R. Brown, N. Keath and T. Wong, Transitioning to Water Sensitive Cities: Historical, Current and Future Transition States, in Proceedings of 11th International Conference on Urban Drainage, Edinburgh 31st August-5th September 2008.
5. B. Mitchell, Integrated water resource management, institutional arrangements and land-use planning, *Environ. Planning A*, 2005, **37**, 1335–1352.
6. A. K. Biswas, Integrated water resources management: A reassessment, *Water Int.*, 2004, **29**, 248–256.
7. C. S. Holling, Two cultures of ecology, *Conservation Ecology* (online), 1998, URL: http://www.consecol.org/vol2/iss2/art4/.
8. J. Carter, Spatial planning, water and the Water Framework Directive: insights from theory and practice, *Geograph. J.*, 2007, **173**, 330–342.
9. S. Kidd and D. Shaw, Integrated water resource management and institutional integration: realizing the potential of spatial planning in England, *Geograph. J.*, 2007, **173**, 312–329.
10. Council of the European Community (CEC) 2000. Directive 2000/60/EC of the European Parliament and of the Council of 23 October 2000 establishing a framework for Community action in the field of water policy. *Official J. Eur. Communities*, 2000, L327/1.
11. B. Moss, The Water Framework Directive: Total environment or political compromise?, *Sci. Total Environ.*, 2008, **400**, 32–41.
12. C. Folke, T. Hahn, P. Olsson and J. Norberg, Adaptive governance of social-ecological systems, *Annu. Rev. Environ. Resources*, 2005, **30**, 441–473.
13. Millennium Ecosystem Assessment (2005) – *Synthesis Report*, Island Press, Washington D.C., 2005.

14. R. Costanza, R. d'Arge, R. De Groot, S. Farber, M. Grasso, B. Hannon, K. Limburg, S. Naeem, R. V. O'Neill, J. Paruelo, R. G. Raskin, P. Sutton and M. van den Belt, The value of the world's ecosystem services and natural capital, *Nature*, 1997, **387**, 253–260.
15. G. C. Daily, *Nature's Services. Societal Dependence on Natural Ecosystems*, Island Press, Washington D.C., 1997.
16. J. Boyd and S. Banzhaf, What are ecosystem services? The need for standardized environmental accounting units, *Ecol. Economics*, 2007, **63**, 616–626.
17. R. S. De Groot, M. A. Wilson and R. M. J. Boumans, A typology for the classification, description and valuation of ecosystem functions, goods and services, *Ecol. Economics*, 2002, **41**, 393–408.
18. R. K. Turner and G. C. Daily, The ecosystem services framework and natural capital conservation, *Environ. Resource Economics*, 2008, **39**, 25–35.
19. R. K. Turner, S. Georgiou and B. Fisher, *Valuing Ecosystem Services. The Case of Multi-functional Wetlands*, Earthscan, London, 2008.
20. A. Castelletti and R. Soncini-Sessa, Bayesian Networks and participatory modelling in water resource management, *Environ. Model. Software*, 2007, **22**, 1075–1088.
21. P. Reichert, M. Borsuk, S. Hostmann, S. Schweizer, C. Sporri, K. Tockner and B. Truffer, Concepts of decision support for river rehabilitation, *Environ. Model. Software*, 2007, **22**, 188–201.
22. R. K. Turner, S. Georgiou, R. Brouwer, I. J. Bateman and I. J. Langford, Towards an integrated environmental assessment for wetland and catchment management, *Geograph. J.*, 2003, **169**, 99–116.
23. R. Grimble and K. Wellard, Stakeholder methodologies in natural resources management: a review of principles, contexts, experiences and opportunities, *Agric. Syst.*, 1997, **55**, 173–193.
24. M. S. Reed, Stakeholder participation for environmental management: a literature review, *Biol. Conservation*, 2008, **141**, 2417–2431.
25. C. Prell, K. Hubacek and M. S. Reed, Social network analysis and stakeholder analysis for natural resource management. *Society Nat. Resources*, 2009, **22**, 501–518.
26. S. Cinderby, *How communities can use geographical information systems*, in *Towards Understanding: Community People and Places.*, ed. C. Clay, M. Madden and L. Potts, Palgrave Macmillan, UK, 2007.
27. C. Prell, K. Hubacek, M. Reed, C. Quinn, N. Jin, J. Holden, T. Burt, M. Kirby and J. Sendzimir, If you have a hammer everything looks like a nail: traditional versus participatory model building, *Interdisciplinary Sci. Rev.*, 2007, **32**, 263–282.
28. C. Pahl-Wostl, The implications of complexity for integrated resources management, *Environ. Model. Software*, 2007, **22**, 561–569.
29. HarmoniCOP, Learning together to manage together: improving participation in water management, 2005, www.harmonicop.uos.de/harmonicophandbook.pdf.

30. C. Pahl-Wostl, M. Craps, A. Dewulf, E. Mostert, D. Tabara and T. Taillieu, Social learning and water resources management, *Ecol. Soc.*, 2007, **12**, 5. URL: http://www.ecologyandsociety.org/vol12/iss2/art5/.
31. A. P. Kinzig, Bridging disciplinary divides to address environmental and intellectual challenges, *Ecosystems*, 2001, **4**, 709–715.
32. D. G. Raffaelli, P. C. L. White, A. Renwick, J. C. R. Smart and C. Perrings, *The health of ecosystems: the Ythan estuary case study*, in *Handbook of Indicators for Assessment of Ecosystem Health*, ed. S. E. Jørgensen, R. Costanza and F. L. Xu, CRC Press, Boca Raton, FL, 2004.
33. M. S. Reed, A. J. Doughill and T. R. Baker, Participatory indicator development: what can ecologists and local communities learn from each other?, *Ecol. Appl.*, 2008, **18**, 1253–1269.
34. L. Maltby, A. Paetzold and P. Warren, Sustaining industrial activity and ecological quality: the potential role of an ecosystem services approach, in *Ecology of Industrial Pollution: Remediation, Restoration and Preservation*, ed. L. C. Batty, K. Hallberg and A. Jarvis, Blackwell, UK, in press (due June 2010).
35. E. Beinat and P. Nijkamp, *Multicriteria Analysis for Land-Use Management*, Kluwer Academic Publishers, Dordrecht, The Netherlands, 1998.
36. I. Bateman and K. G. Willis, *Valuing Environmental Preferences: Theory and Practice of the Contingent Valuation Method in the US, EU and Developing Countries*, Oxford University Press, Oxford, UK, 1999.
37. N. Hanley, R. E. Wright and V. Adamowicz, Using choice experiments to value the environment, *Environ. Resource Economics*, 1998, **11**, 413–428.
38. R. K. Turner, D. Burgess, D. Hadley, E. Coombes and N. Jackson, A cost-benefit appraisal of coastal managed realignment policy, *Global Environ. Change*, 2007, **17**, 397–407.
39. M. Svendsen, P. Wester and F. Molle, *Managing river basins: an institutional perspective*, in *Irrigation and River Basin Management: Options for Governance and Institutions*, ed. M. Svendsen, CABI Publishing, Wallingford, 2005, pp. 1–18.
40. S. R. Carpenter and L. H. Gunderson, Coping with collapse: ecological and social dynamics in ecosystem management, *BioScience*, 2001, **6**, 451–457.
41. Ph. Quevauviller, P. Balabanis, C. Fragakis, M. Weydert, M. Oliver, A. Kaschl, G. Arnold, A. Kroll, L. Galbiati, J. M. Zaldivar and G. Bidoglio, Science-policy integration needs in support of the implementation of the EU Water Framework Directive, *Environ. Sci. Policy*, 2005, **8**, 203–211.
42. Y. Liu, H. Gupta, E. Springer and T. Wagner, Linking science with environmental decision making: Experiences from an integrated modeling approach to supporting sustainable water resources management, *Environ. Model. Software*, 2008, **23**, 846–858.
43. M. Gibbons, C. Limoges, H. Nowotny, S. Schwartzmann, P. Scott and M. Trow, *The New Production of Knowledge*, Sage Publications, London, 1994.
44. J. Horgan, *The End of Science*, Abacus, Little, Brown and Co., London, 1996, pp. 324.

45. M. P. Hare and C. Pahl-Wostl, Stakeholder categorisation in participatory integrated assessment processes, *Integrated Assesss.*, 2002, **3**, 50–62.
46. S. Connelly and T. Richardson, Exclusion: the necessary difference between ideal and practical consensus, *J. Environ. Planning Manage.*, 2004, **47**, 3–17.
47. M. S. Reed, A. Graves, N. Dandy, H. Posthumus, K. Hubacek, J. Morris, C. Prell, C. H. Quinn and L. C. Stinger, Who's in and why? A typology of stakeholder analysis methods for natural resource management, *J. Environ. Manage.*, 2009, **90**, 1933–1949.
48. G. Harris, This is not the end of limnology (or of science): the world may well be a lot simpler than we think, *Freshwater Biol.*, 1999, **42**, 689–706.
49. G. Harris, Integrated assessment and modelling: an essential way of doing science, *Environ. Model. Software*, 2002, **17**, 201–207.
50. B. Walker, C. S. Holling, S. R. Carpenter and A. Kinzig, Resilience, adaptability and transformability in social-ecological systems. *Ecol. Soc.* (online). URL: http://www.ecologyandsociety.org/vol9/iss2/art5/ (2004).
51. C. Pahl-Wostl, Towards sustainability in the water sector – The importance of human actors and processes of social learning, *Aquatic Sci.*, 2002, **64**, 394–411.
52. C. S. Holling, *Adaptive Environmental Assessment and Management*, John Wiley & Sons, Inc., New York, 1978.
53. J. C. Refsgaard, J. P. van der Sluijs, A. L. Hojberg and P. A. Vanrolleghem, Uncertainty in the environmental modelling process – A framework and guidance, *Environ. Model. Software*, 2007, **22**, 1543–1556.
54. J. Alcamo, Scenarios as tools for international environmental assessments, European Environment Agency, Copenhagen, *Environmental Issue Report*, Experts Corner Report, Prospects and Scenarios No. 5, 2001.
55. M. Matthies, C. Giupponi and B. Ostendorf, Environmental decision support systems: Current issues, methods and tools, *Environ. Model. Software*, 2007, **22**, 123–127.
56. P. Parker, R. Letcher, A. Jakeman, M. B. Beck, G. Harris, R. M. Argent, M. Hare, C. Pahl-Wostl, A. Voinov, M. Janssen, P. Sullivan, M. Scoccimarro, A. Friend, M. Sonnenshein, D. Barker, L. Matejicek, D. Odulaja, P. Deadman, K. Lim, G. Larocque, P. Tarikhi, C. Fletcher, A. Put, T. Maxwell, A. Charles, H. Breeze, N. Nakatani, S. Mudgal, W. Naito, O. Osidele, I. Eriksson, U. Kautsky, E. Kautzky, B. Naeslund, L. Kumblad, R. Park, S. Maltagliati, P. Girardin, A. Rizzoli, D. Mauriello, R. Hoch, D. Pelletier, J. Reilly, R. Olafsdottir and S. Bin, Progress in integrated assessment and modelling, *Environ. Model. Software*, 2002, **17**, 209–217.
57. A. J. Jakeman, M. B. Beck and M. J. McAleer, (eds), *Modelling, Change in Environmental Systems*, John Wiley & Sons Ltd, Chichester, 1993.
58. S. M. Cuddy and C. Gandolfi, Integrated catchment modeling and decision support, *Environ. Model. Software*, 2004, **19**, 989–990.
59. S. Moss, T. Downing and C. Pahl-Wostl, Agent based integrated assessment modeling, *Integrated Assesss.*, 2001, **2**, 17–30.

60. S. Moss, Agent based modeling for integrated assessment, *Integrated Assess.*, 2002, **3**, 63–77.
61. R. S. Sojda, Empirical evaluation of decision support systems: Needs, definitions, potential methods, and an example pertaining to waterfowl management, *Environ. Model. Software*, 2007, **22**, 269–277.
62. (a) J. C. Refsgaard, H. J. Henriksen, W. G. Harrar, H. Scholten and A. Kassahun, Quality assurance in model based water management – review of existing practice and outline of new approaches, *Environ. Model. Software*, 2005, **20**, 1201–1215; (b). W. E. Walker, P. Harremoes, J. Rotmans, J. P. van der Sluijs, M. B. A van Asselt, P. Janssen and M. P. Krayer von Krauss Defining uncertainty: a conceptual basis for uncertainty management in model-based decision support, *Integrated Assess.*, 2003, **4**, 5–17.
63. A. Dewulf, M. Craps, R. Bouwen, T. Taillieu and C. Pahl-Wostl, Integrated management of natural resources: dealing with ambiguous issues, multiple actors and diverging frames, *Water Sci. Technol.*, 2005, **52**, 115–124.
64. P. Kloprogge and J. van der Sluijs, The inclusion of stakeholder knowledge and perspectives in integrated assessment of climate change, *Climatic Change*, 2006, **75**, 259–289.
65. F. Morillo, M. Bordons and I. Gomez, Interdisciplinarity in science: A tentative typology of disciplines and research areas, *J. Am. Soc. Information Sci. Technol.*, 2003, **54**, 1237–1249.
66. HM Treasury, Department for Education and Skills and Department of Trade and Industry, *Science and Innovation Investment Framework 2004–2014*, HMSO, Norwich, 2004.
67. J. Tait, R. Williams, A. Bruce, C. Lyall, E. Gravalos, P. Rodriquez, E. Jolivet, U. Jorgensen and J. Laessoe, Interdisciplinary integration in the fifth framework programmes (II-FP5) Final report to European Commission, Contract No. SEAC-1999-00034 (2002). http://www.supra.ed.ac.uk/publications/FINAL REPORT.pdf
68. P. Lowe and J. Phillipson, Reflexive interdisciplinary research: The making of a research programme on the Rural Economy and Land Use, *J. Agric. Economics*, 2006, **57**, 165–184.
69. Office of Science and Technology, Quinquennial review of the grant awarding research councils, Report of the review team, Stage 2. Office of Science and Technology and DTI, London, 2001. http://www.rcuk.ac.uk/cmsweb/downleads/rcuk/publications/
70. D. Sarewitz and R. A. Pielke, The neglected heart of science policy: reconciling supply of and demand for science, *Environ. Sci. Policy*, 2007, **10**, 5–16.
71. G. A. Bradshaw, M. Bekoff, Ecology and social responsibility: the re-embodiment of science, *Trends Ecol. Evolution*, **16**, 460–465.
72. C. S. Holling and G. K. Meffe, Command and control and the pathology of natural resource management, *Conservation Biol.*, 1996, **10**, 328–337.

73. F. Clark and D. L. Illman, Dimensions of civic science, *Sci. Commun.* (online), 2001, URL: http://scx.sagepub.com/cgi/content/abstract/23/1/5.
74. H. W. J. Rittel and M. M. Webber, Dilemmas in a general theory of planning, *Policy Sci.*, 1973, **4**, 155–169.
75. J. Qin, F. W. Lancaster and B. Allen, Types and levels of collaboration in interdisciplinary research in the sciences, *J. Am. Soc. Information Sci.*, 1997, **48**, 893–916.
76. D. Stuart, M. Thelwall and G. Harries, UK academic web links and collaboration – an exploratory study, *J. Information Sci.*, 2007, **33**, 231–246.
77. P. C. L. White, A. de Bruin, S. Cinderby, A. Holt, D. Raffaelli and M. Huby, Enhancing the effectiveness of policy-relevant integrative research in rural areas, *Area*, in press.
78. B. W. J. Surridge and R. C. Harris, Science-driven Integrated River Basin Management: a mirage?, *Interdisciplinary Sci. Rev.*, 2007, **32**, 298–312.
79. O. R. Young, Evaluating the success of international environmental regimes: Where are we now?, *Global Environ. Change*, 2002, **12**, 73–77.
80. D. W. Cash, W. C. Clark, F. Alcock, N. M. Dickson, N. Eckley, D. H. Guston, J. Jager and R. B. Mitchell, Knowledge systems for sustainable development, *Proc. Natl. Acad. Sci. USA*, 2003, **100**, 8086–8091.
81. D. J. Roux, K. H. Rogers, H. C. Biggs, P. A. Ashton and A. Sergeant, Bridging the science-management divide: moving from unidirectional knowledge transfer to knowledge interfacing and sharing, *Ecol. Soc.* (online), 2006. URL: http://www.ecologyandsociety.org/vol11/iss1/art4/.
82. M. C. Lemos and B. J. Morehouse, The co-production of science and policy in integrated climate assessments., *Global Environ. Change*, 2005, **15**, 57–68.
83. R. J. Cabin, Science-driven restoration: A square grid on a round earth? *Restoration Ecol.*, 2007, **15**, 1–7.

CHAPTER 1.6

Analysis of EC Framework Programme and LIFE Projects for their Relevance to the Water Framework Directive

KEES J. M. KRAMER[a] AND XENIA SCHNEIDER[b]

[a] Mermayde, P.O. Box 109, NL-1860 AC Bergen, The Netherlands; [b] XPRO Consulting Ltd, 7 Nelson St., CY-2021 Strovolos, Lefkosia, Cyprus

1.6.1 Introduction

This chapter discusses the issues of disseminating scientific results in order to capitalise on the created knowledge to achieve a sustainable environment. The chapter identifies and analyses challenges and recommendations concerning the dissemination of scientific knowledge generated by projects related to the Water Framework Directive.

A sustainable environment is of vital importance; hence the European Union has funded research in the environmental area for several decades and has invested millions of euros in supporting scientific development. The results from these investments must be utilised and applied to achieve the benefits that they promised. Today there appears to be a gap between generating, disseminating and utilising scientific results by stakeholders other than scientists. Stakeholders may be the public, policy makers/implementers, industry and other practitioners. 'Science is too difficult to understand' and 'Scientific reports are written by scientists mainly for scientists' are opinions that are often

Water System Science and Policy Interfacing
Edited by Philippe Quevauviller

Published by the Royal Society of Chemistry, www.rsc.org

heard from citizens and policy makers alike. Whether these are true or not, there is nevertheless a challenge for scientists to disseminate to the non-scientific world what their work is about, what is their drive for this work and what is the importance of their results.

Furthermore, the scientific community should not only have a drive to disseminate their results to the scientific community, *e.g. via* publication in reviewed scientific journals. Especially, research that is intended to be scientific support to policy (SSP) should be communicated to policy makers in terms of, for example, how their scientific results may be used for the implementation of European Commission Directives such as the Water Framework Directive, the WFD.[1] Scientists usually speak and report in scientific terminology, a 'language' not always understood by the policy field (Chapter 1.3).

But are they the only group that is to blame? Policy makers and policy implementers also speak their own 'language' that is not easily understood by scientists, even by those that cater for SSP. Where the policy area would require more input from the scientific community, in return this scientific community may greatly benefit from understanding the policy needs. In essence, both areas would largely benefit from better and improved communication.

1.6.1.1 Research Dissemination and Communication

Since the adoption of the Water Framework Directive by the Member States it was recognised that a common approach to some of the water challenges is necessary through sharing of knowledge, expertise and information for facilitating, as far as possible, a coherent and harmonious implementation of the Directive. EC environmental research has two main funding components: Research and Technological Development (RTD), funded through the Framework Programmes (FP) of DG Research, and the environmental research demonstration projects funded by DG Environment, notably in the LIFE Programmes[2] (*see* Chapter 1.4 for further details). Viewing the dissemination–communication interactions, as displayed in Figure 1.6.1, it becomes clear that there is one more stakeholder to be considered: the public audience. They also require and need to be informed about the research outcomes, not only because they have, through taxes, paid for the public funded research.

The interactions in Figure 1.6.1 summarise the challenges as well as the needs:

1: RTD primary output by researchers is traditionally *via* (fundamental) research papers in recognised scientific journals and (applied) research reports, the latter including those to the European Community. These dissemination products are of a technically high level but unfortunately full of jargon and intended for the scientific 'specialists'. These dissemination products mostly stay within the research sector.

2: Policy makers/implementers equally have their own publications and reports, full of jargon, and intended for the policy 'specialists'. These dissemination products also mostly stay within the policy sector.

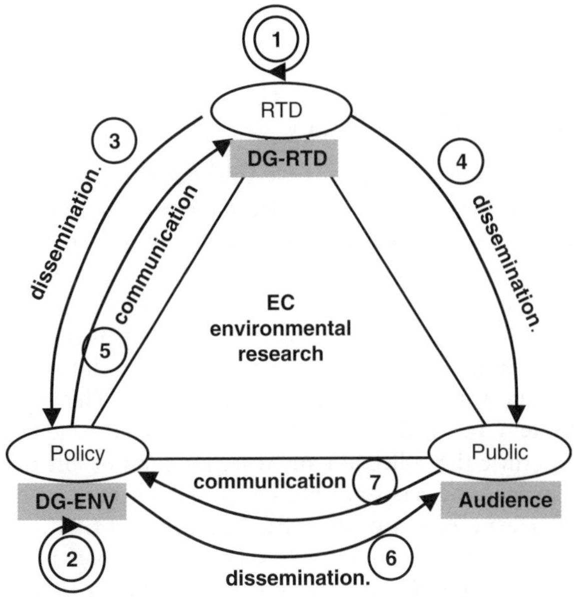

Figure 1.6.1 Interacting links in dissemination and communication of environmental research.

3: There is a need for making (relevant) RTD results known to the policy sector. This requires a different 'language' and an adapted level of technical detail.
5: *Vice versa*, the (EC) policy sector should disseminate the outcome of environmental research demonstration projects (and their needs for research) to the RTD sector; true communication should be bi-directional and thus 'bridging the gap'.
4 and 6: The RTD and Policy sectors should disseminate their environmental research results to the public, while public communication to policy usually (7) goes *via* the democratic process.

In the water area, for example, science–policy integration seems to be one of the most complex challenges that scientific and policy-making communities are facing.[3,4] Repeatedly it has been discussed that better means for supporting a continuous exchange between scientists, policy makers and practitioners need to be identified and established (Chapters 1.2 and 5.3).[5]

All stakeholders have the right to know where and why the research budgets are spent, what has been achieved and discovered in the European Research Area and how the European Community (EC) will benefit from these results. Likewise, scientists, possibly from other sectors of the scientific universe, may largely benefit from proper dissemination of RTD results, if they knew where to find them. To whatever audience at whatever level, communication and dissemination are key factors in our information society for achieving a sustainable environment.

1.6.1.2 Bridging Science–Policy

The EU bodies (DG ENV, JRC, ESTAT and EEA) and the then EU25 Member States, Norway, Bulgaria and Romania (at the time candidate countries) believed that there was a need for streamlining the reporting process, gathering more useful and relevant information and making the exchange process as efficient as possible using modern technology for the information on water. At their Water Directors' meeting in November 2003 they agreed on the: 'Development of a new, comprehensive, European data and information management system for water, including river basins'. This system, the Water Information System for Europe (WISE) was thus launched. WISE was an initiative to modernise the collection and dissemination of information on water policy across Europe, and has since become a major source of information.[6]

Under the EC Harmoni-CA project (EC FP5 contract EVK1-2001-00192; see Chapter 2.4) an information system has been developed to bridge science–policy for the topic of the EC Water Framework Directive: a 'portal for science & technology transfer to policy making & implementation in integrated water resources management', in short the 'WISE-RTD web portal' at http://www.wise-rtd.info/ (Chapter 4.2).[7]

Under the EC SPI-Water project (EC contract 044357;[8]) this portal was further developed into an easy to use tool for both dissemination and retrieval of RTD results. Part of this project's objectives included the collection of EC research projects relevant for the WFD implementation, to 'translate' the abstract into non-scientific language, and to upload this information on research output (such as documents, tools, case studies and experiences) to the web-portal.[9] This chapter discusses the problems and challenges encountered in the collection of information and the (apparent) limited 'shelf-life' of some research projects and their deliverables, thus hampering the bridging of the science–policy interface. Several options are provided in the recommendations, which may help to overcome limitations in communicating science to policy.

1.6.2 Collection of EC Research Project Information

The strategy on what information to include in the WISE-RTD web portal consisted of several steps, in particular the collection of as many as possible RTD projects, the 'long list'. From this long-list a selection of the most relevant projects was made, based on a set of criteria developed beforehand, leading to a 'short list'. One important criterion was whether the project results had a direct link to the WFD tasks, *i.e.* to the common implementation strategy (CIS Guidelines); if there was only a 'potential link' they were classified as 'may be' and treated with a lower priority.

1.6.2.1 EC Projects Considered

The 'long list' of EC funded projects was based on the following EC programmes:

- RTD projects (DG Research), mainly consisting of FP5 and FP6 projects (Chapter 1.4);
- research demonstration projects from the LIFE Programme (DG Environment), notably LIFE Environment (LIFE-ENV), LIFE Nature (LIFE-NAT) and LIFE Third Countries (LIFE-TCY);
- RTD results from other sources (notably COST and INTERREG Programmes).

1.6.2.2 Nature of the Information

The main objective was to make an inventory of research projects for easier dissemination that may be of interest for the implementation of the Water Framework Directive. The following project details were collected: acronym, full title, funding programme, contract code, start and end dates, name coordinator, e-mail coordinator and project website, and language(s) of information. Furthermore, projects were classified as to whether they were relevant for implementation of the WFD.

In addition, information on the RTD outcome and the results of the projects were collected. This related to several types of project deliverables:

- Project 'brochures', 'fliers' or (for LIFE) 'Laymen's report', describing the project in a short text (2–4 pages), and where attempts were made to explain the project in non-scientific English.
- 'Documents' that were defined as deliverables in the project. Explicit focus was on deliverables that would be of interest to users of WISE-RTD and in support of WFD implementation. Example documents that were considered of interest included policy and technical (RTD) guidance documents, case studies and experiences.
- 'Tools', defined as deliverables in the projects, which may relate to, for example, mathematical models, socio-economic support, monitoring, databases of relevant information, data management, quality assurance of data, uncertainty analysis, decision support or training products.

Project and/or project deliverables information uploaded to the WISE-RTD web portal included for each a short abstract, about 15 lines of laymen's text, definition(s) of the relevance to the implementation of the WFD, such as information on WFD eco-region, river basin, river or lake, country where demonstrated/implemented, and, finally, a set of keywords that reflected the WFD tasks.

Although a document repository was implemented in WISE-RTD, it was agreed to limits in principles the uploading of documents to a flier/brochure, and to use hyperlinks to forward users of the portal to the specific project's website.

1.6.2.3 Sources of Information

A prime source for DG-Research RTD projects was the information obtained *via* the public DG Research project information website and database 'CORDIS' (http://www.Cordis.europa.eu). Nearly all projects could be identified in CORDIS, and the basic information such as acronym, full title, funding programme, contract code, start and end date could be found. However, the database did not contain all information or was not kept up-to-date at the time of the search. Other sources included existing compilations and databases such as IWRM-NET (Integrated Water Resource Management, part of ERA-Net; http://www.iwrm-net.org/, *see* Chapter 3.2), NeWater (New methods for adaptive water management under uncertainty; http://www.newater.info/, see Chapter 2.6), and EMWIS (Euro-Mediterranean Information System on the know-how in the Water Sector; http://www.emwis.net).

The main database used to gather information about LIFE projects that are possibly of interest for WFD implementation was the 'Butler' database, managed by the LIFE Unit of DG-Environment (non-public database used as internal tool for DG Environment). Publicly accessible LIFE actions' websites were scanned for demonstration projects: LIFE Environment (now at http://ec.europa.eu/environment/life/project/Projects/index.cfm).

Projects funded by the European cooperation in the field of scientific and technical research (COST) were searched *via* the CORDIS-COST website (http://cost.Cordis.lu/src/home.cfm).

Interreg is a Community initiative that aims to stimulate interregional cooperation in the EU. It is financed under the European Regional Development Fund (ERDF). Focus was on INTERREG III projects. Projects were located *via* http://ec.europa.eu/regional_policy/interreg3/index_en.htm, the dedicated Commission website.

Unfortunately, in many cases these sources did not provide all the details required for uploading information to the WISE-RTD portal. However, having obtained a project acronym and full title from these sources, by using a web browser (Google) it was usually possible to locate project websites and other relevant information.

1.6.3 Project Information Uploaded to the WISE-RTD Portal

1.6.3.1 Research Projects with WFD Relevance

Table 1.6.1 summarises the information on projects and their deliverables (based on the situation at 1 May 2008). It can be concluded that many EC funded projects may have a direct or indirect impact on the implementation of the WFD. This may, obviously, be all results of a given project or limited information that is contained in only one project deliverable. Projects with WFD relevance are spread over the different EC funding programmes from

Table 1.6.1 Projects and project deliverables with WFD relevance from different EC funding programmes that were uploaded to the WISE-RTD database (by 1 May 2008).

	Total no projects	*Projects entered*	*Documents entered*	*Tools entered*
RTD	167	138	131	40
LIFE-ENV	88	85	121	15
LIFE-NAT	115	115	9	3
LIFE-TCY	22	22	25	19
COST	8	8	1	0
INTERREG	29	27	29	0
sum	429	395	316	77

Table 1.6.2 Overview of the number of projects where information has been uploaded to the WISE-RTD web portal sorted by EC funding programmes and instruments (status 1 May 2008).

				Funding instrument		
Framework programme	*RTD funding instrument*	*Projects*	*LIFE programme*	*ENV*	*NAT*	*TCY*
FP5	EESD	84	LIFE00	23	21	3
	GROWTH	2	LIFE02	18	19	3
	HP	1	LIFE03	13	21	3
	INCO2	14	LIFE04	10	24	2
	IST	5	LIFE05	13	14	5
			LIFE06	11	16	6
FP6	COORDINATION	1				
	INCO	6				
	INNOVATION	1				
	JRC	4				
	MOBILITY	5				
	POLICIES	7				
	SME	3				
	SUSTDEV	20				

DG Research and DG Environment (Table 1.6.2). For example, FP5 projects were identified under five RTD funding instruments (106 projects), with a dominance for FP5-EESD (84). Additional projects were identified under the FPs 3 and 4; WFD relevant information about 15 'may be' projects were entered as well. FP6 projects were identified under eight RTD funding instruments (47 projects) with a higher contribution by FP6-SUSTDEV (20).

The different LIFE Programmes showed also, within the sectors ENV (88 projects), NAT (115) and TCY (22), an almost even distribution over the years. Some 105 'may be' LIFE projects were identified.

For the INTERREG IIIa, b and c funded projects most were found for the IIIb instrument (26 out of 29 projects). For COST there is no subdivision for different funding instruments.

Table 1.6.3 Types of WFD relevant information entered in the WISE-RTD web portal (status 22 May 2008).

WFD relevant information entries	
Policies	78
Policy guidance documents	828
Policy implementation projects	37
Policy cases and reports	720
Technical guidance documents	217
RTD cases and experiences	521
Tools	124
No type specified	14

1.6.3.2 WFD Relevant Information

Identification of the projects was the first step. Probably even more important was the identification of information that is available, either as background document or as project deliverable, notably as guidance documents, case studies and tools. The database is expanding continuously with new projects and new WFD-relevant information. Table 1.6.3 gives an overview of the diverse types of information contained (situation of 22 May 2008).

The highest number of information entries is for policies and policy guidance documents. They relate to the WFD itself, the Common Implementation Strategies (CIS) and their related (guidance) documents. Also, the number of policy implementation projects, policy case studies and reports are well represented. The number of RTD technical guidance documents, case studies, experiences and tools highlights the large potential of the research deliverables for use in the implementation of the WFD.

Contained in this heap of information were the identification of or links to in total 262 rivers and lakes. The top-10 of rivers with inputs of various nature are the rivers Danube, Scheldt, Odense, Sebou, Litani, Júcar, Rhine & Lake Constance, Rhone, Lake Geneva and Thames. These include the largest river basins in Europe. It is no surprise to find high inputs for the rivers Danube, Scheldt, Sebou and Litani as they have been specially highlighted in the various project workshops of the SPI-Water project.

1.6.3.3 Shelf Life of RTD Results

For the past decade, EC funded projects have required the preparation of a project website. Unfortunately, no detailed instructions or criteria were provided, so these sites largely vary in content and quality. Some have extensive websites detailing the project and making available relevant deliverables of the project in a downloadable format. In a WWW oriented society it is expected that websites become more and more important for the dissemination of scientific information, including project deliverables. Unfortunately, at present too many projects fail to disseminate their project deliverables to a wider audience.

The basis for the WISE-RTD web portal is not to become a database of documents, but to be a switchboard that links WFD related questions to answers that are contained on the World Wide Web. The WISE-RTD portal in principle only contains few documents such as fliers/brochures/'laymen's reports' of projects and their deliverables. The power of WISE-RTD lies in the hyperlinking to the original information. This implies that the information shall be accessible *via* the World Wide Web. Without a live project website that contains information about the project and its deliverables that are research results in the form of guidance documents, case studies, tools, *etc.* information seems to be lost!

1.6.3.4 RTD Information Lost?

While executing the collection of project information it appeared that not all projects have a website or had an expired website. However, even if a project website does exist, their deliverables were frequently not present or they were reserved in a partners-only closed section. In these situations access to information about projects and their deliverables was seriously hampered and no hyperlinks could be defined in the WISE-RTD portal. Thus, valuable information on EC-sponsored research, of potential interest to the implementation of the WFD, has become less accessible for those interested in the research outcome. If a team of research specialists from SPI-Water project is unable to locate this type of information, one may fear that the less experienced policy maker has even less chance of finding and accessing relevant information, if at all.

As is demonstrated in Table 1.6.4, for most EC programmes a considerable number of projects do not have a working website. For example, in April 2007, for 31 out of 168 RTD projects no working website could be identified to provide useful information about the project, either *via* the Commission database CORDIS, web browser (Google) or other source of information. This means that 18.6% of the RTD projects either never had a website or their

Table 1.6.4 Overview of the number of projects where no (functional) website could be identified (status 1 May 2008).

		No website		*No longer have a website (or not functional)*	
EC funding programme	*Total no. of projects*	*no.*	*%*	*no.*	*%*
RTD	167	31	18.6	53	31.7
LIFE-ENV	88	8	9.1	14	15.9
LIFE-NAT	115	15	13	35	30.4
LIFE-TCY	22	4	18.2	4	18.2
COST	8	0	0	2	25
INTERREG	29	2	6.9	5	17.2

website was closed down. The situation seems somewhat better for the LIFE, COST and INTERREG projects.

When the SPI-Water team, by 1 May 2008, performed a quality check and scanned all project websites for proper access they found that an additional number of websites had become obsolete, as indicated in the last column of Table 1.6.4. Within one year of SPI-Water activities RTD projects 'lost' an additional 22 project websites, the LIFE programme 26, COST 2, and INTERREG: 3.

1.6.3.5 Maintenance

Even when a project website was identified, it often appeared that the site was poorly maintained and updated. Usually, the introductory texts were simply a copy of the project proposal text, even when the project had been completed. Many projects fail to place the project deliverables on the project website, or have them accessible to project members only. Almost never, near the completion of a project, is a website updated with all project results. Once a project is completed, there is apparently less interest within the consortium to keep the site updated or even alive. Costs of outsourced hosting of the website, although marginal, were mentioned as a reason.

One would expect that the older the project, the less interest there is in keeping a website alive, *e.g.* because the 'shelf life' of the results are considered limited. Indeed, there is a tendency for projects completed long ago to no longer have a site, but this trend is not very strong. Notably, there are former project consortia that manage to keep their website alive, thus continuing to disseminate the project deliverables to a wider audience.

1.6.4 Challenges and Recommendations for Bridging Science and Policy

As already discussed, there is a gap between generating scientific results and the effective dissemination to non-scientific stakeholders for their use and implementation. This is also evident from the 2008 FP7 calls for project proposals (Chapter 1.4) that:

aim is to elaborate innovative strategies, practices, and products to communicate results and activities of EU environment research activities to different target audiences (policy makers, scientific stakeholders, general public, educational purposes, etc.). Particular emphasis is placed on the development of knowledge transfer, strategies, and tools involving close interactions with stakeholders and other users of science and technology.

As summarised above, in several RTD FP Programmes and LIFE Programmes many research projects have received EC funding for addressing the issues raised by WFD and for facilitating its adoption by the Member States.

The SPI-Water Project, recognising the need for dissemination of scientific results to the WFD-related stakeholders, has further developed a web-based knowledge-sharing platform: WISE-RTD (Chapter 4.2). During the project several challenges have surfaced, which are analysed below, and a set of recommendations is provided.

1.6.4.1 Challenge 1: Preserving the Project's Generated Knowledge is a Necessity

Many projects had at least some form of a website. As noted Section 1.6.3.5, any website is often poorly maintained. Frequently, projects have two sections on their website, a project-partner section and a public section. Nevertheless, projects rarely publish their results in the public section as expected, and even if they do the projects frequently fail to include the project deliverables on their website in a downloadable format. Unfortunately, a project's website is almost never updated with the project's results when the project is nearing completion. Obviously, this hampers the dissemination of scientific results to a wider audience.

Project results usually exist in the form of flyers/brochures, documents/guidance reports/case studies and tools and may be reached *via* hyperlinks. However, without a live project website, these hyperlinks are unreachable. If the hosting of a project website had been outsourced, the website link might still exist but the website did not.

Finally, project websites are very diverse in nature: in lay-out, contents, accessibility of information, language, level of maintenance and quality of information.

The reasons for all this may be that there is no incentive to keep a website updated or even alive after project completion. Scientists apparently see no need to disseminate their research results beyond the (obligatory) dissemination products (reports) to the European Commission, and to publication in scientific journals.

During the SPI-Water project, Cordis, Butler and LIFE databases have been used in identifying projects relevant to the WFD. Yet, these databases were often incomplete, did not always have latest project updates and thus did not facilitate the retrieval of project information.

Recommendations

- Project consortia should take the necessary provisions to use the project website as a tool for dissemination, to the Commission and scientific and non-scientific audiences. The website should be well maintained and updated, and kept alive for a reasonable time after project has been completed. Hosting a website nowadays is not an expensive affair and costs should not hamper continued access.

- A template for standardising minimum project website contents would be useful for the diverse audiences to tap into the project's results in an easy manner without having to spend too much time in understanding the website's functionality.
- It is suggested that the EC keeps project related databases (Cordis, Butler and LIFE) regularly updated for facilitating the dissemination of project information.
- Standardisation across EC programmes regarding the dissemination of research results would be beneficial (*e.g.* a 'Laymen's report' is compulsory for LIFE projects, RTD projects should deliver a dedicated project website).
- At present there seems no overall strategy on how to preserve and disseminate research results. A structure for sustainable storage of RTD results and knowledge is necessary. An EU RTD overall strategy would be beneficial for preserving and making available research results. The web portal WISE-RTD is an example; however, advancement in its functionality and sustainability would be necessary.

1.6.4.2 Challenge 2: Tapping into the Knowledge of WFD Related Projects through a Single Source

During the SPI-Water project, we investigated the RTD, LIFE, COST and INTERREG programmes for projects and project results that relate to WFD implementation. Even though most of the projects that were uploaded to WISE-RTD portal did not have WFD implementation as a key objective, they frequently had a direct link to water issues, such as sustainable/integrated water management, water quality, monitoring, stakeholder participation, database, GIS and mapping, *etc.* Furthermore, trials performed in river-basins with water managers have shown that tapping into one consistent knowledge based information system that merges the information from different origins (Programmes) generates new possibilities and answers diverse questions.

Recommendations

- Tapping into the knowledge generated by research projects through a single interface, such as the WISE-RTD portal, allows an easy access and the combination of scientific results that before was probably not accessible other than by sheer chance.
- Apart from a project flyer (Layman's report), for most LIFE projects there are relatively few deliverables (documents, experiences, tools) available *via* the project websites. The addition of such deliverables, for better dissemination of the results, is encouraged.

1.6.4.3 Challenge 3: Simplifying Communication with Non-scientific Audiences

The challenge is to create a successful and continuing dialogue between science and policy. However, it is difficult to create such a dialogue because different stakeholders have different interests and agendas. Some of the communication problems in terms of water and environmental issues emerge due to different interests and different communication manners between the stakeholders. How is it then possible to communicate scientific results to diverse audiences? The use of suitable language for each audience will facilitate the dialogue and the transfer of scientific knowledge. Translation mechanisms that would allow the relevant information to be passed on to other stakeholders or policy makers are missing. Often, researchers do not have the time or the capacity to translate research results into policy and in a language that is more 'popular' (*see* Chapter 5.2 for further discussion about 'translation' aspects). During the SPI-Water project we have identified very few RTD projects that had created a project brief and newsletter written in non-scientific language, avoiding technical jargon. In our experience such an approach would be very effective for transferring the project's knowledge to other non-scientific audiences.

Language may also pose a barrier in terms of dissemination of knowledge. Information provided in English has become common practice, and most websites studied have been filled with English text, some sites were bi-lingual (including English). Sometimes one of the languages is dominant, and the second language is only used for the website. Not all stakeholders that search for information will master the English language sufficiently well, but with the help of (automated) translation facilities, they will be able to grasp the contents of the project/project deliverables. If a website (and the project deliverables) is in a language other than English, dissemination will be less effective, especially when the language is not one of the dominant languages of Europe (next to English: German, French and Spanish).

The LIFE programme, with demonstration projects, operates much more nationally/regionally and allows websites (and dissemination products) to be in the national language(s). As a result, quite a large number of the LIFE project websites are in national languages only, which will impair the wide use of their results.

Recommendations

- To attract non-scientific audiences to look at and evaluate the potential use, functionality and practicality of a project's results it is necessary to also communicate the project's aim and results in simpler language, avoiding jargon that any non-scientific group will not understand.
- The language of communication should be carefully taken into account when preparing the project's dissemination strategy and the common

language of the most important audiences should be considered. For all projects and their dissemination products at least an executive summary in the English language will stimulate better dissemination and use.

1.6.4.4 Challenge 4: Dissemination of Research into Practice and Change Management are Lagging Behind

For many years environmental research activities have increasingly paid attention to incorporating environmental policy relevant topics in their research agendas. For example, current water-related RTD projects have already established operational links with practitioners that implement the WFD in several catchments and river basins, allowing the needs of policymakers to be taken into account. Furthermore, dedicated advisory and/or governing boards with the presence of diverse decision makers are being established in RTD projects with the purpose of discussing project results and their significance for policy implementation. However, experience has shown that success in transferring scientific findings from the research community to operational use by the environmental managers is lacking.

Analysis of the process of research uptake in the field of water management has shown the need for efforts in translating research outcomes into management practice. This translation is generally undertaken by technical advisors rather than by water managers. Nevertheless, the technical advisors do not take part in the projects' boards.

Another issue is the acceptance of new methods and tools by the practitioners, who tend to 'stick' to the known and widely accepted ways of managing water issues. In our experience, suggestions for changes in management and application are either non-existent or insufficiently addressed in the dissemination strategies of research projects.

Recommendations

- In the dissemination of research into practice, the technical advisors play a crucial role and therefore should be included as a focus group for a project's communication activities and participate in the project's boards.
- The project dissemination strategy should also address the changes that new methods and tools might bring, in support of easier acceptance by practitioners.

1.6.4.5 Challenge 5: Having a Continuous Dialogue and Adopting a Participatory Approach

There appears to be several inter-related reasons for the science–public and the science–policy interfacing problems, such as an insufficient dialogue among the scientific and public or policy communities. Moreover, in EC funded projects very often scientists view the Commission as the 'legitimate' client and end-user for their research results and dissemination products. They thus miss the point that the Commission is funding the research but that the beneficiaries of the results will be the science community and other stakeholders, including the policy area and the general public.

Environmental and water issues have a very broad impact on many different actors from diverse backgrounds and with diverse interests. To improve communication between researchers, policy-makers and other groups a continuous dialogue and a participatory approach is necessary. During the SPI-Water project we have identified many projects that mentioned in their project objectives that communication with diverse audiences was planned in the form of focus groups for disseminating results and through case studies.

Recommendations

- The media are usually an unexplored option by project consortia to disseminate their results. Media can help improve communication and the dissemination of environmental research results to diverse audiences, and can be used in popularising research results.
- The participatory approach may help scientists spread their message and findings. The participation of diverse groups during project conferences and especially through the use of facilitators/'translators' can improve the communication and dissemination in environmental issues. These facilitators should have a scientific background and an understanding of the multi-disciplinarity of the different audiences.

Acknowledgements

The authors express many thanks for the excellent cooperation of their colleagues that participated in the SPI-Water (Work Package 1) exercise: Mike Briers and Steve Fletcher (UKEA, United Kingdom), Ildi Ipolyi and Yael Meroz (QualityConsult, Italy), Remi Laane (Deltares, The Netherlands) and Klio Monokrousou (NTUA, Greece).

References

1. European Commission, Directive 2000/60/EC of the European Parliament and of the Council of 23 October 2000 establishing a framework for Community action in the field of water policy, *Official J. Eur. Community*, L 327: 1–72 (2000).
2. M. Oliver, L. Sorkin and A. Kaschl, Science and policy interface: the LIFE programme and its links to the EU Water Framework Directive, *Environ. Sci. Policy*, 2005, **8**, 253–257.

3. Ph. Quevauviller, P. Balabanis, C. Fragakis, M. Weydert, M. Oliver, A. Kaschl, G. Arnold, A. Kroll, L. Galbiati, J. M. Zaldivar and G. Bidoglio, Science-policy integration needs in support of the implementation of the EU Water Framework Directive, *Environ. Sci. Policy*, 2005, **8**, 203–211.
4. Ph. Quevauviller, B. Harris and P. Vervier, *Needs for an operational science-policy mechanism in support of WFD monitoring – National and regional examples*, in *The Water Framework Directive. Ecological and Chemical Status Monitoring*, ed. Ph. Quevauviller, U. Borchers, K. C. Thompson and T. Simonart, John Wiley & Sons, Ltd., Chichester, 2008, pp. 411–444.
5. A. Scott, J. Holmes, G. Steyn, S. Wickham and J. Murlis, Science meets policy in Europe, Report for the science-meets-policy conference, London, November 2005, The Knowledge Bridge, Lewes, UK, 22 pp (see http://www.sciencemeetspolicy.org/site/11.asp).
6. WISE, see http://ec.europa.eu/environment/water/index_en.htm (visited: 29 June 2009).
7. WISE-RTD, see http://www.wise-rtd.info/ (visited: 29 June 2009).
8. SPI-Water, see http://www.spi-water.eu/ (visited: 29 June 2009).
9. K.J.M. Kramer, M. Briers, I. Ipolyi, Y. Meroz, K. Monokrousou, R. Laane and X. Schneider, Project evaluation and access to project information: recommendations, in Deliverable D26, EC-Project Science-Policy interfacing in support of the water framework directive implementation (SPI-Water), EC contract 044357, 2008, 37 pp (http://www.spi-water.eu/).

CHAPTER 1.7

WFD Economics and the Science–Policy Interface: Status and Perspectives

EDUARD INTERWIES[a] AND ILKE BOROWSKI[b]

[a] Intersus – Sustainability Services, Chodowieckistr. 2, 10405 Berlin, Germany; [b] University of Osnabrueck, Institute of Environmental Systems Research, Barbarastrasse 12, 49076 Osnabrueck, Germany

1.7.1 Introduction

The Water Framework Directive (WFD) requires Member States to integrate economics in various stages of the implementation process. Based on the economic analysis conducted in the context of Article 5 that was due 2004, Member States are currently working intensely on the establishment of the first River Basin Management Plans by March 2010. While this work is ongoing, it already is clear that the economic aspects of the WFD pose significant challenges for water administrations at different institutional levels, as also can be seen in the assessment by the European Commission of the economic analysis reports.[1] This concerns both the policy side (since, in most countries, the integration of economic considerations has not been systematically conducted so far when taking water management decisions) and the methodological requirements concerning the use of economic methods and tools in this work. Here, an important question is how results of academic/scientific projects are considered in practical implementation and how this link of science and policy can be improved.

Water System Science and Policy Interfacing
Edited by Philippe Quevauviller

Published by the Royal Society of Chemistry, www.rsc.org

Towards this background, a workshop took place in June 2007 in Berlin to better understand the current situation regarding the use of economic methods and tools as well as to develop elements of a future research agenda to better implement the economic aspects of the WFD. This workshop took place within the context of the Harmoni-CA concerted action (Chapter 2.4) and raised huge interest among water managers, researchers and consultants working at the science–policy interface on the issue of WFD-economics. Fifty people took part, coming from diverse backgrounds and positions: these included policy makers from very different institutions (local level 'WFD-implementers' up to two Water Directors), scientists working very close to policy making but also with a rather academic direction in their work. The discussions of the workshop built on the state of WFD implementation of economic elements that was elicited through a series of interviews with relevant experts from several Member States (Spain, Germany, Netherlands and France). This chapter presents both this information collected for the background of this workshop[2] as well as the results of the workshop itself.[3]

Notably:

1. Numerous activities are taking place on WFD-economics implementation, so this chapter can not give a full picture of these activities.
2. The information on the implementation approaches in the different Member States does not present an official description of the WFD-economics implementation status in the respective Member State. This is due to the complexity of the implementation on different administrative levels and the status of work, which is ongoing.
3. As the workshop that forms the basis of this chapter took place in June 2007, additional work both at the European level for clarifying certain aspects of WFD-economics (*e.g.* on the understanding of 'disproportionality of costs') and within Member States on practical implementation has taken place since then.

This chapter first briefly indicates the main fields of work for the implementation of the WFD-economic aspects, identifies the most crucial/problematic implementation issues and gives an overview of the current implementation work (Section 1.7.2).

Section 1.7.3 then moves on to the current usage of academic research and the recent/current EU-funded research projects supporting the socio-economic aspects of water management. In addition, some insights developed in the context of the CIS-process on the integration of research into WFD-implementation in general are presented. Following this, the current use and awareness regarding EU-funded socio-economic research as well as the difficulties described on integration of academic knowledge is depicted.

Finally, Section 1.7.4 concludes and identifies elements for a future research agenda, both regarding the structure of the policy–science interface and the research content.

1.7.2 Main Fields of Work for the Implementation of the Economic WFD-aspects: Gaps and Current Activities

In the WFD, economics play a big role in the overall implementation of the Directive and especially for the development of the RBMP. The main points of the implementation cycle are:

- To estimate the cost-effectiveness of measures and sets of measures at different scales to reach the WFD-objectives (Art. 11).
- To assess the proportionality/disproportionality of costs associated to proposed measures to justify potential exemptions from the WFD-environmental objective of good surface water status by 2015 (Art.4), which can be sought on different grounds, they include: time derogation (Article 4.4 WFD) involving an extension of the timeframe in which the objectives have to be reached (beyond 2015); less stringent environmental objectives (Article 4.5 WFD) due to unfeasibility or disproportionate costs of the measures that would be required to reach good water status; derogation obtained for new (hydromorphological) modifications and new sustainable economic activities that lead to a deterioration in water body status (Article 4.7 WFD). In addition, disproportionate costs play a role for the designation of Heavily Modified Water Bodies (HMWB) & Artificial Water Bodies (AWB) (according to Article 4.3 WFD). For each of these derogations, several 'derogation tests' have to be applied as input into decision making.
- To assess and improve the cost recovery level of water services (including environmental and resource costs) as well as the adequate contribution of different water uses/service users to these costs (Art. 9).

Further information on the understanding of these economic WFD-elements has been provided by several documents produced in the context of the Common Implementation Strategy, *e.g.* the WATer ECOnomics-document[4] as well as different information sheets covering specific topics (on cost-effectiveness analysis, environmental and resource costs, *etc.*). In addition, almost all Member States have developed specific analysis and guidance documents clarifying the main concepts of interest regarding WFD-economics. Thus, no detailed description of the economic elements of the WFD is given in this chapter.

In this context, and providing a short overview of the current situation (status: 2007), the implementation work in the Member States is focussed on the selection of measures, while the methodological basis for assessing disproportionality of costs is currently being laid. Limited work is being conducted on the issue of cost recovery.

An important issue to mention here is that due to the federal structure of most Member States and the WFD-river basin authorities not coinciding with the central state level, some general methodological advice is prepared at the Member State level to reach consistent approaches taken. At the same time, the application of these methodologies is taking/will take place at a more decentralised level (depending on the responsibilities regarding water management: federal states, river basin agencies, *etc*., as well as their more local institutions).

During the workshop mentioned above, participants where asked to identify the *most crucial/problematic* issues related to WFD-economics implementation. More than half of the issues mentioned were linked directly to the overall importance and role of economics within WFD implementation: will economics be able to support sustainable water management (*e.g*. by preventing the exemptions of Article 4 being used too often based on a solid analysis) or will economics be misused to legitimize political decisions? While many decisions with the WFD implementation are political (*i.e*. they have to consider policy fields beyond water management), economics and especially the scientific economic understanding do have a role to play.

The lack of data/information, as well as the use of available data (and at which scale), is seen as a significant difficulty for a sound implementation of the WFD objectives. Unclear definitions (*e.g*. of water services, disproportionality of costs, *etc*.) as well as a lack of expertise on economics at all implementation levels hinder a smooth implementation of the WFD-economics elements.

Another prevailing theme is uncertainty, which was identified as originating from gaps in expertise concerning economic methods and tools and lack of trust in them. In this context, the accumulation of uncertainty is central.

In general, the need for negotiation and agreement was also strongly highlighted, going beyond the context of unclear financing structures. These structures concern mostly the financing of measures but also of monitoring programmes and the establishment of a transparent approach to dealing with financial issues. Other topics raised were related to affordability of measures, cost allocation to different user groups/the state and especially in the transboundary context as well as the establishment of appropriate, incentive cost recovery structures (and the reaction of consumers to it).

Based on the three main areas of WFD-economics mentioned above, a short overview on current work is given in the following sub-sections, based on information by the interviewed experts and other additional available information.

1.7.2.1 Selection of Measures, Cost-effectiveness Analysis (CEA), Integrated (Hydro-economic) Modelling

The main bulk of economic work and input has so far taken place on the selection of measures and the assessment of cost-effectiveness. Considerable efforts have been undertaken on achieving the link between the potential measures and their costs with their effects on the pressures and impacts affecting the status of water bodies.

Notably, in most countries, the process is largely based on and driven by the more traditional water management expertise having a good understanding of the hydrological cycle and pressures and impacts of water uses/potential measures, but in which economics play a more limited role. With the WFD, economic issues are shifted to a much more prominent place, thus requiring an expansion of the hydrological and ecological expertise of water managers.

Various pilot studies and projects have taken place across Europe to better understand how the selection of measures according to the WFD can practically be performed, including a better understanding of existing data/information and capacities within the relevant authorities.

In many countries, a general approach/methodological guidance has been developed to support the cost-effectiveness analysis, while currently this approach is being either further refined or already practically implemented at a more decentralized level, in most cases including a strong element of stakeholder participation. These more local first proposals of measures are then complied to assess the strategic implications of these proposals (at a sectoral level, *e.g.* for agriculture) and to identify potential needs for changes in national policies, but also to identify shortcomings in these first drafts and to improve them. Thus, a combination of top-down and bottom-up approaches is implemented.

A main aim of the approaches developed is screening/prioritisation, *i.e.* identifying the cases for which higher planning efforts are needed for the selection of measures (in cases for which this additional effort is needed/leads to a more informed evaluation of cost-effectiveness of sets of measures) and when a simpler approach leads to sufficiently stable assessments (avoiding planning costs). In addition, information on standardised measures is in many cases collected to simplify the decision-making on the selection of measures. This includes costs of measures and to some extent information on the effectiveness of measures, which is more difficult to standardize.

The consideration of baseline scenarios/main future developments affecting the status of water bodies (*e.g.* climate change, changes in agricultural policies) and thus the measures needed seems to so far be limited.

Overall, a main challenge is linking biophysical and socio-economic information to provide a base for assessing different measures/management options for reaching the WFD-objectives. While there have been efforts to use integrated hydro-economic models in this work, the practical limitations of these approaches have become apparent. The main difficulties with having an overall, catchment scale approach evaluating different measures options are the upstream–downstream interrelations of effects of measures, difficulties concerning the assessment of measures concerning hydromorphology and diffuse pollution.

Nevertheless, linking existing hydrological knowledge on the effectiveness of measures with their related costs to support the selection of cost-effective sets of measures is a main part of current work using, for example, some modelling approaches/specific supporting tools for the effects of possible measures, even if not based on one specific, overall integrated model.

1.7.2.2 Cost–Benefit Analysis (CBA), Exemptions, Environmental Benefits

Work regarding these crucial issues for the success of WFD-implementation is underway in most Member States, but not as advanced as on the selection of measures. The approach to be taken is not finalized in most countries on how and at which level to assess the disproportionality criteria and in how far to use cost–benefit analyses supporting these assessments. When considering the costs and benefits of reaching good environmental status, the benefits assessment seems to pose the most difficulties, leading to various approaches, both regarding the scale of analysis but also the specific approach for the estimation of benefits. A commonality among the Member States so far seems to be to first consider a time exemption before looking at an exemption lowering the environmental objectives due to disproportionality of costs.

While the exemptions criteria need to be assessed at the water body level, aggregation approaches seem to be prominent, where the decision on the disproportionality is taken at a sub-basin/basin scale or even at national scale, considering the consequences on economic sectors in an aggregated manner. In some cases, guidance documents/guiding principles are currently being prepared at the national level on how to deal with the assessment of disproportionality/exemptions.

Some form of screening/step-wise approach is sought in many States to focus resources and capacities to the more 'difficult' cases. Thus, a practicable approach is looked for under the principle of providing as much additional information needed in support of taking a specific policy decision, while the initial screening steps for the disproportionality assessment are often centred on affordability issues. Overall, no converging picture is yet available on how the issue of disproportionality will finally be decided, one reason for this being the highly political nature of the exemptions issue. This might lead to difficulties regarding comparable approaches in transboundary river basins.

At the same time, the fact that no uniform procedure has been accorded on the issue of disproportionality is considered by some to simply reflect the decentralized decision making processes in place in many EU-countries as well as the different political settings (including different interests) across Europe. To reach the required consistency/comparability of approaches, information exchange between Member States sharing a River Basin as well as transparency regarding the different approaches is considered to be sufficient.

1.7.2.3 Water Pricing, Cost Recovery/Polluter-pays Principle, Economic Instruments

While the issue of the definition of water services has not been finally resolved yet at the European level, most counties so far consider only drinking water supply and wastewater treatment as water services.

Overall, only limited work is being conducted on cost recovery in most Member States beyond what has been done for the Article 5 analysis. The main reason is that the current cost recovery level is in many cases considered to be adequate, based on existing legislation requiring financial cost recovery, permit systems and charges and fees in place for internalisation of environmental and resource costs. Some additional work is being carried out (especially in countries lacking detailed information) to improve the data situation concerning financial cost recovery and a more in-depth understanding of certain issues (*e.g.* financial flows/subsidies); however, there is an example of this work not being considered useful, since it implies considerable effort with only a very limited better understanding/additional benefits (based on an already good statistical overview).

Regarding the environmental and resource costs of water services, a widespread opinion is that most/all of the costs are already internalized though permit systems as well as charges and fees established by democratically elected institutions. In cases in which good environmental status is not reached in a water body due to a specific water service, the environmental and resource costs of that service are assumed often to be as high as the costs of the measures that would be needed to reach the good status (abatement cost approach). This approach is considered to be more feasible for this first implementation cycle in most countries than conducting a specific assessment of these costs based on methods of 'forgone benefits' estimation.

Limited efforts are also currently underway on the adequacy of the contribution of water uses to the costs of water services. It is important here to note different interpretations of who needs to contribute to the costs of the water services: the main interpretation of Article 9 (as also described in WATECO-document) sees the 'water users' contributing to the costs of water services; an alternative view (from the UK) considers only the 'water service users' having to contribute to these costs. For a more detailed description of these two views, their rationale and consequences see Interwies/Dworak/Görlach/Best (2006).[5] Overall, the current contributions are generally considered be adequate, that is socially acceptable contributions by the different sectors.

Several countries are considering changes in their current economic instruments or the introduction of new ones (tradable permits systems *etc.*) since these might be a more cost-effective solution for reaching the WFD-objectives, foster the cost recovery of water services and better reflect the user/polluter-pays-principle.

Owing to the long time needed for required changes in legislation in case some of these instruments are considered to be more efficient, it does not seem probable that these will be implemented in the first implementation cycle of the WFD.

Finally, there is the awareness that potential changes in pricing structures/economic instruments in the water sector will need to be integrated into the RBMPs and to 'loop back' to the selection of measures for the implementation of the WFD, since they will affect the basis on which these measures have been selected, thus potentially leading to some measures no longer being necessary or different sets of measures being the cost-effective solutions.

1.7.3 Use of Scientific Economic Models and Methods

While several economic methods and models exist in academia and research that might be able support the implementation work, the practical use of such approaches in WFD implementation has been limited so far. On the contrary, the economic issues at hand for WFD implementation are often dealt with by non-economists, using much more limited, but implementable, approaches like expert judgements and so on.

Some main reasons for this, so far, limited use of economic methods and models are:

- the novelty of the WFD requirements (many water administrations are not used in applying economic methods and do not have many water economists on their staff);
- the complexity of issues at stake in the context of an integrated river-basin approach;
- the need to apply economic methods and models at a very broad scale (entire river basins), while most available models have been developed and tested at a limited scale (sub-basin at the most);
- the difficulties of estimating the impact of measures on the ecological status at the water body scale that also needs to be correctly presented in the models/methods used;
- financial restrictions, hindering an in-depth investigation in all cases where such work would be needed.

Since the more general integration of economic considerations in water management process has so far been limited in most EU countries, this clearly also affects the specific use of economic methods and tools provided by academia. But, going beyond that, there seems to be a gap between what is currently available from scientific side and the urgent need of policy makers to have simple-to-use, broadly applicable tools and methods that can be used based on the existing information and capacities within the water administrations in implementing the economic requirements of the WFD.

This has also been acknowledged within the Common Implementation Strategy (CIS)-activities. The Activity on 'Information Exchange and Research Needs' carried out within the CIS aimed 'to identify and prioritise issues arising from the WFD Article 5 activity, which in turn require an EU-wide approach, and to identify blank spots in research'.[6]

Based on a large-scale questionnaire of activity with the Member States authorities, the issue category 'Economics' was reported as relatively poorly covered by research. More specifically, the economic topics identified by the Member States as meriting a common understanding at EU-level of main principles, objectives and methods are related to cost–benefit and cost recovery topics and to building common understandings concerning what are 'economic instruments' and 'economic measures' (ref. 6 p. 14).

The gap between research and water management has led to several specific activities to develop practicable approaches and methods/models for implementing the WFD-economics requirements, based on available scientific knowledge/models and adapted to the existing information/data:

- The most common approach is to commission specific projects (either to consultancies from practically-oriented research institutions) that investigate, summarize and transform the scientific methodologies and results into practicable approaches for WFD implementation.
- Another option is to set-up a specific programme for finding ways to integrate existing scientific knowledge into the practical implementation.
- Finally, an alternative way to deal with the issue is to create an 'economic analysis group' within the environment ministry, bringing together scientists from different universities and from public sector technical assistance companies.

1.7.3.1 EU-funded Research on Socio-economic aspects of Water Management

Already, the 4th and 5th European Research Framework Programme has put significant efforts into water-relevant research; this has been continued in the closed 6th Framework Programme and will continue in the 7th Framework Programme. While socio-economic aspects have been specifically asked for in many of the calls and included in almost all research projects funded by FP 5 & 6, there is so far limited research conducted specifically on economic methods; the focus has been rather on general socio-economic aspects of water management.

Thus, research regarding the specific economic requirements of the WFD and oriented towards their broad implementation has also been limited (with the AQUAMONEY-project being a notable exception). One reason for this is that the WFD only went into force in 2000; consequently, many older projects could not build on the specific requirements (which were some of the last to be decided in the negotiation phase of the WFD, most notably the content of Article 9). Another reason is that even if economic research was conducted, this was more case-study specific and not oriented towards a broad-scale implementation as needed for the WFD.

Nevertheless, several EU-funded research projects have worked on socio-economic aspects of water management that can provide some insights of relevance also for the WFD-economics implementation. In addition, specific activities have been funded to improve the communication of European research to potential users of the project results (see also below). This better linkage of research and policy is also supported by the WISE-RTD portal (Chapter 4.2). During the interviews, WISE-RTD was mentioned as a useful potential 'clearing house of available scientific knowledge' when it has become fully operational.

In the background note to the workshop, several specific European research projects from the 4th, 5th and 6th framework programme are mentioned that have a component of socio-economic aspects in their research, based *inter alia* on a survey document prepared by the European Commission/DG Research in 2002 on 'The WFD: Socio-Economic aspects in Research Projects' as well as a search through the DG Research-project database and the WISE-RTD portal. Most of the websites of the projects mentioned there can be found at www.wise-rtd.info.

On the linkage of science and policy, two projects need to be specifically mentioned since they do deal explicitly with socio-economic aspects of water management and this link: the large-scale concerted action Harmoni-CA (of the 5th framework program) and the SPI-Water project, a 6th framework program project in the context of Specific Support to Policies, aiming to enhance the interfacing between science and policy with respect to the Water Framework Directive Implementation.

Overall and based on the clear need for streamlining of information from, for example, the scientific community to policy decision-makers, efforts are on-going in the framework on various initiatives to examine how an efficient and operational 'science–policy interface' could be developed in support of the implementation of the WFD. This development is being undertaken in the framework of the Common Implementation Strategy of the WFD and in close cooperation with the Harmoni-CA concerted action that initiated the WISE-RTD Web Portal. The ultimate aim is to develop an interface that could meet the demand of different levels of users and other stakeholders, ensuring an efficient dissemination and use of research results and experiences.

1.7.3.2 Awareness and use of EU-funded Research

To get an indication of the applicability of the research conducted at European level, the experts interviewed prior to the workshop were asked on if they currently use the results of EU-funded projects. Besides the repeated mention of the AQUAMONEY-project basically no EU-funded research projects are directly used in the implementation of WFD-economics.

During the interviews, it was repeatedly mentioned that content and results of EU-funded projects are not well known to the policy sphere due to communication difficulties (both from the projects to the policy makers, but also concerning the dissemination of information within the policy institutions) and time constraints.

At the same time, the experts supporting the practical implementation work on economics (consultants, practically-oriented researchers, economic analysis group members, *etc.*) are assumed to know and use all available scientific knowledge, both from European and also national research projects.

1.7.3.3 Main Difficulties in Current Implementation Regarding the use of Scientific Tools and Methods

Besides their use of EU research projects, the experts and workshop participants exchanged views on the main difficulties when trying to integrate scientific knowledge into practical implementation. This can be helpful for a better understanding of the main issues that could be improved, both for the utilisation of current and the content of future research, and also for giving indications of the kind of difficulties faced with this integration, going beyond content. Some of the topics relevant to the economic aspects of the WFD have also been discussed in other contexts of the WFD, such as the support of the implementation in general with regards to the use of model-based tools. In this context, data availability and constraints are not only a major issue in water management itself but have also been addressed by the research community. Harmoni-CA/WP4 worked on bringing new perspectives into the joint use of monitoring and modelling and to bridging the possible gaps that might exist between the monitoring and the modelling communities (see www.harmoni-*ca*.info for relevant reports).

Overall, the need to have practicable approaches (regarding data/information requirements and capacities/resources needed) for WFD-implementation seems to be a crucial need of policy-makers that the current scientific basis can only satisfy to a limited extent. Some of the main specific issues mentioned are given below.

1.7.3.3.1 Selection of Measures, Cost-effectiveness Analysis (CEA), Integrated (Hydro-economic) Modelling

Under this heading, issues were:

- dealing with existing data/information constraints, since collecting and analyzing such information, especially regarding economics, do not have a tradition in water management; existing scientific projects/models are often based on additional data collection in specific case studies; while such analysis provides useful insights, such an approach would imply very costly new data collection if these approaches were implemented at the broad scale;
- limitations on the possible transferability/application of scientific approaches developed and tested for specific case/pilot studies to broad-scale implementation;
- closely related, general issues of scale regarding the methodology for the selection of measures;
- linking of biophysical and socio-economic information to have a base for assessing different measures/management options for reaching the WFD-objectives. Integrated hydro-economic models have practical limitations, limiting a possible overall, catchment scale approach evaluating different measure options due to complex upstream–downstream interrelations of

effects of measures, difficulties concerning the assessment of measures concerning hydromorphology and diffuse pollution (*e.g.* linking agri-environmental measures to the effects on the status of water bodies);
- analyzing and integrating possible measures that have impacts on different WFD-objectives at the same time (especially since complex modelling approaches that might deal with this issue have proven to be too difficult to implement at a practical level due to the data requirements but also the capacities available at more regional levels regarding economics);
- limited usability of Decision Support Systems (DSS), *e.g.* due to the greater effort needed compared to the additional benefits through potentially improved, practicable results.

1.7.3.3.2 Cost–Benefit Analysis (CBA), Exemptions, Environmental Benefits

In this context the issues raised were:

- the challenge of determining the required use of CBA and the appropriate level(s) for its application in the context of the exemptions issue;
- in general, methodological and data constraints regarding the evaluation of benefits of an improved water status;
- doubts regarding the practicability of using valuation of non-market based benefits within the disproportionality assessment, regarding the methodological soundness of valuation study results, their transferability (through benefit transfer: not enough reference values, uncertainties of transferability), their acceptance by stakeholders (*e.g.* since accepting and using results of CBA has not been common so far) as well as to the costs associated with such approaches;
- accumulation of uncertainty through possible benefits evaluation, since the uncertainties regarding such an evaluation are added to the effects of measures that can not be precisely estimated;
- methodological difficulties/reliability of results specifically as related to stated preference approaches, as well as issues regarding the communication of the results;
- capacity issues related to the regional authorities that will need to implement the approach for disproportionate cost assessment.

1.7.3.3.3 Water Pricing, Cost Recovery/Polluter-pays Principle, Economic Instruments

Specific issues mentioned in this regard were:

- the main difficulties of estimating cost recovery levels are related to data constraints (*e.g.* on estimating investments needed for maintaining the quality of the water services, data on subsidies);

- how to estimate resource costs and especially in a way avoiding double-counting with environmental costs estimates;
- better understanding needed of different possible pathways of improving cost recovery, better understanding of current incentive structures and ways of improving them;
- moving the fiscal system towards the establishment of 'green taxes' is a complicated process that needs to be better investigated to better understand potential effects of such changes (socio-economic and ecological impacts, changes in the incentive structures, *etc.*), also based on existing experiences with the implementation of green taxation;
- more analysis needed on the institutional and economic opportunities and challenges connected with the setting up of tradable permit systems for pollution/emission rights, which could provide interesting and flexible solutions; for example, on the elasticity to pollution price, since this would provide a better understanding of the potential effects of such tradable pollution permit systems.

1.7.4 Summary and Conclusions: Elements for a Future Research Agenda

Overall, significant room for improvement seems to exist to reach a better understanding of the reasons behind the limited use so far of EU-funded research of WFD-economics implementation. Such a better understanding can guide the way to improving both the current use and future set-up of research projects.

Demand for additional support on the economic aspects of the WFD is large. While there is a general openness towards 'external input', including research, based on the acknowledgement that scientific work might be able to support the selection of approaches for practical WFD-implementation, the willingness to use specific EU-funded research remains low.

Both constraints in time and resources do not allow policy makers to maintain an overview that enables them to assess the usefulness of the different research programs. Water managers are looking for and setting up their own tailor-made projects that fit directly their interests and needs. At the same time, the limited focus of research so far specifically on economic methods related to WFD-economics, as well as the content/scope of the economic work done, leaves in many cases the research results serving as 'idea givers' to the most.

The workshop identified several research topics. These were divided into content-related research needs as well as needs to improve the structure of the research–policy interface. The results show that, when focusing on economic issues, the challenges for the policy–science interface are often of a rather general nature. They already begin during the definition of the problem for which water managers need solutions and on which scientists should do what research on. The subtleties of the related challenges have been discussed more deeply in Borowski and Hare.[7]

Content related issues of a future research agenda: Demand for more research on economic issues was broad and diverse. However, more research on integration and integrated models was valued highest, directly followed by more research on economic methods and tools in general and more clarity on definitions of main terms regarding WFD-economics.

Within the continuum of economic science 'interference' with the political decision making, three levels of this interference were identified: At a first level, economic science can inform the political decision-making process, while also the limitations of the approaches used need to be made transparent to reduce the risk of 'backfiring'/misuse of economic assessment. Using economics for examining/assessing problem-solving approaches (regarding effectiveness but also distribution) needs to be dealt with cautiously, since the risk of misuse here is given. At a second level, beyond informing decisions transparently by scientific approaches applied in the WFD implementation, transparency was also repeatedly discussed in the context of public participation and negotiation with/between different stakeholders. As discussed in other chapters of this book, scientists were also expected to act as a 'transmitter' of knowledge beyond policy and as a facilitator to improve processes of joint implementation and social learning. At a third level, economics should not be used for legitimizing already taken decisions; which would indicate a process failure in the overall WFD implementation.

The results on the content issues might indicate two things: Methodological issues both on integrated modelling and management, and on economics, are of high importance. The diversity of the content of the research demanded (*e.g.* on scales, uncertainty or case studies) indicated that, especially, the content of work to be done depends often more strongly on the specific situation, not only in terms of basin specifications and national/ regional situation but also in terms of available capacity. Although not highly ranked in the prioritization exercise, data availability and accessibility, including the challenges of appropriate scale, was a persistent theme throughout the workshop. Harmoni-CA has explored this issue of data intensively by looking at the good potential practice for joint use of modelling and monitoring.[8,9]

In the following, the main research topics identified from the discussion are listed, organized along the main WFD-economics issues.

Research demand on the selection of measures, cost-effectiveness-analysis and integrated (hydro-economic) modelling:

- Approaches supporting systematic assessments of uncertainty throughout the dose–response chain are needed, including environmental, social and economic system responses to measures;
- understanding of the value of extra information to reduce uncertainty in decision-making (through Bayesian approaches) should be improved;
- integrated-extended CEA approaches (*versus* CBA and MCA), including weighting procedures for multiple objective 'satisfying' problems/policy contexts, should be tested;

- integrated modelling, better linking biophysical to economic models for different purposes at different scales (*e.g.* adaptation–mitigation), should be improved;
- development of standards for non-market benefit analysis (based on the AQUAMONEY-project).

Research demand on cost–benefit analysis, exemptions and environmental benefits:

- better understanding of 'disproportionality of costs' (at strategic/ national level; local implementation) and to what extent a CBA is needed;
- Research in required on the role of non-monetary benefits: what and how to assess?
- increase the applicability of valuation approaches: more specific non-market valuation studies (of specific WFD-relevance); methodological work on benefit transfer;
- work on how to better conduct market-based benefit evaluation (*e.g.* at a bigger scale);
- better understanding and assessment of opportunity costs;
- analysis of stakeholder acceptance and exploring ways to increase it;
- role of monetary valuation/CBA in general in decision making.

Research demand on water pricing, cost recovery/polluter-pays principle and economic instruments:

- More case studies are needed: to learn from approaches in other countries but also to test new instruments and approaches coming from research;
- more studies needed on a broader definition of water services including *e.g.* hydropower, navigation, *etc.*;
- more research is needed to find the appropriate criteria for cost recovery, polluters-pays-principle and cost distribution, including aspects such as affordability, hidden costs (*e.g.* subsidies), spatial differentiation of approaches and justification of exemptions;
- How to take the results from small scale studies and use them on a larger scale?
- environmental resource costs: more research is needed on defining appropriate criteria and procedures for the justification of exceptions from cost recovery, for the application (and the limits) of the polluters-pays-principle in the practical implementation, process, as well as for dealing with affordability issues, hidden costs, subsidies, and the need for spatially differentiated approaches;
- environmental and resource costs: research on practical implementation and on quantification needed, providing guidance to water managers;
- continue research on tradable markets, *e.g.* nutrient trading.

Issues for improving the structure of the science–policy interface: The workshop identified the communication of the results of finalized and ongoing projects as most important. In addition, more continuous support from researchers for the management process was strongly appreciated. These results showed a strong trust that the available scientific results and the scientists can offer a lot of support without changing the basic approach towards European research.

In the following, the main suggestions for improving the structure of the science–policy interface are listed, indicating a strong interest in closer interaction and exchange between scientists and water managers. This included:

- More focused communication on research/projects taking place and their results in a transparent and accessible way. For example, WISE-RTD (*see* Chapter 4.2) should be used more (*i.e.* filled with research results from the scientist) or regular 'show case events' presenting research results should be established.
- At national level, 'focal points' were suggested where experience from management and research at regional level in a country could be brought together to facilitate exchange and national harmonisation where appropriate. This is in line with chapters of Section 3 of this book (Links to Water National or Regional Research, Policies and Management).
- In addition to the provision of project-based research results, continued support from the scientific perspective is needed. Here, repeatedly, the call came for improving the authorities' capacity by engaging more economists.
- In approaching uncertainty, scientists have to be more explicit about probabilities and policy makers have to be more explicit about required confidence levels for decision-making. However, experience from earlier Harmoni-CA-workshops show that uncertainty is not only a matter of uncertainty analysis and confidence levels that are expressed in different numbers. Uncertainty analysis quantifies only a small part of the uncertainties involved in decision making. Building confidence and trust into decision making goes far beyond it. More insights into the challenges of uncertainty and options how to approach it can be found in the literature.[10,11]
- Both sides need to work on their communication. Water managers need to phrase the right and specific questions in terms of the problems to be solved. Researchers need to reflect better on these questions and adapt their research towards them or need to be more transparent about the boundaries of their research. Also, research should be better linked to the existing system of water managers.
- During the development and implementation of European research projects, closer interaction of policy makers and researchers was considered to improve the applicability of research results in practical water management.
- A shared commitment for the implementation of the WFD and its economic aspects is necessary for a good exchange between not only science and policy but also with other stakeholders.

- The interface between science and policy should also be strengthened by researchers acting as facilitator or mediator between different stakeholders in participatory processes.
- Integrated hydro-economic models were seen as a potential tool to facilitate interaction and public participation. However, the complexity of such tools and the involved uncertainties might at the same time limit their application for improving the understanding of the system. Also, complex models are often approached with less trust. Simpler models were expected to serve well for communication purposes of results.
- Scientists need to focus first on compatible approaches to the existing system that are easier to implement (*e.g.* financing mechanisms). Also, to apply research results, a more pragmatic approach is often necessary. This might also ask for changes in the incentive structure for scientists: Pragmatic and case study approaches often involve a lower level of scientific innovation and lead to less scientific acknowledgement (*i.e.* high-ranking publications).
- Research results should more often provide guidance and give synthesis of the state of the art.
- The value of case study approaches could be improved by a combination with broad-scale applicability (and testing the added value of analysis). Pilot implementations of new instruments and the replication of 'good practice' examples were also needed.

Finally, notably, a thorough and early analysis of the first RBMP due March 2010 from a scientific point of view and especially regarding economics would be extremely useful. Such an analysis would give an even clearer picture of the main implementation gaps regarding WFD-economics, helping to pinpoint the future research needed to obtain better results for the second implementation cycle of the WFD.

Acknowledgements

The present work is mainly based on the workshop 'Using Economic Methods and Models for the Implementation of the Water Framework Directive: Status, Options and Challenges for a better integration of water economic research into policy implementation' (June 2007, Berlin) and its results. This workshop was part of the European Concerted Action Harmoni-CA (Contract no [EVK1-CT1-2002-00192]), especially Work Package 5 dealing with Integrated Assessment and the Science Policy Interface and the Human Dimension more generally. We thank the European Commission for the financial support. We would also like to thank all participants for their contributions and comments before, during and after the workshop. Special thank goes to the presenters for initiating the discussion.

References

1. European Commission (2007), Communication from the Commission to the European Parliament and the Council: 'Towards sustainable water management in the European Union – First stage in the implementation of the Water Framework Directive 2000/60/EC', COM (2007) 128 final. http://eur-lex.europa.eu/LexUniServ/LexUniServ.do?un=CELEX:620070C0128:EN:NOT.
2. E. Interwies and I. Borowski, Current use of Economic Methods and Models for the Implementation of the WFD in selected EU-Member States: the role of European research, Background note for the 4th Harmoni-CA/WP5 Policy Workshop 'Using Economic Methods and Models for the Implementation of the Water Framework Directive: Status, Options and Challenges for a better integration of water economic research into policy implementation', 14–15 June, 2007, Berlin.
3. E. Interwies and I. Borowski (2007): *Workshop Report*, Osnabrack for the 4th Harmoni-CA/WP5 Policy Workshop 'Using Economic Methods and Models for the Implementation of the Water Framework Directive: Status, Options and Challenges for a better integration of water economic research into policy implementation', 14–15 June, 2007, Berlin..
4. WATECO, Common Implementation Strategy for the Water Framework Directive (2000/60/EC), *Guidance Document No 1.: Economics and the Environment*, 2003 (see http://circa.europa.eu/Public/irc/env/wfd/library.
5. E. Interwies, T. Dworak, B. Görlach and A. Best, *Incentive water pricing and cost recovery in the WFD: Elements for linking EU Agricultural and Water Policies*, Paper endorsed by the EU Water Directors in May 2006, pp. 15–16.
6. M. de Rooy et al., Key issues and research needs under the WFD, 2005. See: http://circa.europa.eu/Public/irc/env/wfd/library > public library > other thematic information > research.
7. I. Borowski and M. Hare, Exploring the gap between water managers and researchers: Difficulties of model-based tools to support practical water management, *J. Water Resources Manage.*, 2007, **21** (No. 7 – Special Issue 'Methods for participatory integrated water resources management') 1049–1074.
8. J. C. Refsgaard, L. F. Jørgensen and A. L. Højberg, Data availability and accessibility: State of the art on existing data required for modelling for research purposes and for the implementation of the Water Framework Directive, Harmoni-CA, 2007. http://www.wise-itcl.info.
9. A. L. Højberg, J. C. Refsgaard, L. F. Jørgensen, F. v. Geer, I. Zsuffa and V. Vandenberghe, Catchment Modelling Guidance – Good practise in joint use of monitoring and modelling Harmoni-CA, http://www.wise-itcl.info, GEUS, 2007.
10. M. Brugnach, A. Tagg, F. Keil and W. de Lange, Uncertainty matters: computer models at the science–policy interface, *J. Water Resources Manage.*, 2007, **21** (No. 7 – Special Issue 'Methods for participatory integrated water resources management'), 1075–1096.
11. J. C. Refsgaard, J. P. van der Sluijs, A. L. Højberg and P. Vanrolleghem, *Harmoni-CA Guidance Uncertainty Analysis*, European Commission, Brussels, 2005.

Section 2: Interfacing Science and Policy in the Context of Selected RTD Projects

CHAPTER 2.1

SKEP Network: Facilitating Improvements in Science into Policy Process

SIMON GARDNER,[a] EEVA FURMAN,[b] PAULA KIVIMAA,[b] PIRJO KUUPPO,[b] HANNA MELA,[b] PÄIVI KORPINEN,[b] ERIK FELLENIUS,[c] JENNIE SAVGÅRD[c] AND JOHN HOLMES[d]

[a] Science Strategy, Environmental Protection Directorate, Environment Agency for England and Wales, Block 1, Government Buildings, Burghill Road, Westbury-on-Trym, BS10 6BF, Bristol, United Kingdom; [b] Finnish Environment Institute, Mechelininkatu 34a, P.O. Box 140, FI-00251 Helsinki, Finland; [c] Swedish Environmental Protection Agency, SE-106 48, Stockholm, Sweden; [d] University of Oxford, Department of Earth Sciences, Parks Road, Oxford OX1 3PR, United Kingdom

2.1.1 Introduction

The SKEP (Scientific Knowledge for Environmental Protection) network[1] is a European Commission 6th Framework Programme (FP6) funded initiative[2] that was launched on the 1st June 2005. It is one of a cluster of environmental European Research Area Networks (ERA-Nets).[3] The network has the twin objectives of ' . . . facilitating improvements in science into policy processes, and supporting the evidence-base for better regulation . . . '.

The SKEP network has brought together a consortium of key European public-sector funders of environmental research programmes to generate

Water System Science and Policy Interfacing
Edited by Philippe Quevauviller

Published by the Royal Society of Chemistry, www.rsc.org

knowledge in collaboration and improve links with high level policy makers. Consortium members are divided between environment ministries, environmental protection agencies and research institutes. This blend of experience has allowed the SKEP network to study the science-into-policy process from a holistic perspective. This has resulted in the development of a series of reports outlining good practice in research management, evaluation and dissemination designed to enhance the value of research in meeting the needs of policy makers and those who implement environmental regulations.

By taking a progressive, strategic overview of the research needs of environmental policy-makers and regulators, SKEP has coordinated the effective deployment of research resources to develop innovative research and meet European Union (EU) environmental challenges. One of the enduring legacies of the SKEP network will be the creation of a lasting structure for research co-ordination and co-operation between SKEP participants, which will provide for effective alignment of national programmes.

SKEP is also a cross-cutting research initiative, with the objective of '. . . developing a formal network to . . . close the gap between high-level policy makers, policy-based science funders and relevant information from scientists in a structured, coordinated way across member states . . . '.[4] There are therefore close ties between the SKEP network and the Science Meets Policy (SMP) initiative,[5] the Bridging the Gap (BtG) initiative,[6] and the Network of Heads of European EPAs (EPANet).[7] This has been reflected by SKEP representation at the Helsinki meeting of the Network of Heads of European Environmental Protection Agencies in April 2007, and at the 4th Bridging the Gap conference in May 2008 in Slovenia.[8]

2.1.2 Science–Policy Aspects of the Network

The SKEP network has deliberately drawn together a blend of environment ministries, environmental protection agencies and research institutes and councils, to allow it to study the science into policy process in a holistic and inclusive manner, and link funding streams for environmental regulation (Figure 2.1.1).

In the first four years of its existence, the SKEP network has focused on three key areas of the science into policy process. These are concerned with the management of research programmes, the mid-post and ex-post evaluation of research, and the dissemination and implementation of research.

2.1.2.1 European Overview of Research Management Approaches in the Field of Environment Protection

The SKEP network's studies on research management have been undertaken by the Finnish Environment Institute (SYKE), in liaison with the Finnish Ministry of the Environment.[9]

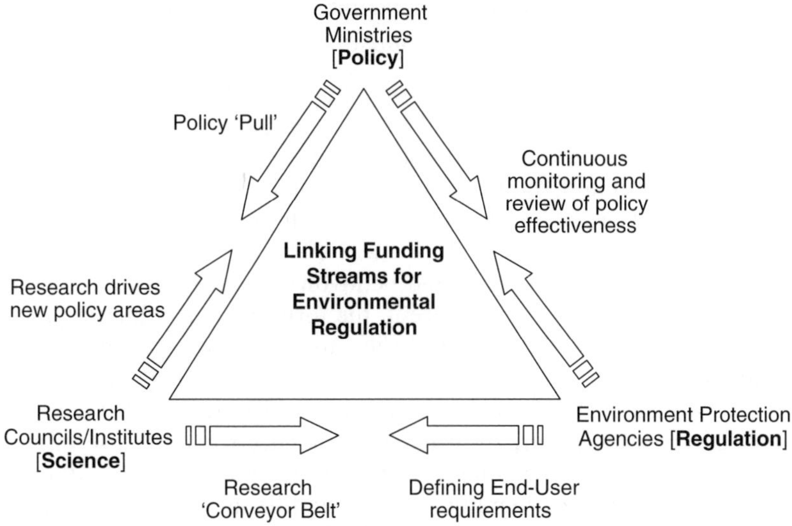

Figure 2.1.1 Interdependent relationship between environment ministries, environment protection agencies and research councils.

In November 2005, a, detailed *Good Practice Guidelines*, questionnaire, together with an invitation note to a workshop, was distributed to external experts within each Member State containing a participating organisation. Thirty-three completed questionnaires were returned, from which a sub-set of 16 practitioners in seven European countries were selected for more in-depth interviews. These interviews included a wide range of thematic areas, including water, soil, air, biodiversity, global change, environmental hazards and risks, human health and the environment, the urban environment, energy, environmental technology and environmental policy. A proportion of those interviewed subsequently attended a SKEP workshop in Helsinki on the 27th–28th April 2006. This workshop was also attended by representatives of the EC-sponsored 'Science Meets Policy' initiative (Chapter 1.5), private industry and external stakeholders from outside of the EU, including the United States Department of Energy. The workshop focused on the following areas:

- a review of environmental research management case-studies;
- a discussion of how to best match programme management practices to research strategy;
- sharing experiences from public and private sector programme managers;
- programme initiation (including scope, funding and evaluation of proposals);
- an assessment of how effective programme management can facilitate collaboration between projects.

The workshop resulted in the production of a report that made several recommendations in the following key areas.[10]

2.1.2.1.1 Research Funding Initiation (Including Scope, Funding and Evaluation of Proposals)

The report concluded that research programmes should have *clear boundaries and scope*; it should have realistic and specific objectives, and clear criteria that are communicated to the applicants and proposal evaluators. It also ascertained that programmes should be defined broadly enough to enable meaningful research projects for academics but still specific enough to have a *common perspective* and to enable a successful monitoring of progress. In addition, different *stakeholders* should be involved in scoping the programme already at an early stage, for instance in the form of stakeholder seminars or consultations. The involvement of policy makers, industry and NGOs is important in relevant programmes. The inclusion of these different stakeholders was found to add valuable transparency. The report also agreed that *seminars* and *workshops* are transparent and effective tools for discussing organisations' interest and commitment to fund research, and for outlining the scope and objectives of a programme. It was additionally recommended that a *pre-assessment* of the research area can be beneficial in setting up the programme scope and objectives. Whilst it was found that a comprehensive assessment can generate a lot of information for the applicants, equally it represents a fairly onerous approach towards setting up a programme. Finally, the report stated that the *duration* of a programme should be sufficiently long to generate meaningful results. At the same time, budget and time limitations should be taken into account in programme planning, and over-ambitious goals should be avoided.

2.1.2.1.2 Calls for Project Proposals: their Evaluation Processes and Criteria for Evaluation

Several criteria were considered as being of relevance by the interviewees and questionnaire respondents (Figure 2.1.2). The most relevant of these was adjudged to be the scientific quality of the proposal (and the competence of the research team), the presence of international (or national) networks of researchers within projects, and the policy relevance of the proposal.

The study illustrated several methods used to assess project proposals in different policy contexts. These ranged from the use of in-house experts, the use of international experts and the employment of thematic evaluation teams. The blend of stakeholders used to evaluate project proposals varied markedly with the nature of the selection criteria used (Figure 2.1.3).

2.1.2.1.3 Links between Setting up Programme Objectives and Evaluating Proposals

The report highlighted that research programmes should be socially responsible and include both objectives and criteria related to *social or public relevance*.

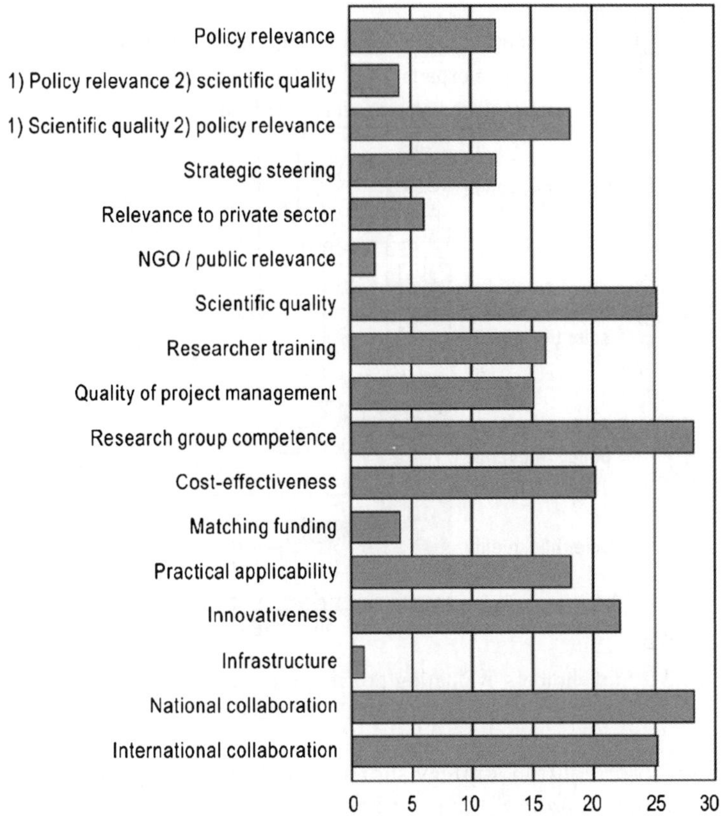

Figure 2.1.2 Criteria used in assessing project proposals.

It was recommended that the *policy relevance* and/or applicability of results, or at least their feasibility as objectives and criteria, should be considered in the start-up phase of each programme. Despite including policy-makers in programme planning, project selection criteria regarding policy relevance are generally not self-evident. Therefore, objectives related to policy relevance should be reflected in project evaluation criteria. Finally, the report recommended that a *combination* of science and policy relevance is important both in setting the objectives and project selection criteria. For example, a research programme on biological diversity has the capacity to focus both on ecosystems and on the management of biodiversity.

2.1.2.1.4 Research Programme Co-ordination and Collaboration between Projects

The findings of the report were split into several different sections. In terms of *overall* programme management structure, it was highlighted that a robust

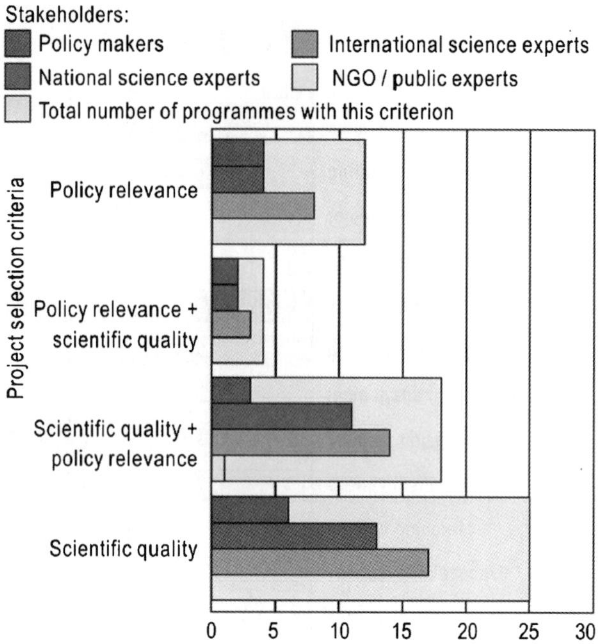

Figure 2.1.3 Stakeholders evaluating project proposals in programmes with criteria for scientific quality and/or policy effectiveness.

programme co-ordination strategy should be determined by the *scope*, *budget*, *duration* and the *purpose* of the programme. The presence of a *key person* was adjudged essential to the programme management process; someone who should be supported by a core team. The responsibility for the various *tasks* of programme management should include communication with projects and programme board, ensuring the dissemination of information and results to different stakeholders, monitoring progress and technical quality, solving problems and balancing risks, handling finance issues and ensuring delivery according to the agreed time schedule. A premium was put on selecting board representatives who were motivated and aligned with the aims of the programme, and (where required) combining this management body with the use of a steering committee. Also, various stakeholders, including funding agencies, scientific experts and end-users should ideally be employed in the process.

With respect to recommendations for best practice in *tracking the progress* of projects within research programmes, it was advised that progress reports should be submitted by projects to programme management periodically (*e.g.* every six months or annually). It was also highlighted that research programme management could use intelligent and innovative reporting formats by issuing guidelines to projects that include specific themes or questions. In addition, the periodic progress reporting could be matched with the payment of funds, so that projects get the payment of funds for each next period only after submitting the progress report, and that the system of processing progress reports should be transparent, especially when connected to the payment of funds. The responsibility of reviewing of progress reports should be specifically assigned to someone (*e.g.* the programme coordinator, programme board, external experts or personnel of the funding organisation). In addition, the funding agency can clearly benefit from assigning the responsibility of reviewing progress reports to in-house people. In this way, detailed information and knowledge from the research projects is effectively transferred to the funders. Annual programme seminars and smaller seminars involving the programme board can also be used as an aid to review progress, when projects present their results.

Collaboration between researchers adds clear value to research funding programmes. In terms of management approaches that *promote information exchange* between researchers it was found that collaboration (both nationally and internationally) can be encouraged through the careful selection of project funding criteria, and the merging of closely related project proposals. However, the focus on collaboration should be in proportion to the size of the programme, and not place too heavy a burden on projects or on programme management (for instance, the main effort should be only focused on one aspect, either collaboration within projects or between projects). During the progress of a programme, collaboration and information exchange can be encouraged through annual conferences, thematic small group seminars, joint publications and the internet. Informal meetings are also useful for generating true collaboration: to facilitate this process a specific person, usually in the core team or programme board, can be assigned to strategically plan and co-ordinate collaboration activities.

In relation to issues concerning *motivation* of researchers it was felt that this could be raised by providing possibilities for researcher training, mobility and career development (over and above the provision of monetary incentives). One component of motivation to staff includes the awareness of *gender issues*, which should be dealt with throughout the programme life-cycle using various mechanisms.

2.1.2.2 Linking Research and Policy through Evaluation: A European Overview of Approaches and Practices in Mid-term and Ex-post Evaluation of Environmental Research Programmes

'Policy relevance' is an integral part of discussions in relation to the science–policy interface and a key criterion in public research funding. However, it is both difficult to define and measure in general terms. One reason for this is that the links between research and political decision-making are complex, and policy relevance needs to be defined case-specifically for each research programme or project. In addition, research may have long-term relevance for policy-makers that cannot be detected or realised until some years after a programme has ended.

In June 2007 the SKEP network produced a report[11] that focused on the planning and selection of programme evaluation techniques, the selection of environmental foci and the utilisation of research results. Specific attention was paid in this report towards evaluation at the EU-level and, in particular, how to evaluate joint-funded research programmes with respect to the ERA-Net scheme.

2.1.2.2.1 Planning the Evaluation

Planning of evaluations is an important part of programme management. The context of the programme, partly arising during the implementation of the programme, defines the focus and methods that would be beneficial for evaluation. However, certain key areas, such as the general need for, the budget of, and the main purpose, of the evaluations can be decided upon at the programme initiation stage.

Several actors may take part in planning the evaluation, including the funding agency, the programme preparation committee, external experts and programme management. In some programmes, evaluators are given an active role in planning the evaluation, while in others the funding agency may set strict requirements for the evaluation that is undertaken by external evaluators. It is often difficult to estimate to what extent the programme has influenced a given outcome. This is commonly referred to as the '*attribution problem*'. An assessment of the operational context of the programme can be included in the planning phase of the evaluation. This would support the recognition of which changes can be directly attributed to the programme, and which are due to changes in the operational context.

2.1.2.2.2 Categories of Evaluation

Programme evaluations can be divided into four categories: programme evaluations, thematic evaluations, and evaluations of knowledge systems and of research fields. Programme evaluation is the most common type. Mid-term and continuous evaluations typically belong to this group, as they concentrate on improving a certain ongoing research programme. Thematic evaluations and evaluations of knowledge systems can be seen as alternative or complementing types of evaluation. Thematic evaluations examine a group of programmes from the point of view of a given goal or outcome, while evaluations of knowledge systems concentrate on the social configurations around a research programme, emphasizing the idea that scientific knowledge is produced as a result of interaction and co-operation between several actors.

2.1.2.2.3 Focus of the Evaluation

Programme evaluations may focus on different issues. Some of the most typical foci include evaluating processes within the programme itself, both in terms of its effectiveness and its outcomes. Outcomes can be divided into immediate and final outcomes. Evaluations tend to focus on the immediate, often purely scientific outcomes of the programme, while it is less common (but often more valuable) to evaluate outcomes that are seen sometimes a long time after the programme has ended, *e.g*. impacts on policy, environment, society or economy.

2.1.2.2.4 Evaluation Methods

Methods for evaluating research programmes can be categorized on the basis of the nature of the evaluator, or the techniques actually used, for carrying out the evaluation. Evaluation may be carried out by, for example, a panel of experts, external consultants or other professional evaluators, or it can be done internally by the funding agency. Self-evaluation, in turn, means that people participating in the programme evaluate themselves. These methods are useful for different purposes. Peer panels are suitable especially for evaluating the scientific results and quality of the programme, while professional evaluators have more competence in evaluating the management or other-than scientific outcomes of the programme. Internal evaluations may be especially suitable for undertaking mid-term evaluations, as they require commitment from the programme owners. Different techniques for carrying out evaluations include for instance bibliometric methods, stakeholder surveys, economic evaluation and benchmarking. One method is rarely used alone but, rather, is supplemented by other methods.

2.1.2.2.5 Evaluating the Environmental Dimension

Evaluating the environmental outcomes of a research programme is challenging because a number of societal (and other) factors influence the actual uptake of the results or recommendations of a research programme. Therefore, a programme may have the potential to result in several positive environmental outcomes, but their realization may depend on factors outside the programme. One way to address this problem is to separate the evaluation of the potential impacts of the research programme from its realized, or actual, impacts. In addition, an analysis of the operational context of the programme should be included in the evaluation. This supports the identification of both actors and activities, which are influenced by the programme, and are relevant from an environmental policy perspective.

2.1.2.2.6 Utilisation of Evaluation Results

The utilization of evaluation results is very important from the point of view of improving learning processes and the implementation of lessons learnt into the programme management process. It concerns several groups, *e.g*. programme and project managers, researchers, the funding agency, policymakers and the private sector, who can all learn from the evaluation and use its results for different purposes. One factor that may hinder the use of evaluation results by the programme's funding agency is that evaluations can be experienced (or perceived) as criticism instead of a possibility to learn and improve future programmes. There may also not be enough time and resources to go through in detail, and take onboard, the recommendations of the evaluation, especially after the programme has come to its end. This is one reason why mid-term evaluations are of particular importance.

2.1.2.2.7 Evaluation Linked with Supra-national Collaboration

The SKEP report sought to discover previous experiences in evaluating research programmes at the EU-level, both the Commission's Framework programmes and ERA-NET activities, to identify specific elements related to jointly funded research programmes on a supra-national level. Some important elements to consider in evaluating jointly funded research programmes were identified. These included the following:

- Evaluations need to take into account the settings in the various funding countries. At the same time, a common evaluation protocol, or uniform terms of reference, is needed to guarantee fair and consistent evaluations.
- The composition of evaluation panels, if used, is important. There needs to be a balance in panel members, so that each funding country or organization is equally represented.
- Modified peer panels, combining scientific and other experts, are often regarded as important, because policy and business level outcomes also need to be evaluated.
- Evaluations should especially focus on the added value of joint funding. Regional and cultural differences and conflicting national interests are also important focuses of supra-national evaluations.

As the ERA-Net scheme is a new kind of networking instrument, it can be anticipated that new evaluation approaches will also be needed. Some innovative tools for mid-term or continuous evaluation have been developed, such as PROTEE,[12] but their use for evaluating jointly funded programmes has not been rigorously tested. Evaluation processes as well as lessons learned from ongoing and finalized funding programmes of various ERA-Nets should be collected, analysed and synthesized as these networks progress. In addition, the pilot research funding programmes of ERA-Nets should also be important sources of information for the development and testing of evaluation processes for supra-national research programmes.

2.1.3 Assessing the Dissemination and Implementation of Research

In addition to conducting studies on 'front-end' science into policy issues, the network has also been considering some 'back-end' issues in relation to science communication. In the autumn of 2006, a study was initiated by SKEP to study differing European approaches to the dissemination and implementation of research in SKEP member organisations, and to produce a series of guidelines.[13] The specific aims of the study were to:

1. compare and contrast approaches to dissemination and implementation of research in SKEP member organisations;
2. identify what works (and what does not) and why;
3. develop findings and pointers towards 'good practice'.

Face-to-face interviews with staff from SKEP member organisations and associated bodies were the main mechanism for exploring the above areas. Using a carefully-prepared set of questions, 95 people from 33 organisations were interviewed. In most cases, the interviews were carried out on a one-to-one basis, but occasionally small groups of people were interviewed. In one or two cases the interviews were carried out by telephone.

Interviews were conducted with staff from 14 SKEP member organizations, and also with staff from associated sponsoring and funding bodies, research institutes and groups, subsidiary agencies, sister organizations and other relevant initiatives. Interviewees included researchers, research users, interpreters and intermediaries, funders and commissioners of research.

The report[14] describes their approaches to, and experience of, research dissemination and implementation. Each organization described constituted a 'case study' in its own right, but, in addition, particular programmes, projects and initiatives are presented as case studies to illustrate specific issues. The ways in which approaches to research dissemination and implementation corresponded to different factors and '*settings*' was also examined.

While there are some differences across SKEP member organizations, the study revealed that SKEP members possessed much in common in terms of their approaches, experiences of what works and what does not, and in recognizing remaining challenges that need to be addressed to improve the effectiveness of their research dissemination and implementation processes.

Key conclusions were summarized as follows for the five areas of investigation of the study.

The *planning and management* of research programmes and projects is critical to successful dissemination and implementation. If research is to be used in policy-making and environmental management, users should be involved throughout the planning and execution stages to ensure the continuing coherence of the research questions and the answers that are needed. The dissemination and implementation of research needs to be properly thought through at the planning stage, and adequate resources and time allocated in project budgets and schedules.

With regard to the *communication of results,* the channel and content need to be tailored to the audience: one size does not fit all. An understanding of the audience should be developed, preferably through interactions during the research phase, so that messages can be conveyed in a way that is readily assimilated. In an age of information overload, succinct messages in clear language are required. Wherever possible, an opportunity should be provided for face-to-face interaction between researchers and users so that issues of interpretation can be resolved.

Interpreters and intermediaries can play an important role in synthesizing results into a useful form, and in providing a balanced overview where there are competing claims to the 'truth'. They need to put the science into context and in proportion, describing uncertainties in a way that is helpful to the users but true to the science. Interpreters need to develop good relationships with both users and researchers, understanding both and able to see the world through their eyes. Good social skills, a breadth of view, and the ability to synthesize information and communicate it clearly are all key skills for interpreters.

SKEP network members are putting increasing emphasis on effective *engagement with stakeholders*: the wider group of organizations and people, including the general public, with an interest in the research beyond the direct users. The motivation is to ensure that they have the information that they need to be informed participants in robust debates about policy and environmental management decisions, and that those decisions are informed by a better appreciation of stakeholder views. The media will inevitably play a key intermediary role in communications with the public and need to be courted as valued partners in stakeholder engagement.

Finally, *evaluation* of research impact and of the effectiveness of dissemination processes is recognized as important but is, on the whole, a neglected area. There are some significant methodological difficulties involved in evaluation. However, where it is carried out systematically, it has proved to be a useful management tool. The approach needs to engender the active participation of users and researchers in the evaluation process, encourage honesty in responses, and ensure that lessons are taken on board in future research management activities.

2.1.4 SKEP Joint Calls: Maintaining a Focus on the Practical Application of Science into Policy Processes

The core objective of the ERA-Net scheme has been to consolidate the European research area by creating a new trans-national tier of research funding. The second SKEP joint call was launched in February 2008.[15] This research call, which employed a true-common-pot architecture, and engaged co-funding organisations across six European countries, was intended to build upon the science–policy assessments undertaken over the previous two years with reference to comparative studies of approaches towards research management ('Experiences in the management of research funding programmes for environmental protection') and overviews of research evaluation ('Approaches and practices in mid-term and ex-post evaluation of environmental research programmes').

As part of the development of a framework for the co-ordination of trans-national research, the SKEP network invited the submission of proposals for the following science–policy areas.

2.1.4.1 Assessment of Current Approaches towards the Evaluation of the Uptake, and Impact, of Research Projects and Programmes by Environmental Policy-makers

This section of the call focused on the commissioning of research on the evaluation, uptake, implementation and impact on the policy-making community – widely acknowledged to be an important, but neglected, science–policy area. It was designed to provide essential feedback on the relevance and quality of evaluation approaches undertaken, and their contribution to the development of evidence-led policy.

There are several methodological challenges intrinsically associated with this type of trans-national, evaluation-based research:

- commonly used measures such as citations or patents are often inappropriate for policy relevant research;
- it is difficult to trace the uptake of research in policy-making and regulatory decision taking. The research output from an individual project will be just one of the considerations taken into account, and it may be the integrated results from several projects that has the greater influence;
- it can be some time after the completion of a research project before the impact is realized.

Consequently, a lot of research is aimed at building conceptual understanding rather than at instrumental use, which is generally easier to evaluate. In addition, the relevance of a project or programme may either be reviewed against its starting conditions, or the context pertaining when it is completed; programme and project objectives often tend to be imprecisely defined, making the achievement of objectives difficult to evaluate, and studies might look for examples of where research evaluation has been done (potentially outside the environmental arena) as the basis for drawing up some guidelines on how to undertake this.

The practical objective of the research call was to commission work that would ultimately propose guidelines for the evaluation of the implementation and uptake of environmental research. These guidelines would then be trialed and embedded into the management practices of SKEP network members. Key aspects of the call included:

- a theoretical model of research evaluation that defines the idealised pathways for gauging uptake and impact;

- comparative study of approaches taken towards evaluating the impact of environmental research to support policy development and regulatory activities;
- subsequent development of an applied model for research evaluation and an implementation plan for trial application by SKEP network participants.

This work was designed to pave the way for an improved flow of information between network members, an enhanced ability to track research effectiveness over the lifecycle of projects listed in the SKEP on-line research database, and the potential for external (and hence more objective) assessments of research effectiveness (by one SKEP network member of another SKEP network member).

2.1.4.2 Assessment of Lessons Learnt in the Communication and Dissemination of Emerging Scientific Issues to Environmental Policy-makers

This section of the research call focused on the transition from the detection within research communities of weak signals of environmental concern to the formal recognition by governing institutions of an environmental problem, and ultimately to the initiation of political action – a long and complex process. A good example of this is climate change, which is now occupying centre-stage on the political agenda, approximately three decades after the problem of climate change was first reported by scientists. A similar pattern was evident in early warnings of *Waldsterben* in Germany and central Europe.

Many key factors come into play when attempting to communicate scientific issues, and to bridge any perceived governance gaps, including:

- the importance of two-way communication,
- the communication of uncertainty, including how to prioritize between new issues,
- the role local or lay knowledge may play in detecting and establishing the sensitivity of emerging issues,
- the building of critical mass (within scientific and policy communities) for change,
- the importance of context in how messages are picked up and acted upon,
- the diffusion of ideas through multiple institutions,
- the role of society in catalyzing political action,
- the ability of environmental policy-makers to influence those in charge of economic, trade and industry agendas.

The European Environment Agency's report on 'Late lessons from early warnings: the precautionary principle 1896–2000' (2001)[16] addressed several these issues, and drew 12 key conclusions, which included the need for more integrated science, the need to accurately assess environmental costs and benefits, and the need to use lay and local knowledge when appraising emerging environmental concerns.

The SKEP network launched this research call to assess the lessons learnt in the communication and dissemination of emerging scientific issues to environmental policy-makers. The objectives of the call were to provide perspectives on the following key areas:

1. With reference to case-studies, how can we learn from past attempts to influence (either at an EU-level, a Member State level or a regional level)? How could these findings be used in other contexts (*e.g.* as applied to biodiversity loss, water security or converging technologies)?
2. What are the requirements for scientific advice on emerging issues to be taken-up by policy-makers?
3. How can this process be catalysed – what could be done to accelerate the political decision-making process?

In responding to these objectives, research proposals were encouraged to consider the following ancillary questions:

- How do the communication of risk, uncertainty and precautionary considerations influence decision-making processes?
- How can we analyse how contextual factors and policy-windows impact on the preferred approach to communication?
- How can institutional obstacles to learning and action be reduced by effective communication?
- Can we distil examples of best practice from previous case-studies?
- How can communication and dissemination approaches be used to bridge parallel gaps between research and society, and policy and society?
- Can a generic route-map be constructed for dissemination and communication of emerging issues to improve influencing strategies?

The progress of research resulting from this joint call can be monitored on the SKEP website at www.skep-era.net or by contacting the SKEP co-ordination team (SKEP@environment-agency.gov.uk).

References

1. www.skep-era.net.
2. http://cordis.europa.eu/fp6/fp6_glance.htm.
3. http://cordis.europa.eu/coordination/era-net.htm.
4. SKEP Description of Work, p. 6.
5. http://www.sciencemeetspolicy.org/site/11.asp.
6. http://www.sciencemeetspolicy.org/site/24.asp.
7. http://epanet.ew.eea.europa.eu/.

8. http://www.bridgingthegap.si.
9. http://www.skep-era.net/site/79.asp.
10. E. Furman, P. Kivimaa, P. Kuuppo, M. Nykänen, P. Väänänen, H. Mela and P. Korpinen, *Experiences in the Management of Research Funding Programmes for Environmental Protection – Including Recommendations for Best Practice*, Finnish Environment Institute (SYKE), on behalf of the SKEP network, June 2006.
11. P. Kivimaa, H. Mela and E. Furman, *Approaches and Practices in Mid-term and Ex-post Evaluation of Environmental Research Programmes*, Finnish Environment Institute (SYKE), on behalf of the SKEP network. June 2007.
12. http://cordis.europa.eu/transport/src/protee.htm.
13. http://www.skep-era.net/site/82.asp.
14. J. Holmes, *A Study of Dissemination and Implementation of Research in SKEP Member Organisations*, University of Oxford (on behalf on behalf of the Swedish EPA, and the SKEP network), June 2007.
15. http://www.skep-era.net/site/193.asp.
16. http://reports.eea.europa.eu/environmental_issue_report_2001_22/en.

CHAPTER 2.2

Evolution of Methods to Link Science and Policy: The Experience of EUPOL

ANTONY CHAPMAN,[a] ADRIAAN F.L. SLOB,[b] MARC RIJNVELD[b] AND CORINNE MERLY[c]

[a] wca environment ltd., Brunel House, Volunteer Way, Faringdon, Oxfordshire SN7 7YR, United Kingdom; [b] TNO Innovation & Environment, Van Mourik Broekmanweg 6, PO Box 49, 2600 AA Delft, The Netherlands; [c] BRGM, Avenue C. Guillemin, 45060 Orléans, France

2.2.1 Introduction

The issue of how science can be used to inform policy and political decisions, whether it is underused in the policy environment and whether links between science and policy could be more effective is complex and increasingly significant at both national and European level.[1] If science is used well, it can be a positive asset to policy making;[2] however, it appears that a large body of scientific work is not being effectively used outside its original sphere of influence. There have been many models developed to describe the process of transferring scientific information to the policy community, including the linear relationship,[3] the 'two communities theory',[4] push models, pull models, dissemination models and interaction models.[5] The fact that so many descriptions of this relationship have been developed illustrates its complexity, yet to date none of

Water System Science and Policy Interfacing
Edited by Philippe Quevauviller

Published by the Royal Society of Chemistry, www.rsc.org

these models has been able to completely describe it and further investigation is required in order to recommend new approaches.

AquaTerra was an Integrated Project in the EU sixth Framework Programme. Its aim was to produce a range of coordinated scientific research to assist in river basin management in the context of climate change through the provision of tools, knowledge and information. It consisted of a wide range of scientific sub-projects investigating air, soil, water and groundwater processes at a range of scales from laboratory studies through field investigation to large-scale climate modelling. It also contained socio-economic elements, focused on how the scientific work relates to wider stakeholders in river basins and the social and economic impacts of river basin management in the context of climate change. It began in 2004 and was completed in May 2009.

The EUPOL sub-project was the main element of AquaTerra, aiming to link available science with the needs of policy in a constructive way in order for policy makers to gain maximum use from the various AquaTerra science sub-projects. As such EUPOL needed to have an understanding of the stakeholders who may have use of the information, as well as their specific requirements and the nature of AquaTerra science and what it is capable of providing. In this context, EUPOL considered stakeholders as policy makers and managers with a connection to the implementation of the Water Framework Directive (WFD). EUPOL was closely linked with the INTEGRATOR sub-project within AquaTerra, which aimed to synthesise the varied scientific output of the whole project into a format that makes the information more accessible to a wider audience.

Although EUPOL was a relatively small sub-project within AquaTerra, the work it has undertaken to try and bring science and policy closer together is significant in a wider context. This chapter explores the evolution of EUPOL by examining the various practical approaches that have been used to link science and policy and identifying the lessons that other projects and stakeholders can learn from the EUPOL experience.

2.2.2 Scientific Framework

The first approach adopted by EUPOL was a Scientific Framework to link the rationale behind the AquaTerra science projects to the policy and stakeholder needs.[6] The concept of the framework was to provide a means of identifying how the very specific, technical knowledge of the natural system generated by AquaTerra relates to the wider ranging current and likely future needs of policy makers and stakeholders. The Scientific Framework is effectively a matrix that allows a two-way movement from the initial rationale of each sub-project, through their individual deliverables (supply) to the needs of stakeholders and policy makers (demand) and vice versa; thus the framework facilitates matching the supply of technical information to policy demands. Figure 2.2.1 illustrates the concept.

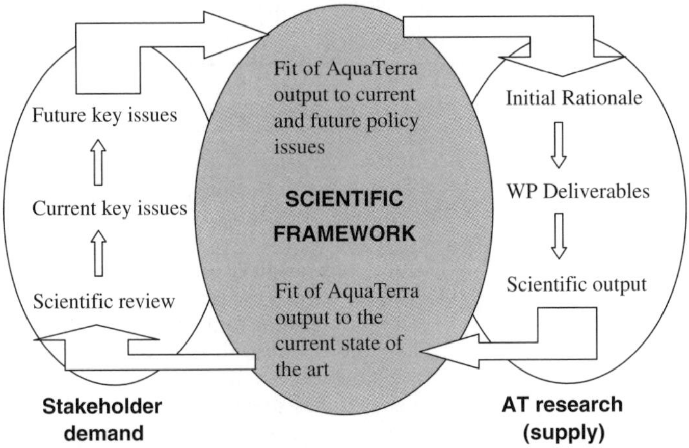

Figure 2.2.1 Linking role of the EUPOL Scientific Framework.

2.2.2.1 Inventory of Policy Questions

Populating the Scientific Framework involved defining both policy demand and the available supply of scientific information. In the case of policy demand, the EUPOL team consulted representatives involved in policy development and implementation through telephone interviews and brainstorm sessions to obtain an understanding of current and future policy needs. Brainstorm sessions were held in Amsterdam, Brussels, Budapest and Paris; in total 25 stakeholders were interviewed in either the brainstorm sessions or the telephone interviews. The brainstorm sessions took place over a three to four hour period and involved the stakeholders being invited to give their individual view on what, in their opinion, the most important environmental issues facing their countries and/or Europe were, both now and in the future, followed by an open forum discussion. The results of the sessions provided a list of questions that could be raised by policy makers now or in the future. The questions were hypothetical, being based on the discussion rather than on direct questions from individual participants. In this way a balance was achieved between making the questions subject or person-specific and reflecting the genuine issues of concern of a wide range of policy makers and stakeholders from various European backgrounds. In total the brainstorming sessions yielded 54 policy questions. The range and nature of the questions is neither exhaustive (for example, a recent study,[7] identified 100 policy questions in a subsequent paper) nor can it be assumed that all policy issues have been highlighted. These policy questions were used as the means to link policy issues to AquaTerra *via* the Scientific Framework. First they were linked to one of six broad policy areas: three (climate change, land use and agriculture) were identified as key driving forces influencing policy decisions during a session with policy makers, the fourth was the overarching theme of river basin management, and the remaining two (soil and water, with a sub-category of soil/water for questions

Table 2.2.1 Functions and uses of the natural environment for the purposes of the scientific framework.

Functions of natural resources/ ecosystems	*Uses (with users – liable parties – land planners)*
Biomass production (food, feed, raw material, energy)	Natural vegetation & animals, agriculture & forestry, fisheries & fish farming, biofuels
Storing, filtering, transformation, including CO_2 sink	Agriculture & forestry
Habitat and gene pool	Natural vegetation & animals
Physical & cultural environment for mankind	Urban infrastructure (residential/commercial/traffic), recreation/leisure/tourism, transport (including navigation), cultural heritage (relic industrial/historic landscape/buildings)
Source of raw materials (other than biomass)	Industry/mines/storage of wastes and residue, agriculture (water abstraction, soil nutrients/growing media), municipal water use (abstraction/reservoir storage), energy (hydroelectricity, geothermic energy, fossil fuels)

that relate to both) were specific to particular issues. Once the questions were linked to a policy area they could be further defined by being linked to a more specific subtopic, defined as either a function that the environment is required to serve or a use to which the environment, or products from the environment such as food crops, are put (Table 2.2.1). This sub-division was based on the draft classification of the European Soil Strategy, as recommended by the participants of the first brainstorm session: it was developed further to take into account the functions and uses of the water system, which is also of relevance to AquaTerra. This subdivision by function and use helps to define the location of links between policy demands and the supply of scientific data, which are not always clear.

2.2.2.2 Linking Demand to Supply

To match the supply of scientific information from AquaTerra to the framework the work packages were linked provisionally to the functions and uses in Table 2.2.1 through the deliverables scheduled for each work package between months one and thirty of the project. These were linked initially by the EUPOL team on the basis of information in the Description of Work (DoW). Individual deliverables were categorized as being of high, medium, low or no relevance to each policy question. As a means of quality control this initial assessment was sent to the WP leaders in AquaTerra to allow them the opportunity to comment on how their science work had been linked to policy demands *via* the EUPOL Scientific Framework. This was, firstly, to ensure that the science as portrayed in the DoW matched the perceptions of it by the EUPOL team linking it to policy demands and, secondly, to identify any further linkages between the two as identified by the WP leaders themselves.

Having made the link between policy questions and AquaTerra deliverables, the degree of relevance of AquaTerra science as a whole to each question was assessed 'as a whole'. The relationship between AquaTerra deliverables and policy questions was expressed in the form of an index. The three categories of relevance into which the deliverables are divided (*i.e.* high, medium and low) were weighted by a factor of 1, 0.5 and 0.1, respectively, and then a percentage value was calculated using the formula below; 0% indicates AquaTerra can be of no assistance in addressing the issue and 100% means that the entire output of AquaTerra is directly relevant to the issue:

$$R = \left[\frac{(n(a) \times 1) + (n(b) \times 0.5) + (n(c) \times 0.1)}{N}\right] \times 100\%$$

where R = relevance index (%), $n(a)$ = number of deliverables of high relevance, $n(b)$ = number of deliverables of medium relevance, $n(c)$ = number of deliverables of low relevance and N = total number of deliverables (246 in the case of AquaTerra).

As there were a large number of deliverables within AquaTerra, most of which will not be directly relevant to any individual policy question, the percentages derived by this calculation are very low and this may give a misleading indication of the value of AquaTerra science to policy. To overcome any undeserved negative interpretations the figures for each question were normalized to a value between 0 and 1 and subsequently ranked as high, medium and low relevance (0–0.19 = low; 0.2–0.65 = medium; >0.65 = high). Note that these figures are a calculation of the relevance of what AquaTerra as a whole can contribute to the policy question rather than the extent to which the question can be answered by AquaTerra. In other words a figure of 100% indicates that all the pertinent AquaTerra deliverables are directly relevant to the question rather than indicating that the question can be answered in its entirety by AquaTerra.

The scientific framework methodology provided an effective means of linking the broad nature of policy issues to the more specialized output of a large integrated science project such as AquaTerra. The methodology is also transferable, it could be applied to other projects or other policy questions could be derived and linked to AquaTerra. However, there are limitations as well as positive aspects to this approach. The main limitation is that it is a subjective approach, both in terms of the responses and views of the stakeholders interviewed and in the interpretation of how the available scientific information relates to the policy questions generated. As a result some policy issues may not have been raised by the stakeholders and some connections may not have been realized by the assessors. Thus any end user should exercise their personal judgment when using the assessment, using it as a guide to the linkages between project and policy demand rather than a definitive assessment. A further issue is that the assessment of the output deliverables was undertaken before the deliverables were completed, so it is possible that the outcome perceived from the DoW may not match the output of the project as delivered. This was unavoidable if the assessment was to be undertaken with the lifespan of the project.

Despite the successful development and application of the scientific framework in the early stages of the project, it was not pursued further as there were several practical limitations within the project. There was a limit both to the number of brainstorms and interviews it was possible to conduct and to the number of questions it was possible to analyze in the depth required by the methodology within the time constraints of the EUPOL project. The scientific framework is a time consuming methodology to undertake and requires a researcher with broad-based knowledge rather than a specialized one to make effective links between elements of the wide variety of information at their disposal. These are not limitations of the methodology *per se*; now the methodology has been developed it could be applied to other questions (*e.g*. ref. 7) or to other science projects. However, within the context of EUPOL, the development of the project stopped at this stage (see ref. 6 for a comprehensive report) and the focus of EUPOL moved to other methods of linking science and policy.

2.2.3 DPSIR

The AquaTerra project as a whole used the DPSIR (Drivers, Pressures, State, Impact, Response) model as a framework to link the various aspects of the project together into a coherent structure. This model has been widely used at a European level[8] and is a useful alternative to the Scientific Framework as it connects physical science and social and economic systems. In this model *driving forces* refer to large scale changes, such as climate change or changes in land use, which lay *pressures* on the use and functions of the natural system, for example increased quantities of diffuse pollution brought about by changes in land use will place pressure on the buffering capacity of soil and water systems. This pressure may in turn alter the *state* of the system as a whole, which may have *impacts* now and/or in the future. *Responses* are the actions that might be undertaken to remediate the *impact* produced by the *state* of the particular component of the system (water or soil for example), with the aim of relieving the *pressure* on the system as a whole in a sustainable manner, so as not to create further pressures elsewhere. Responses may come in the form of policy changes or direct practical measures. In general, policy measures affect the larger scale drivers and pressures (such as legislation to mitigate carbon emissions), whereas practical measures tend to be applied to remediate or alleviate the state of a system or the impacts caused by it (such as limiting nitrate fertilizer applications or increasing controls on air and water emissions from a chemical plant).

Most of the scientific work within AquaTerra focuses on the state of the different systems within the river basin (air-soil-water-groundwater-sediment) and the impact of the state of the system, with a lesser emphasis on driving forces and pressures. Responses are not dealt with; this would imply that the information is being provided within the project for stakeholders to make a decision on the best form of response based on scientific evidence. This interpretation of DPSIR and AquaTerra suggests that the connection between science and policy lies in linking DPS and I to R. DPSIR is an effective concept for describing the interrelationship between different physical and political elements of a river basin system; it helps to show where AquaTerra (and other) science fits into the wider river basin system and how it relates to other physical and social aspects of the system. This has been demonstrated at a practical level within the INTEGRATOR sub-project of AquaTerra, which has close links with EUPOL. The DPSIR model was used as the basis for a series of socio-economic models developed at a range of scales in the AquaTerra study river basins (The Ebro, Meuse and Danube) which were designed to assess the relationship between specific economic activities on specific aspects the soil-water-sediment-groundwater system in each case.[9] The aim of these models was to improve understanding of the functionality of the system.

2.2.4 Resilience

The concept of resilience has been used in the wider sphere of environmental management as a means of understanding the functionality of complex systems and identifying the key influences on their stability.[10]

The core of resilience thinking as applied to ecosystems is the notion that a system has a finite capacity to absorb pressures while maintaining its functionality. Resilience is defined as 'the capacity of an ecosystem to tolerate disturbance without collapsing into a qualitatively different state that is controlled by a different set of processes'.[10] If a pressure is maintained on the system, eventually its capacity to resist the effect of this pressure will be exhausted and an apparently instant, potentially catastrophic change in the system's functionality will occur. If the pressure is removed at this point the system may have the capacity to revert the impact of the pressure or it may not. A system may have more or less resilience to a pressure and its resilience will vary with different forms of pressure. From this description it can be seen that resilience is a dynamic concept suited to the improved understanding of complex systems. One example is the application of resilience models to the assessment of sustainability, which is usually described as a balance between economic, social and environmental elements that interact with each other and vary within themselves according to outside influences. Thus sustainability can be seen as a complex and dynamic system.

The EUPOL project considered the possibility of using the resilience of ecosystems as a framework both to unify and understand the scientific output of AquaTerra and to aid the translation of this output to a policy orientated context. Limited progress was made in these investigations and it was not ultimately taken up within the project. However, if the notion of resilience were considered in a project from the outset it may provide a useful tool to guide science–policy interaction.

2.2.5 Connecting with Scientists and Policy Makers: Workshops and Interviews

A core approach within EUPOL has been direct contact with both the scientific and policy communities through telephone interviews, surveys and workshops. Communication and dialogue with stakeholders is a fundamental part of the Water Framework Directive (WFD) and one of the aspects of the WFD that sets it apart from earlier European legislation. The experiences from the communication exercises within EUPOL (listed in Table 2.2.2) can provide some useful insight both into the nature and extent of the current links between science and policy and how these might be improved.

One of the widely accepted aspects of science–policy integration is that good quality interaction between scientists, stakeholders and policy makers greatly enhances the uptake of science beyond its source community.[11] The sooner this is undertaken within a project, the more successful the uptake tends to be. In large projects such as AquaTerra, such communication also helps scientists from widely differing backgrounds to communicate more effectively with each other. It would be more accurate to describe this interaction as multiple community rather than two-community.[11]

The workshop sessions that the EUPOL project has produced during the course of AquaTerra have resulted in lively debates that demonstrate the interest this topic currently generates. However, the fact that relatively few people have actively participated in EUPOL activities suggests that the field of science–policy communication is not at the core of many people's agendas. This does not imply that it is not of interest, but that it is a low priority for many stakeholders who have limited time to devote to all of their activities. Many also have the view that adapting science to the needs of policy and making policy requirements explicit to the scientific community is seen as a role for someone outside both fields. A future issue will be to try and identify who, if anyone, should take responsibility for linking science and policy and how this could best be achieved.

Table 2.2.2 Communication and information gathering activities of the EUPOL project during the course of AquaTerra.

Description of activity	*Date*
Internet and telephone survey of policy makers to identify important issues	2004
Brainstorm sessions with policy makers and stakeholders (Amsterdam, Brussels, Paris, Budapest)	2005–2006
EUPOL session at AquaTerra General Assembly, Bordeaux	2005
EUPOL session at AquaTerra General Assembly, Menorca	2007
Combined EUPOL/RISKBASE workshop, Venice	2007
EUPOL special session at ConSoil 2008, Milan	2008

2.2.6 Future Plans

Having explored the potential connection between AquaTerra science and policy demand through various means, a final task for the EUPOL project was to investigate the connections between the science, policy and political spheres. Two more workshops were undertaken prior to the end of the project. One focused on what AquaTerra can give at a management level by communicating with those directly involved with the implementation of River Basin Management Plans, while the second involved politicians to establish the kind of links that they have with science and the use that they make of scientific information.

The nature of environmental policy has been changing over the past ten years; it is becoming more holistic and less about thresholds and limits related to specific substances and discrete environmental compartments such as lakes or a stretch of river. While such assessments of environmental quality remain, the consideration of the quality of the environment as a whole is becoming the means by which the success of environmental action is assessed. Science, as well as the interaction between the different communities, should reflect this holistic view. Both science and policy groups have made significant progress in this respect; however, more could still be done. From the experience of the EUPOL project the development of boundary spanning approaches to link different areas of experience and expertise should be adopted now and in the future. Boundary spanning, in its broadest sense, is the active transfer and integration of knowledge from discrete sources that do not typically interact. Boundary spanning might involve a person, an object or an activity, or a combination of these.[12] Boundary spanning is also considered as a two-stage process, namely the gathering of information from numerous sources and secondly the digestion and dissemination of this information within the internal group from whom the boundary spanning originated. Clearly this process may already occur at a subconscious or small scale: a future goal of EUPOL team members will be to promote the conscious use of boundary spanning to improve knowledge and understanding between science groups policy groups and stakeholders. This activity will not form a part of EUPOL itself, but will be an ongoing activity.

2.2.7 Lessons Learned From the EUPOL Experience

EUPOL has developed and implemented a range of methods to connect policy and science more effectively. From this work a range of conclusions and recommendations can be made for future work to improve the connection between scientific investigation and the wider community who could make use of this information:

- Effective use should be made of scientific information to inform policy and management decisions where appropriate.
- Making effective use of the information that is already available will help to focus further scientific activity into areas where most research is needed.

- In science projects that have an applied approach, particularly those that aim to provide support to major initiatives, such as the WFD, more effort should be made to translate the outputs of these results into valuable information for stakeholders beyond the immediate scientific community.
- Not all science is geared towards providing answers for policy and this should be recognised; there should always be a role for pure science.
- The role of exporting scientific information is a specific one and should be undertaken by someone suited to the role rather than someone who is available; this would not necessarily be someone from within either the scientific or policy communities, but they should have knowledge of both.
- The DPSIR framework used in the AquaTerra project is a useful tool for explaining the interrelationships between science and policy, and between the environment and society. It could also be used as a practical means to link science and policy where there is a clear objective and there has been close science–policy interaction from the start of a project.
- The Scientific Framework developed within the EUPOL project is a useful tool for providing a direct link between the supply of specific scientific information from a large Integrated Project such as AquaTerra and policy demands as defined by direct communication with stakeholders. However, it is a subjective assessment and should only be used as a guide to finding relevant information. While it is practical and effective, it is also time consuming.
- The concept of resilience has been used in other environmental fields to understand the functionality of complex systems and identify influences on their stability. The EUPOL project considered using the resilience of ecosystems as a framework both to unify and understand the scientific output of AquaTerra and to aid the translation of this output to a policy orientated context.
- Direct personal communication through interviews and workshops has proved to be a very effective means of canvassing opinions and gathering information. However, more effort should be made by both the policy and science communities to increase the amount of such activities and the degree of involvement on both sides.
- To help achieve more effective communication between science and policy stakeholders, it would help if the goals of the science–policy interaction process were more clearly defined, both from the science and the policy perspective.

2.2.8 Conclusions

A large quantity of scientific information is generated throughout the world each year. An effective infrastructure of publications, conferences, websites and interest groups exists to facilitate the exchange of science within the scientific community, but this infrastructure is less well adapted to communicating scientific information beyond its immediate sphere of interest and into other areas or groups of people where the information would be of interest and value, such as policy makers, politicians and regulators responsible for River Basin Management Plans in the WFD. It may be that scientific information would have to be adapted to suit this purpose; it may also be the case that different lines of communication between the science and policy communities need to be identified and established to facilitate the process. If the policy community wishes to make the most of such communication activities they need to be actively involved with them to provide a dialogue with scientists and help to define the aims and objectives of the discussion. In this way each group can learn more about how the other functions and help to create a more effective link between the two.

Acknowledgements

EUPOL was a sub-project of the EU FP6 Integrated Project AquaTerra (project no GOCE 505428), which was undertaken under the thematic priority: sustainable development, global change and ecosystems.

References

1. P. Quevauviller, P. Balabanis, C. Fragakis, M. Weydert, M. Oliver, A. Kaschl, G. Arnold, A. Kroll, L. Galbiati, J.-M. Zaldivar and G. Bidoglio, Science-policy integration needs in support of the implementation of the Water Framework Directive, *Environ. Sci. Policy*, 2005, **8**(3), 203–211.
2. B. Ballentine, *Enhancing the Role of Science in the Decision-making of the European Union*, EPC Working Paper 17, European Policy Centre, Brussels, Belgium, 2005.
3. C. H. Weiss, The many meanings of research utilisation, in *Social Science and Social Policy*, ed. M. Bulmer, Allen & Unwin, London, 1986, 31–40.
4. M. Wingens, Toward a general utilisation theory: a systems theory reformulation of the two communities metaphor, *Knowledge: Creation, Diffusion, Utilisation*, 1990, **12**, 27–42.
5. R. Landry, N. Amara and M. Lamari, Utilisation of social science research knowledge in Canada, *Res. Policy*, 2001, **30**, 333–349.
6. A. S. Chapman, A. Slob, M. Rijnveld, P. Menger, M. Lansiart, P. Strosser and R. P. Bardos, The EUPOL Scientific Framework; development and refinement. AquaTerra deliverable E1.1c, February 2005; http://www.attempto-projects.de/aquaterra/.
7. W. J. Sutherland, S. Armstrong-Brown, P. R. Armsworth, T. Brereton, J. Brickland, C. D. Campbell, D. E. Chamberlain, A. I. Cooke, N. K. Dulvy, N. R. Dusic, M. Fitton, R. P. Freckleton, H. Charles, J. Godfray, N. Grout, H. J. Harvey, C. Hedley, J. H. Hopkins, N. B. Kift, J. Kirkby, W. E. Kunin, D. W. MacDonald, B. Marker, M. Naura, A. R. Neale, T. Oliver, D. Osborn, A. S. Pullin, M. E. A. Shardlow, D. A. Showler,

P. L. Smith, R. J. Smithers, J. -L. Solandt, J. Spencer, C. J. Spray, C. D. Thomas, J. Thompson, S. E. Webb, D. W. Yalden and A. R. Watkinson, The identification of 100 ecological questions of high policy relevance in the UK, *J. Appl. Ecol.*, 2006, **43**, 617–627.

8. European Environment Agency, EEA, *Environmental Indicators: Typology and Overview*, EEA Technical Report No 25, EEA, Copenhagen, Denmark, 1999. http://reports.eea.europa.eu/TEC25/en/tech_25_text.pdf.
9. C. Herivaux, M. Bouzit, N. Graveline, E. Ansink and P. Strosser, Synthetic report on the economic analysis including a sensitivity analysis for each selected study areas, and the link with the conceptual model. AquaTerra deliverable I2.4, November 2006; http://www.attempto-projects.de/aquaterra/.
10. S. Jansen, I. Immink, A. Slob, M. Rijnveld and J. Brils, Resilience and Water Management; a literature review. Discussion document produced by AquaTerra Netherlands, November 2007.
11. A. Slob, M. Rijnveld, A. S. Chapman and P. Strosser, Challenges of linking scientific knowledge to River Basin Management; AquaTerra as a case study, *Environ. Pollution*, 2007, **148**, 867–874.
12. M. Duijn, W. St-Amour, I. Bogenrieder and M. Rijnveld, An integrative approach to knowledge transfer and integration: spanning boundaries through objects, people and processes, OLKC 2008 International Conference on Organisational Learning, Knowledge and Capabilities: The many Senses of Organization and Learning, 28–30 April 2008, Copenhagen, Denmark, 2008.

CHAPTER 2.3

Science–Policy Interfacing in the CA Project RISKBASE

JOS BRILS

Deltares, Daltonlaan 400, 3584 BK, Utrecht, The Netherlands

2.3.1 Introduction

The European Commission (EC), under its 6th Framework Programme (FP6), funded the Coordination Action (CA) RISKBASE. This project aims to bring together and synthesise the state-of-the-art in risk-assessment and risk-based management of European river-basins. This synthesis leads to recommendations in this field for policy, management and research. The background and objectives of RISKBASE are further described in Section 2.3.2. RISKBASE has its roots in the biophysical sciences domain. However, already at the design phase of the project, it was realised that to achieve its goal it is essential to have a strategy on how to bridge the gap between biophysical science on one side and policy and management at the other side. Five pillars are the basis to Science–Policy Interfacing (SPI) in RISKBASE:

1. a clear idea about the primary target group (science for whom?);
2. a clear idea about the risk objective (risk to what?);
3. a transparent, SPI supporting project structure;
4. actual engagement with actors from policy, science and practice at the European level;
5. a common language, needed to enable this engagement – or better – to enable communication.

Water System Science and Policy Interfacing
Edited by Philippe Quevauviller

Published by the Royal Society of Chemistry, www.rsc.org

These five SPI pillars are further described in Section 2.3.3. Furthermore, there are three different types of engagement, and thus also three different types of SPI:

1. Science–Science Interfacing (SSI);
2. Science–Policy Interfacing (SPI);
3. Policy–Policy Interfacing (PPI).

The specific SPI activities in RISKBASE are categorised and briefly described under these three types of interfacing, along with a brief summary of the preliminary results (the project is due to be completed at the end of 2009), in Sections 2.3.4–2.3.6. Finally, this chapter concludes with a brief summary of the SPI lessons learned in RISKBASE (Section 2.3.7).

2.3.2 RISKBASE

In the last decade several EC RTD Framework Programme (FP) projects (FP4-FP6) and other major research initiatives have addressed and promoted issues related to risk-assessment based management. Most of these initiatives focused on quality and management aspects of one specific compartment such as surface- and groundwater (BRIDGE), soil and groundwater (INCORE, JOINT, WELCOME, CORONA, WATCH), sediment (SEDNET) or contaminated land (CARACAS, CLARINET, NICOLE). However, all these initiatives stressed the importance of an integrative approach for understanding and managing the multi-compartment water-sediment-soil system at the river-basin scale.

To improve the scientific basis for the integrated management of this soil-sediment-water system several EU RTD and Specific Support Action (SSA) initiatives have been funded, *e.g.* HARMONICA in FP5 (Chapter 2.4 of this book), the Integrated Projects AquaTerra (Chapter 2.2) and MODELKEY, the SSA EUROWET and the SSA SoilCritZone in FP6. Europe is also implementing the Environmental Technology Action Plan (ETAP) on the level of mitigation technologies, because Europe needs to increase investment in more innovative environmental protection methods while boosting competitiveness.[1] In relation to risk-based management, the JOINT project concluded that 'accompanying activities are necessary to help networking within the scientific and end-users communities, to evaluate the results and up-date the state of the art, and to focus on the main needs, strategies and future perspectives'. Furthermore, 'an integrated management approach is needed, which considers soil and waters as one system, interacting with other compartments of the environment and with the socioeconomic world through users and functions'.[2]

At the end of EC FP6, there was a clear need to bring all soil-water RTD efforts together through a CA as a stepping stone towards further policy development and implementation whilst addressing the RTD topics highlighted in the FP7 programme. This CA is the FP6 project RISKBASE (acronym), full title: Coordination Action on Risk Based Management of River Basins.

In RISKBASE, leading European scientists and representatives of major, European stakeholder groups review and synthesise the outcome of EC RTD Framework Program projects, and other major initiatives, related to integrated risk assessment based management of the water/sediment/soil system at the river-basin scale. The synthesis will lead to the development of integrated risk assessment-based management approaches enabling the prevention and/or reduction of impacts caused by human activities on that system. RISKBASE aims to deliver: (i) an overarching concept, generic approach and guiding principles to integrated risk based management of river basins; (ii) recommendations towards evolution and implementation of risk based management operations in national and community policies and towards implementation in management; and (iii) a proposal for the European research agenda related to risk based management.

To be able to achieve this, RISKBASE organised numerous workshops throughout the project, which were dedicated to specific issues related to risk based management at the river-basin scale. In addition, RISKBASE organised an annual General Assembly (GA) and made use of an existing web-based information exchange system, EUGRIS (www.eugris.info) (Chapter 4.3). The workshops, GA and the website (www.riskbase.info) were open to all who are interested and willing to contribute to achieving the RISKBASE goals and objectives.

2.3.3 Science–Policy Interface Pillars in RISKBASE

As introduced in the beginning of this chapter, five pillars are the basis to Science–Policy Interfacing (SPI) in RISKBASE: (1) a clear idea about the target group (science for whom?), (2) a clear idea about the risk objective (risk to what?), (3) a transparent, SPI supporting project structure, (4) engagement with actors from policy, science and practice at the European level and (5) a common language to enable the engagement or communication between these actors. Each of these pillars is further described below.

Science to whom? The primary target audience are the drafters and implementers of Water Framework Directive (WFD)[3] River Basin Management Plans (RBMPs). RBMPs will be published at the end of 2009 and so the RISKBASE findings and recommendations will not be in time for this first generation of plans. This was realised already at the beginning of the project. Hence, RISKBASE aims to get its key findings noted for – and hopefully also integrated in – the second generation (*i.e.* the first update) of WFD RBMPs (2015). It is, hence, anticipated that the 'window of opportunity' for the RISKBASE messages will be from early 2012 to mid-2014 (Figure 2.3.1).

Risk to what?: Sustaining ecosystem services (see explanation further below under 'common language') for human well-being is a main EU environmental policy objective.[4] However, RISKBASE selected ecosystem services provided

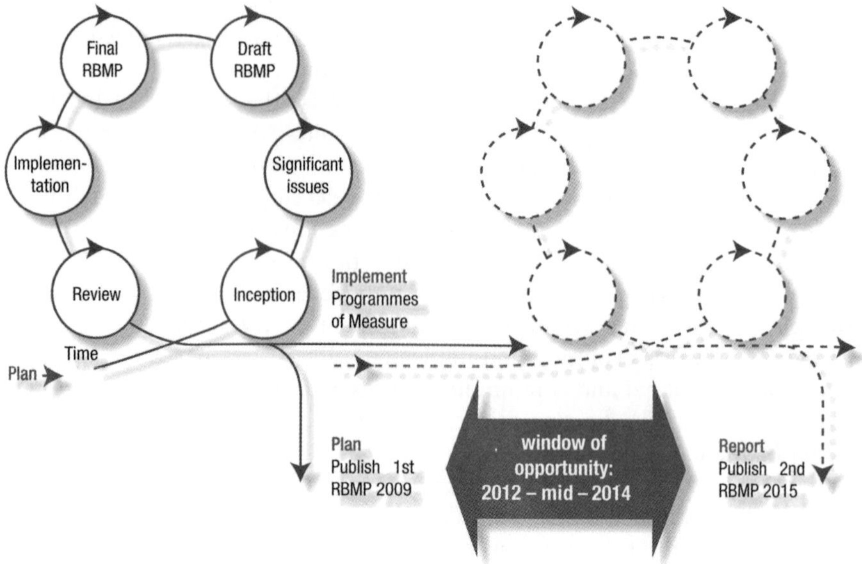

Figure 2.3.1 Window of opportunity for integration of the RISKBASE findings. (Original drawing courtesy of Damian Crilly, presented at RISKBASE Workshop in 2007, Vienna.)

by the biophysical soil-sediment-water system as its main risk objective, because this concept fits closely with the highly dynamic nature of river basins and their adaptive management. Thus RISKBASE aims to develop a risk-based management approach enabling the conservation and restoration of these services.

Supporting project structure: A transparent, SPI supporting structure was provided by the Drivers-Pressures-State-Impact-Response (DPSIR) framework.[5] The different DPSIR components are divided and positioned in a plane that differentiates between the 'biophysical system' and the 'social system' (Figure 2.3.2). The biophysical system includes the natural/biophysical river-basin components with surface- and groundwater, sediment and soil. Related to the input of knowledge and experience – *i.e.* a key prerequisite for the success of RISKBASE – the biophysical system is studied by biophysical related sciences disciplines, such as geology, chemistry, geochemistry, biology, microbiology, ecotoxicology, meteorology, modelling, *etc.* The social system is more the domain of socio-economic, juridical and governance disciplines.

The three different RISKBASE work packages (WP) A, B and C are positioned in such a way that all connecting arrows between the DPSIR components (Figure 2.3.2) are covered by these WPs. WP-A, 'degradation causes, mitigation and remediation', covers in the 'biophysical system' plane the DPSIR connections D-P, P-S, R-P and R-S and the influence of global change on these connections. Thus WP-A covers the identification of the causes of degradation and their actual and future status, and the identification of

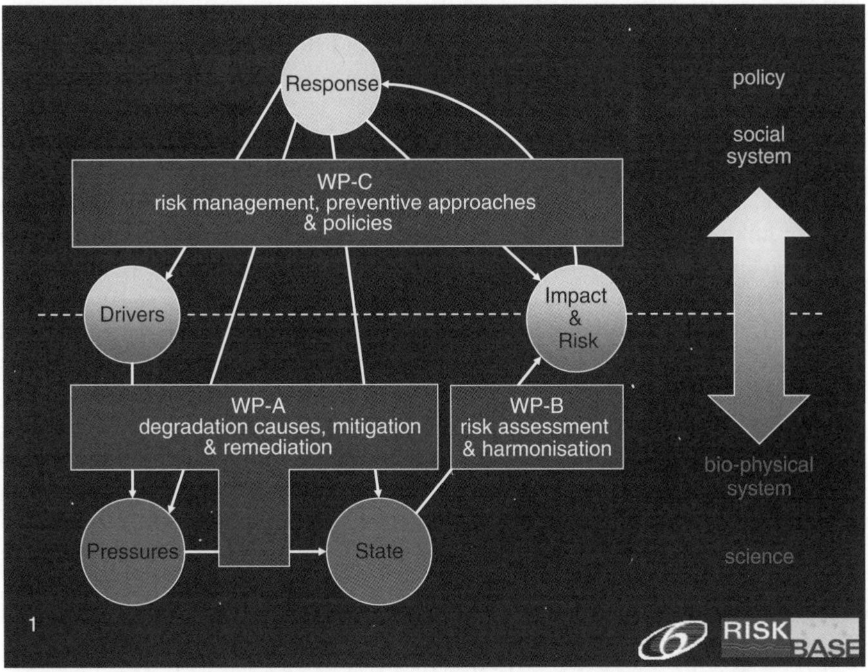

Figure 2.3.2 DPSIR framework at the heart of the RISKBASE project structure.

mitigation and remediation measures. WP-B, 'risk assessment and harmonisation', covers in the 'biophysical system' plane the DPSIR connections S-I, or, with specific focus on risk-based management, S-I & Risk and the influence of global change on this connection. Thus WP-B covers the improvement of risk-assessment methodologies. WP-C, 'risk management, preventive approaches and policies', covers in the 'social system' plane the DPSIR connections R-D, R-P, R-S and R-I & Risk (also in relation to the influence of global change). Thus WP-C covers the development of preventive approaches, development of operational management tools and recommendations and options providing the knowledge base for the evolution and the implementation of community policies.

Although it may appear from Figure 2.3.2 that there is a clear division between the social and biophysical plane, RISKBASE follows the opinion of the Resilience Alliance (www.resalliance.org) that such a clear division cannot be made and that in reality it is one, close-knit social/biophysical, or – in Resilience Alliance (RA) terms – social/ecological system. Figure 2.3.2 combined with this RA viewpoint may result in a conceptual diagram as shown in Figure 2.3.3. A lesson learned so far from contact with stakeholders participating in RISKBASE activities is that many stakeholders find the DPSIR framework still too complex, but seem to have no problems with the 'simple' RA social/ecological system presentation.

Engagement with actors: From the literature it is known that an intensive interaction between scientists, policy makers and stakeholders will promote the uptake of scientific information in policy processes.[6,7] RISKBASE started in a good position to engage with actors from policy, science and practice because it continued to build upon the work of previous Concerted Actions/Thematic Networks/Specific Support Actions (CARACAS, CLARINET, NICOLE, SEDNET, EUROWET) and Accompanying Measures (JOINT) that have actively linked up to policy and environmental management practices. There was a strong link with the contaminated land policy makers (COMMON FORUM) because their (former) secretary is part of the RISKBASE consortium. Furthermore, the other consortium members were also selected based on their involvement in projects and initiatives mentioned above. Thus the consortium represented several decades of experience in research and its application for EU policy making and management in the field of (ground)-water, sediment and soil. In addition, several other existing expert and stakeholder (end-user) networks readily contributed to help to make RISKBASE a success. They delegated relevant experts to participate, on invitation, in selected RISKBASE workshops and/or general assemblies, where they were willing to share their knowledge, expertise and experience in environmental risk assessment and management. To this end, many supporting letters were collated and added to the RISKBASE project proposal.

Common language: In the multidisciplinary and multi-stakeholder processes the language that is being used is very important.[6,8,9] Slob *et al.*[6] have researched the hurdles for science policy interfacing and found that the capability to speak each other's language is a very important factor for that. This observation is similar to that of Totlandsdal *et al.*,[7] who concluded that 'each expert field (whether it is scientific or policy related) has its own set of jargon and expressions that may complicate communication'.[7] Quevauviller *et al.*[10] also stressed the communication difficulties linked to different 'jargon' used in the different communities.

The words River-Basin, Risk, Ecosystem Services, Risk Assessment, Risk Management and Risk Assessment Based (Risk-Based) Management mean different things to different people. RISKBASE has not tried to establish precise definitions of these words. However, consensus on a common perception of these words was developed. This consensus has been described by Brils *et al.*[11] and is summarised briefly below.

River basin: RISKBASE follows the European WFD definition of a river basin. Hence, RISKBASE perceives a river-basin as a natural geographical and hydrological unit rather than looking at administrative or political boundaries.

Risk: Although risk has become a central concept in environmental policy and practice, this does not mean that it is an easy concept to define well. Within RISKBASE risk is related to all agents or sources that may inhibit the achievement of river-basin, ecological quality objectives and specifically the sustainable use of ecosystem services.

Ecosystem services: Societies (present and future generations) depend for their well-being on the goods and services provided by ecosystems[12]

(Figure 2.3.3). Such goods, comprise, *inter alia*, (drinking) water, food, fuel, medicines and building materials. Services are the benefits that society obtains from ecosystems – for instance, life support (*e.g.* biodiversity, fishery, fertile soils for agriculture, water supply and protection against natural hazards), regenerative services (cycling of nutrients) and cleansing services (clean water) that nature provides. Also the well being that nature imparts through enjoyment and recreation is such a service. Biodiversity is seen as metaphor for the health of ecosystems[4,12] and is thus an important measure for the effective functioning of natural processes.

Risk assessment: Risk assessment generally starts with a respective risk agent, or source. It identifies both the potential damage scenarios and their probabilities and then models the potential consequences over time and space. From the early 1980s, there has been an ongoing debate in most developed countries between those who assess risk scientifically and technologically, on the one hand, and social scientists and psychologists, on the other hand, about the measurability or predictability of risk. The debate can, to a large extent, be characterised by two contrasting points of view: the scientific approach (formal risk assessment) versus the risk as perceived by individuals or the general public (intuitive risk assessment). The need for the better integration of these objective and subjective components is becoming increasingly recognised in almost all fields of risk assessment.

Risk management: Risk management relates to the policy driven decisions about risks, which depend on specific time and economic conditions. In more general terms, risk management refers to the creation and evaluation of options for initiating or changing human activities or (natural and artificial) structures with the objective being to increase the net benefit to human society and prevent harm to humans and what they value. Managing of uncertainty – scientific, technical, subjective, needs of society – is an inherent part of risk management. Risk management is a permanent process and interaction of society to organise anthropogenic activities and living with nature[13] (Figure 2.3.3). Thus, it is generally a major task to get a better understanding of how a social system understands the biophysical system and reacts by different means of risk management, preventive approaches and policies. To analyse nature as an ecological system it is a prerequisite to understand political cultures, institutional arrangements, stakeholder involvement, risk perception, risk communication and the how to translate societal values into science related risk management systems.

Risk based management: Risk based management or decision making is only better if the uncertainties associated with the decision are addressed in a transparent way. As a consequence, decision-making can be more efficient in terms of the information needs. The basic question is 'do we have enough information and/or are we certain enough to make a decision'. Risk based decision making usually consists of a tiered approach. If there is not enough information/knowledge/certainty to make a decision, additional investigations are carried out. Following the Risk Based Land Management concept developed by CARACAS and CLARINET[14] the decision making process needs to

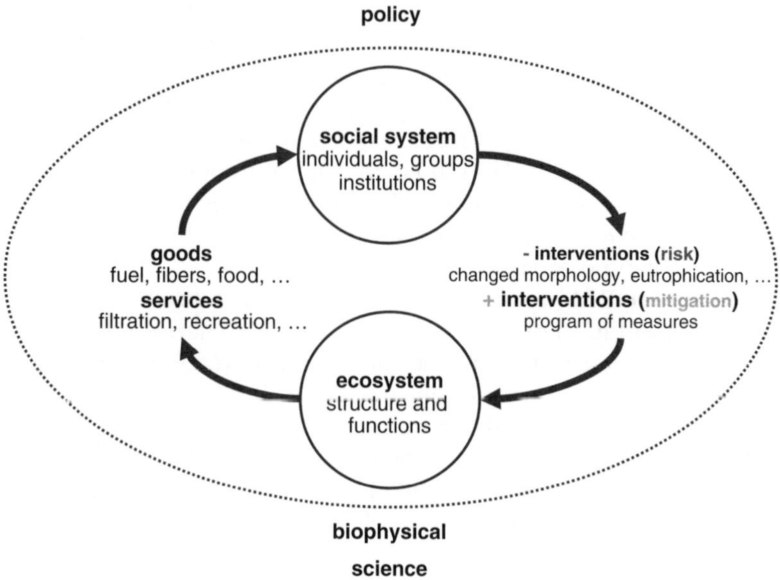

Figure 2.3.3 Resilience Alliance's (www.resalliance.org) social/ecological system from a RISKBASE perspective.

consider three main components that form the core of risk based management: (i) fitness for use, (ii) protection of the environment and (iii) long-term care. The first two components describe goals for the safe use of a certain part of the river basin, including the prevention of harm and resource protection. The third allows for a more rigorous assessment of the manner in which these goals can be achieved sustainably. The three components need to be in balance with each other to achieve an appropriate solution.

2.3.4 Science–Science Interfacing (SSI)

Within RISKBASE it is important to distinguish two sorts of SSI: SSI within the consortium and SSI between consortium and non-consortium scientists.

SSI within the consortium: The RISKBASE consortium consists of scientists with different scientific or disciplinary backgrounds. Although nearly all have a biophysical background, it is an ongoing, stimulating challenge for all to try to speak one common language and share a common perspective on the approach to fulfilling the objectives of the project. In this respect, it is essential to have a clear and transparent Description of Work (DoW) that, whenever disagreement may arise, serves as an ongoing source of reference for the activities to be executed, and is jointly agreed upon. An initial, and considered fruitful by all, activity to further facilitate the speaking of one common language was the joint preparation, right at the start of the project, of a publicly available 'strategy paper'.[15] However, the production of such a paper was not identified as an obligation in the DoW. The aim of this strategy paper was: (i) for external people to connect to the subject area of RISKBASE, (ii) to agree upon a common perspective on that area and (iii) to scope the longer-term focus of the RISKBASE project. It was also agreed that the strategy paper was to be regarded as dynamic, evolving in response to discussions within, and outside, the consortium. This it did, especially in the consortia meetings, which proved very useful for the continuous and still ongoing debate on topics (ii) and (iii). Communication *via* e-mail and phone proves much less effective than the small group, intensive discussions held. An emerging learning point, therefore, is that to further the development and evolution of a shared perspective on any project and its outcome it is necessary to set aside time on the agenda of regular (at least two to three times a year) physical meetings with the *full* project consortium.

SSI between consortium and non consortium scientists: The most prominent SSI activities are the public workshops organised by the biophysical sciences oriented RISKBASE WP-A and WP-B (Figure 2.3.2). Each WP initially organised one workshop independently. The WP-A workshop focussed on 'degradation causes, mitigation and remediation' (Orleans, France, October 2007). The WP-B conference (combined with another EU FP6 project, Modelkey: www.modelkey.org) and held a subsequent workshop focussed on the 'state-of-the-art and future challenges related to risk assessment in European River Basins' (Leipzig, Germany, November 2007). The invited workshop participants had relevant biophysical science backgrounds. Although the project structure (Figure 2.3.2) distinguished between the scope of WP-A and WP-B, it emerged from the discussions at the workshops that there was overlap between them. However, this was perceived positively as it provided a stimulus for better communication between the two work packages. The second scheduled workshop of both WPs was therefore organised as a combined event on the integration and structuring of WP-A and WP-B outcomes, the further identification of overlaps and gaps, suggestions to fill these gaps and the identification of the key findings for river basin management. A lesson learned is that the joint organisation of activities under different WPs can be very fruitful in stimulating SSI.

2.3.5 Science–Policy Interfacing (SPI)

Specific SPI activities within RISKBASE tended to take place initially within the first WP-C (Figure 2.3.2) workshop, which explored the interactions of social and ecological systems (Figure 2.3.3). On top of the three works packages mentioned above, there were two more RISKBASE work packages that organised SPI related activities. WP-D 'River basin cases' organised two river basin case oriented workshops and WP-E 'Communication, dissemination and knowledge management' organised a cross-cutting issues workshop focussed on SPI. Furthermore, an important part of SPI in RISKBASE takes place in meetings with the Advisory Panel. These SPI activities are further described below.

Social/ecological system interactions workshop: The first WP-C workshop (Vienna, Austria, October 2007) was centred on two case study discussions. The 'Dommel-case', a tributary of the river Meuse in the south of The Netherlands, worked on the relationships of contaminated water environments and human health risks. In comparison the 'Stože-case', where catastrophic landslides had been induced by heavy rainfall in Slovenia in 2000, explored risk issues regarding the overall water cycle and natural hazards. Although the topics of the two discussion groups were rather different, the analysis indicated clearly that natural resource management at regional scales and river basins goes beyond natural sciences and engineered or mechanical solutions. Societal values and interests need to be investigated by appropriate cooperation strategies and participative and adaptive approaches are at the heart of future environmental management. These approaches might be necessary through the different levels of government from local to regional. A conclusion was that cultural or political 'top-down' styles often conflict with the idea of natural resource management, where the balancing of natural capacities against economic interests and societal values requires close examination at both local and regional levels.[13]

The workshop also indicated that it is important to make the different societal demands and values visible in the discussion and to deal with the values and worries of stakeholders in the process. The design of participative processes should allow a common systemic understanding so that consensus can develop, decision-making is better informed and risk management solutions are more readily accepted. This understanding of the river basin system needs to go beyond the linked physicochemical and ecological systems and recognise the linkage with the socio-economic system. Joint activities between stakeholders, policy makers and researchers are needed to help the information flow and develop the common understanding. Communication and education (or training) to raise awareness and understanding are regarded as a prerequisite for participative or collaborative approaches.[13]

Adaptive management[16] is also an essential element in these approaches (Chapter 2.6). Adaptive management is a social as well as a scientific process and attempts to use a scientific approach, accompanied by collegial hypotheses testing, to build understanding. Starting from a common system understanding it seeks to anticipate how the system will react on interventions by closely monitoring what happens. Thus, adaptive management should also use interventions to test key hypotheses about the functioning of the system and, by doing this, to understand the system in a progressive way and manage the system at the same time.[13]

River basin cases workshops: The first workshop of RISKBASE WP-D 'River basin cases' focussed on 'Monitoring and assessment of river pollutants: a step towards the implementation of risk based management plans. Case studies of Portuguese river basins' (Lisbon, Portugal, May 2007). Although the main focus was on Portuguese river basins, other European river basin cases were discussed. The delegates included scientists with a large diversity in scientific backgrounds alongside several stakeholder representatives. The participants in this workshop concluded that better management of river systems must incorporate knowledge from many scientific disciplines as well as many practical experiences. However, it is well beyond the scope of individuals in the management, science or practice communities to organise the necessary knowledge transfer. Policy–science interfacing needs to be better recognised and resourced as an important area in its own right, involving dedicated professionals.[17]

The second workshop of WP-D focussed on 'Bridging science, water management and practical experiences in European river basins' (Budapest, Hungary, May 2008). The event was organised in four sessions. Three of them focused on selected river basins across Europe, *i.e.* Llobregat/Ter, Danube/Sava and Dommel river basins, and the fourth one addressed a prognostic case on the impact of bio-fuel production at the EU level. The workshop came to several SPI-related conclusions. The first conclusion was that River basin authorities should focus more on public demand and health issues to improve the quality of life of the inhabitants through adding social values to the river basin. The new goals might be included in the second generation of RBMP. Secondly, it was concluded that although the ecosystem services concept[4,12] is important in a more integrated approach to river basin management it is necessary to develop new practical criteria for its application. For instance, enhancing navigation could decrease other services provided by the river system, such as flood protection, nature conservation, *etc.* From an open discussion held at the workshop on the diverse impacts of an increase supply of bio-energy on humans, ecosystem and the economy it was concluded that our state of knowledge is limited. First, managers, scientists and stakeholder need to thoroughly frame the problem considering the issues at stake. Then scenarios should be built to address the challenges, and to balance the risks against the opportunities and finally suggest alternatives to reduce the risks and to exploit the opportunities.

SPI workshop: The RISKBASE WP-E on 'Communication, dissemination and knowledge management' organised a cross-cutting issues workshop, focussing on the topic of SPI (Venice, Italy, December 2007). The following points were discussed and identified as crucial for successful SPI:[18]

- Objectives need to be clearly identified to decide what kind of knowledge is required in policy.
- Knowledge developed within different national and EU funded research projects should be brought together by coordination, so as not to miss opportunities.
- To bring research results to policy and thus to put them into practice, meetings on both EU level (with EC DG Environment) and on the national level should be held at the end of a research project.
- 'Translators' are needed that (i) are specialised in knowledge transfer/exchange, speaking both 'languages' – *i.e.* being familiar with the

requirements and limitations of both science and policy, (ii) transfer the results and knowledge from the research community to the end users (policy makers and regulators and all other parties involved in putting the policies based on science into practice), but also understand their needs from science and their limitations in applying it (*see* Chapter 5.2 discussing the role of translators). Similar conclusions have been made concerning the revitalisation of brown fields by Broos *et al.*[19] The adaptation of scientific information to policy development and implementation is beginning to be understood to be crucial in the context of river basin management and in its wider environmental setting.

As this 'interface business' was seen as an important step in linking science and policy the rewards for both policy makers joining research discussions as well as for scientists should be much greater. The present systems in most countries mitigate against substantial interaction.

Advisory Panel: The RISKBASE project proposal identified the need to set up an Advisory Panel (ADPAN). In the RISKBASE ADPAN several stakeholder groups/end-user are represented. A stakeholder may be defined as 'Someone who may be affected by, or may affect, a decision that has to be made or its implementation'.[20] In the ADPAN, scientists, national and European policy makers, NGOs and industry are represented. All representatives have a stake in, or are involved in, river basin management within the EU. Through the ADPAN the needs and experiences from these groups are integrated in the RISKBASE activities and resulting deliverables and the raising of their support to these deliverables can be coordinated.

Consultation with the ADPAN proved extremely valuable in the project. The ADPAN was asked to reflect upon the achievements of the project at various stages and to recommend improvements/adaptations for the final phase. Remarkably, SPI-related comments of the ADPAN were 'You should be finished with science by now', 'Move from theory/concepts to practice', and 'Within the second half of the project more emphasis could be given to practical (technical) advice, which can be used for developing second generation WFD RBMPs'. These recommendations were taken on board in a roadmap for the final phase of the project. The focus in the final phase was to try to link better to practice and practitioners. The RISKBASE consortium therefore framed the objective for the last phase in the following way: 'What RISKBASE key-findings are missing in the first WFD RBMPs and could, or should, realistically find their way into the second generation (*i.e.* the first update)?'

2.3.6 Policy–Policy Interfacing (PPI)

The framed objective for the last phase of the project (see above) led to a dedicated PPI workshop being organised. The RISKBASE consortium realised that 'transmission' of the key-findings and related recommendations would be much more effective if the consortium better understood the perspective of the writers of the WFD RBMPs as well as the language they use. Hence a one and a half day workshop has been organised (Barcelona, 4–5 May 2009) by RISKBASE WP-D 'river basin cases' in which the exchange of experiences of, and between, people who have written the first WFD RBMPs is the focus. RBMP writers from river basins with different political as well as biophysical settings participated and shared experiences. The basins were also addressed in previous RISKBASE workshops, *i.e.* Ebro, Meuse, Elbe and Danube river basins. As it is realised by the RISKBASE consortium that several levels of aggregation in RBMP preparation exists, for each basin one representative at three different scales was invited: one for the local scale (*i.e.* water body scale), one for the regional or catchment scale (*i.e.* the scale where people in general still have some feeling of connection to 'their river') and one for the national scale. Also, where the basin is international, a person for the full basin scale (*e.g.* someone from the River Basin Commission) was invited. The workshop structure was designed to facilitate primarily the exchange of experiences between the invited participants, focusing on the following RBMP related questions that are of clear relevance to RISKBASE (NB: some refinement of the questions may take place before the workshop):

1. *Problems?*

   — What problems did you address in your RBMP?
   — How do you know that these are the essential actual problems?
   — How did you select/prioritise these problems?
   — Did you have enough information for the prioritisation?
   — Did scale (spatial and temporal) play a role in the prioritisation of problems?
   — Do you think that there are other problems not yet addressed?

2. *Measures?*

   — How did you select the measures?
   — Did you have enough information to select these measures?
   — Is scale (spatial and temporal) reflected in your measures?
   — Will these measures solve all (major) problems in your river basin?
   — How do you know that these measures will be effective?
   — How will you monitor the effectiveness of these measures?
   — What do you do if these measures appear to be not (enough) effective?

3. *What is needed for next generation RBMP: tools, data, science approaches?*

To facilitate the discussion during the workshop the participants were asked to think over these questions beforehand and draft answers in advance. A benefit for the invited participants was the exchange, in an open atmosphere, of experiences with writing a first generation RBMP with international colleagues, as well as the ability to influence the outcome of RISKBASE. Besides the participation of these invited RBMP writers every RISKBASE consortium member also participated in the workshop, thus ensuring that each had a first-hand opportunity to learn from practitioners and use the experience to further sharpen the key-messages and related recommendations from the RISKBASE activities.

2.3.7 SPI Lessons Learned So Far

Based upon the SPI experiences in RISKBASE so far the lessons learned are summarised below:

- No risk, no glory: SPI is extremely challenging. Only by trying and doing (bringing actors together), and adapting by learning from what went right and wrong, can progress be made in SPI;
- progress can only be made if all the actors involved, either in an SS, SP or PP-interfacing setting, are at least willing to make an attempt to understand each other's language and perspective;
- to this end it is very useful for a project consortium to write a joint and public 'strategy paper' at the start of the project; the paper – after internal discussion – can be refined (on top of your description of work) and a common perspective on the area studied and the common language used to study that area agreed upon;
- a well designed, and professionally moderated, convening of actors facilitates the interfacing;
- SSI, SPI and PPI each need a different approach to be effective;
- objectives must be clearly identified to decide what kind of knowledge is required in policy;
- knowledge developed within different national and EU funded research projects should be brought together by coordination, so as not to miss opportunities;
- to bring research results to policy and thus to put them into practice, meetings on both EU (with EC DG Environment) and national levels should be held at the end of a research project to agree on what can be taken forward into a policy environment;
- SPI must become a new profession: we need 'translators' that (i) are specialised in knowledge transfer/exchange, speaking both 'languages', *i.e.* being familiar with the requirements and limitations of both science and policy, and (ii) transfer the results and knowledge from the research community to the end users (policy makers and regulators and all other parties involved in putting the policies based on science into practice), but also understand their needs from science and their limitations in applying it;
- this adaptation of scientific information to policy requirements is crucial to make sure that it is in a format applicable to the policy end users.

Acknowledgements

The European Commission is greatly acknowledged for funding of the RISKBASE project. Furthermore, the following RISKBASE friends are kindly thanked for their comments and suggestions for additions and improvements to this chapter: Bob Harris (also for improving the English), Adriaan Slob, Thomas Track, Philippe Négrel, Joop Vegter, Wim Salomons and Winfried Blum.

References

1. European Commission, Communication from the Commission to the Council and the European Parliament, 'Stimulating Technologies for Sustainable Development: An Environmental Technologies Action Plan for the European Union', Commission of the European Communities, Brussels, 28.1.2004, COM(2004) 38 final, 2004.
2. JOINT, *Risk Based Management of Contamination and Protection of the Soil System in Urban Environments, JOINT Research Agenda 2005*, Umweltwirtschaft GmbH, Stuttgart, ISBN 3-937750-01-0, 2005.
3. European Commission, *Directive 2000/60/EC of the European Parliament and of the Council establishing a framework for the Community action in the field of Water Policy*, Directive 2000/60/EC, European Commission, Brussels, 2000.
4. European Commission, Communication from the Commission – Halting the loss of biodiversity by 2010 – and beyond, Sustaining ecosystem services for human well–being, Commission of the European Communities, Brussels, 22.5.2006, COM(2006) 216 final, 2006.
5. European Environment Agency, *Europe's Water: An Indicator-based Assessment, Summary*, Luxembourg: Office for Official Publications of the European Communities, ISBN 92-9167-576-8, 2003.
6. A. F. L. Slob, M. Rijnveld, A. S. Chapman and P. Strosser, Challenges of linking scientific knowledge to river basin management policy: AquaTerra as a case study, *Environ. Pollut.*, 2007, **148**, 867–874
7. A. Totlandsdal, N. Fudge, E. Sanderson, L. van Bree and B. Brunekreef, Strengthening the Science–Policy interface: experiences from a European thematic on air pollution and health (AIRNET), *Environ. Sci. Policy*, 2007, **10**, 260–266.
8. P. h. Négrel, D. Darmendrail and A. Slob, Transferring scientific knowledge to societal use: clue from the AquaTerra integrated project, in *Groundwater Science & Policy, an International Overview*, ed. Ph. Quevauviller, Royal Society of Chemistry, Cambridge, 2008, pp. 31–56.
9. European Commission, *Common Implementation Strategy for the Water Framework Directive (2000/60/EC), Guidance Document No 8: Public Participation in Relation to the Water Framework Directive*, ISBN 92-894-5128-9, 2003.

10. P. Quevauviller, P. Balabanis, C. Fragakis, M. Weydert, M. Oliver, A. Kaschl, G. Arnold, A. Kroll, L. Galbiati, J. M. Zaldivar and G. Bidoglio, Science–Policy integration needs in support of the implementation of the EU Water Framework Directive, *Environ. Sci. Policy*, 2005, **8**, 203–211.
11. J. Brils, D. Barceló, W. Blum, W. Brack, B. Harris, D. Müller, P. Négrel, V. Ragnarsdottir, W. Salomons, T. Track and J. Vegter, River basin risk assessment linked to monitoring and management, in: *WFD Ecological and Chemical Status Monitoring.*, ed. P. Quevauviller, John Wiley & Sons, Chichester, 2008, 273–287.
12. Millennium Ecosystem Assessment, *Ecosystems and Human Well-being: Synthesis*, Island Press, Washington, D.C., ISBN 1-59726-040-1, 2005.
13. D. Mueller, Exploring the interactions of social and natural systems, *RISKBASE Newsletter*, May 2008, Issue 2, p. 4, available *via*: www.riskbase.info.
14. J. Vegter, Risk based land management, status and perspectives for policy, in *Soil and Sediment Remediation*, ed. P. Lens, T. Grotenhuis and H. Tabak, IWA Publishing, London, 2004.
15. RISKBASE, 'RISKBASE Strategy towards achieving an Overarching Concept for Risk-based Management of European River Basins', available *via*: www.riskbase.info,Version 9, January 2007.
16. C. Pahl-Wostl, T. Downing, P. Kabat, P. Magnuszewski, J. Meigh, M. Schüter, J. Sendzimir and S. Werners, 'Transition to Adaptive Water Management: The NeWater Project', Water Policy, NeWater Working Paper X, Institute of Environmental Systems Research, University of Osnabrück, 2005.
17. D. Barceló, S. Diaz Cruz and R. M. Darbra, 1st Thematic Workshop of WP 1b – River Basin Cases, *RISKBASE Newsletter*, October 2007, Issue 1, p. 3, available *via* www.riskbase.info.
18. T. Track and K. Wendler, Cross-cutting Issues Workshop, *RISKBASE Newsletter*, May 2008, Issue 2, p. 2, available *via*: www.riskbase.info.
19. L. Broos, T. Ertel, A. Gray, B. Schug and J. Vegter (eds), *European Brownfield Revitalisation Agenda,* Sachverständigen-Bürd Dr. Ertel, Esslingen, 2007, available *via*: www.eubra.eu.
20. United Nations, *World Population Prospects: the 2002 Rev*ision, Population Division of the Department of Economic and Social Affairs of the United Nations Secretariat, Ref ESA/P/WP. 180, 26 February 2003, 36 pp.

CHAPTER 2.4

Narrowing the Science–Policy Gap – Experience from the Harmoni-CA Concerted Action

MICHIEL W. BLIND,[a] JENS CHRISTIAN REFSGAARD,[b] ILKE BOROWSKI[c] AND WILLEM J. DE LANGE[a]

[a] Deltares, PO Box 85467, 3508 AL Utrecht, The Netherlands; [b] Geological Survey of Denmark and Greenland, Øster Voldgade 10, 1350 Copenhagen K, Denmark; [c] University of Osnabrueck, Institute of Environmental Systems Research, Barbarastrasse 12, 49076 Osnabrueck, Germany

2.4.1 Introduction

Improving research integration into the policy-making and implementation process remains one of the major challenges in managing complex environmental problems. Owing to its high ambition to achieve 'good ecological status' of Europe's fresh water resources by 2015, the adoption of the Water Framework Directive (WFD[1]) in Europe has resulted in increased attention to the matter. Prescribing a river basin approach and integration over various disciplines in natural sciences (hydrology, ecology, morphology, *etc.*), disciplines in social sciences, especially economics, and a participatory approach makes the implementation of the WFD a major endeavour. To achieve the goals in a cost-effective and publicly acceptable manner, good use and integration of scientific insights is a prerequisite.

To provide some answers to the challenges of the WFD the European Commission (EC) has invested significant resources in research underpinning

Water System Science and Policy Interfacing
Edited by Philippe Quevauviller

Published by the Royal Society of Chemistry, www.rsc.org

key aspects of the WFD,[2,3] In addition, the EC took the initiative for the large-scale concerted action 'Harmoni-CA' to enhance the science–policy interface (SPI). Harmoni-CA, or 'Harmonised Modelling Tools for Integrated Basin Management,[4,5] was a €3.8 M project running between 2002 and 2007. The overall objective of the large-scale concerted action Harmoni-CA was to create a forum for unambiguous communication, information exchange and harmonisation of the use and development of ICT tools (ICT: information and communication tools, including models) relevant to integrated river basin management, and the implementation of the WFD. Harmoni-CA's popular mission was:

For implementing the Water Framework Directive, developing river basin management plans backed up by high quality ICT-tools, Harmoni-CA will produce guidance and methodologies based on support and consensus achieved by involving science, policy and stakeholders synthesizing available knowledge establishing broad dialogues and tuning projects.

This chapter concerns the results of the project from a science–policy perspective. Section 2.4.2 deals with the rational behind Harmoni-CA, the science–policy interface addressed in the project, methods applied and tangible results. Section 2.4.3 provides an analysis of Harmoni-CA's efforts, including several recommendations. The final section covers conclusions and discussion.

2.4.2 Concerted Action Harmoni-CA

2.4.2.1 Rationale

Over the years increasing emphasis has been put on the applicability of European funded research. Involvement of end-users has been required from the 5th European Research Framework Program onwards. This increase stems from a perceived lack of uptake of research results in everyday practice. Today, several instruments are in place or being developed to improve the situation, the triple-helix approach,[6] combining science with industry with government being one of the buzz concepts pursued. In the 5th Framework Research Programme on Energy, Environment and Sustainable Development (FP5, 2002–2006[7]) this approach was already implied, by promoting both the involvement of end-users and of industry, specifically of small-medium enterprises (SMEs). Nevertheless, in the field of water management, and in particular in modelling and decision support, there was a need to step up the ambition of interaction due to the ambitious WFD. Though not documented, one may safely assume that the fragmentation of research, on one side, and the sheer number of policy implementers (national and regional water managers), on the other side, contributed to the decision that a large-scale concerted action was required to positively impact the implementation of the WFD. Subsequently, the €3.8M concerted action Harmoni-CA[4,5] was funded by the Research

Directorate-General (DG-RTD) of the European Commission under the 5th Research Framework Programme (FP5). It aimed to facilitate specific clustering activities to enhance the collaboration between research projects, but, more importantly, to improve the science–policy interface with respect to modelling and tools to support the WFD implementation.

Though much of the work in Harmoni-CA involved software (computational models and tools), Harmoni-CA was not about software as such: Harmoni-CA was about harmonisation and guidance on proper development and use of software in the light of effective and efficient development of integrated river basin management plans and the implementation of the WFD. Harmoni-CA's point of departure was that transferring newly developed tools to operational use of the water managers had not been fully achieved and the efforts from the different projects were not well coordinated. 'Tools' include both software and methodological guidelines from the research community. The key perceived reasons for this were:

- an insufficient dialogue between the scientific and policy-making and policy-implementing communities;
- a lack of translation of scientific outputs into tools readily applicable to policy-makers and implementers;
- a lack of an organisational framework that connects those responsible for policy development and implementation on one hand to scientists and developers on the other.

2.4.2.2 Science–Policy and Science–Science Interface in Harmoni-CA

The policy life-cycle[8] consists of the following steps: (i) policy preparation, (ii) policy formulation, (iii) policy implementation and (iv) policy evaluation. In the case of the WFD the overarching policy has been in place since 2000. The European legislation is transferred into national legislation and national and regional policy. By the time Harmoni-CA started the policy development occurred on national and regional levels.

The implementation of the WFD is, to a large extent, also a national and a regional activity. On the national and regional levels it is often in the eye of the beholder as to whether somebody is characterized as policy developer or policy implementer/operational manager. 'Science' can have an impact on all stages of the policy cycle. In this chapter 'science' refers mainly to science developed in European funded research projects. During the implementation of the WFD, consultants also play a role. Numerous companies provide advice and tools to the authorities in charge of the WFD implementation. In Harmoni-CA, the four groups addressed were (Figure 2.4.1):

1. 'Policy makers': those in charge of national and regional policy supporting the implementation.

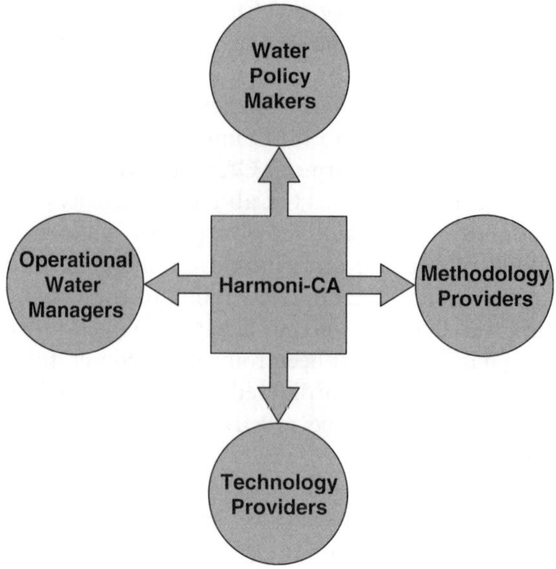

Figure 2.4.1 Position of Harmoni-CA between the different roles in the WFD implementation.

2. 'Operational managers/policy implementers': those in charge of carrying out the implementation.
3. 'Methodology providers': scientists developing and offering the key knowledge.
4. 'Technology providers': consultants providing the key knowledge and tools.

Obviously, reality cannot be that easily grouped and like the double role of some individuals as 'policy makers' and 'operational managers', individuals on the knowledge providing side often have a role both as 'methodology provider' and 'technology provider'. Harmoni-CA's ambition was to narrow the gaps between mainly demand side ('policy makers' and 'operational managers') on one hand and the knowledge and tool offering side ('methodology providers' and 'technology providers') on the other hand. Bringing together the demand and offer side was the key science–policy interfacing carried out within Harmoni-CA.

Harmoni-CA focussed first of all on the more operational side of the WFD, where software tools and methodologies are mostly applied. Furthermore, owing to Harmoni-CA's European dimension, in practice it aimed to involve key representatives involved in the implementation, who coincide approximately with the level of CIS-members. However, Harmoni-CA also reached out to local water managers.

The overall objective of the large-scale concerted action Harmoni-CA was to create a forum for unambiguous communication, information exchange and harmonisation of the use and development of software-tools relevant to integrated river basin management, and the implementation of the WFD.

Of course, Harmoni-CA was only one of the initiatives to facilitate the WFD implementation with a European dimension. To achieve some consistency in the implementation the Common Implementation Strategy (CIS; ref. 9), on a European level more application oriented governmental delegates from Member States joined to develop non-mandatory guidelines. Within the CIS a network of Pilot River Basins (PRB, ref. 10) was set up to test these guidelines and other approaches. Obviously, not every water authority within the European Union was directly involved within CIS and PRB. On national levels the results of these initiatives were disseminated and adapted to local use.

Besides the work on improving the science–policy interface and enhancing the uptake of European Research, Harmoni-CA has played an important role in bringing together the science. Where the CIS aimed to bring together the 'demand-side' of the WFD implementation, Harmoni-CA fulfilled the role of bringing together the 'offer-side', pursuing a 'science–science' interfacing by bringing together scientists working on methodologies and tools for the implementation of the WFD. Within this framework Harmoni-CA focussed on bringing together the scientist working in European projects, in particular those scientists working in the Catchment Modelling Cluster (CatchMod, ref. 11).

2.4.2.3 Approach and Methodology

Being a concerted action, Harmoni-CA's innovation was based on supporting the review, elaboration and synthesis of available knowledge and thus improving its availability, resulting in new insights and achieving a common understanding of problems, methods and solutions. Reflecting on the key perceived reasons for a sub-optimal uptake of research outputs as described in Section 2.4.0 the key methodologies applied were (i) organization of workshops and conferences geared towards the interaction between 'demand side' ('policy makers' and 'operational managers') and 'offer side' ('methodology provider' and 'technology provider') and (ii) active involvement of the 'demand side' in any development of Harmoni-CA products. Since Harmoni-CA also aimed to improve the science–science interface, several workshops were geared towards aligning projects underpinning the implementation of the WFD.

Harmoni-CA's main tasks concerned:

- Task 1: Establishing a communication forum (organisation of large-scale demand–offer conferences), and setting up an open and 'demand driven' communication forum.
- Task 2: Developing a toolbox to provide easy and guided access to approved ICT-tools necessary for the development of River Basin Management Plans.
- Task 3: Developing a General Methodology and Guidance Document for Model-supported Implementation of the WFD.

Task 4: Developing methods and tools to improve the interaction between monitoring and modelling.
Task 5: Developing and strengthening the science–policy interface across sectors and spatial boundaries to establish a dialogue on the requirements for modelling tools and participatory approaches to implement the WFD.
Task 6: Co-ordinating ongoing research activities.

The pivotal aspect of the approach was to focus on involvement of the demand and offer side on a European level. This approach was taken when convening meetings, when developing most of the documents and for the design of some of the software. In workshops and conferences different target groups and stakeholders were invited and actively involved. Through these workshops the available knowledge has been collected and made accessible to non-scientists. They have led to guidelines and synthesis documents, *e.g.* guidelines on (i) the use of models in water management planning processes,[21] (ii) uncertainty analysis[17] and (iii) monitoring and modelling.[22] Other activities targeted some specific issues relevant to the uptake of (modelling) research, *e.g.* (i) synthesis documents on the effectiveness of stakeholder involvement in research,[28] (ii) key factors for successful development of decision support systems[27] and (iii) an exploration of the gap between water managers and researchers.[40]

From the early beginning of the project contacts have been made with many parties in the WFD implementation network, amongst these are the EC Environment Directorate General (DG-ENV), CIS working group 2B on Integrated River Basin Management (CIS-WG2B), the Pilot River Basins (PRBs) and the EC Joint Research Centre Directorate General (DG-JRC). An important activity pursued by Harmoni-CA was to obtain a strong commitment from the European Water Directors. The European Water Directors are the highest policy group on water management within the European Union, consisting of high ranking policy makers in national water and environment ministries. In a joint effort of Harmoni-CA/CatchMod, DG-JRC, CIS-WG2B leaders, DG-ENV and DG-RTD a scope paper[12] was prepared that subsequently was endorsed by the European Water Directors. The scope paper presented tangible actions for the collaboration of the CIS implementation and the research community. By endorsing the scope paper, the Water Directors provided a strong political signal to:

- link WFD requirements to research products;
- build a smart web-portal to link WFD implementation steps with research outputs;
- establish a close cooperation between Pilot River Basins and the European research community.

With this scope paper, Harmoni-CA started the process for cooperation between the 'demand' and 'offer' communities. Elements from the scope paper were incorporated in the WFD-CIS mandates. The development of a smart web-portal was initiated and the CIS groups and PRB were encouraged to involve leading scientist of European funded research.

2.4.2.4 Results

Establishing a dialogue among the scientific and policy-making communities is one of the most important achievements of Harmoni-CA. Workshops, forums and conferences narrowed the gap between both communities and helped in matching the demand for and offer of knowledge and software tools (a.o. models) helpful for implementing the Water Framework Directive (WFD). Issues raised during the conferences were put in the context of the WFD implementation process.

It is beyond the scope of this chapter to provide a complete overview of Harmoni-CA's outputs, including their contributions to improving the uptake of research and improving the science–policy interface. However, the achievements per task can be summarized as follows:

Task 1: Establishing a communication forum (organisation of large-scale conferences, set up an open and 'demand driven' communication forum): The activities regarding the establishment of a communication forum were focussed in facilitating a closer cooperation between the WFD implementation and the research community. Besides the scope paper, important activities were the Harmoni-CA Forums and Conferences and the development of the WISE-RTD web portal (Water Information System Europe – Research and Technology Development;[13] Chapter 4.2), including its introduction in WISE (Water Information System for Europe[14]). The SPI-Water project[15] was initiated by the developers of WISE-RTD. SPI-Water expands WISE-RTD to water related issues beyond the WFD. At the 2007 European Union's Water Directors meeting and at the EU Water Conference the portal was officially launched as part of WISE. To raise further awareness, an animation has been developed that provides a view of the future where research solutions can easily be found by water managers.

Task 2: Develop a toolbox to provide easy and guided access to approved (benchmarked) ICT-tools necessary for the development of River Basin Management Plans: The main outputs of this task concern Catchment Modelling Guidance documents 'model calibration',[16] 'uncertainty analysis'[17] and 'sensitivity analysis' [18] and the document 'Quality Assurance for Model-Based Water Management: Better Modelling Practices'.[19] A web-based statistical tool[20] for model evaluation and analysis with the focus on education has been developed. An extended help function guides the user by examples, links to the guidance documents and other explanatory documents.

Task 3: Develop a General Methodology and Guidance Document for Model-supported Implementation of the WFD: This task resulted in a comprehensive guidance document targeting scientists and practitioners.[21] The guidance formed the basis for the modeller's entry to WISE-RTD.

Task 4: Develop methods and tools to improve the interaction between monitoring and modelling: Based on results from the workshops involving scientists and practitioners a flow chart on WFD implementation has been developed that has been included in the Catchment Modelling Guidance on 'Good practise in joint use of monitoring and modelling'.[22] Furthermore a synthesis report on 'Data availability accessibility – State of the art on existing data required for modelling for research purposes and for the implementation of the WFD'[23] serves the purpose of raising awareness to data-issues in implementing the WFD.

Task 5: Develop and strengthen the science–policy interface across sectors and spatial boundaries to establish a dialogue on the requirements for modelling tools and participatory approaches to implement the WFD: The results of this task clearly call for a distinct perspective on the potential role of models coming from research to support water management.[20] The problems that water managers face can only partly be solved immediately by new research results. The ongoing and cycling nature of water management requires ongoing support and retaining the potential for adaptation. Furthermore, within the framework of this task guidance documents were developed for economic issues,[24] the interaction between agriculture and water management[25] and the integration of the human dimension into modelling.[26] There is considerable interest in more cooperation on new approaches to dealing with uncertainty and a new perspective on policy processes is required to improve stakeholder participation in research projects and in water management that benefit from closer cooperation with the Integrated Assessment community.

Task 6: Co-ordinating ongoing research activities:
The main effort in this task was to achieve more collaboration between research projects formally or informally belonging to the CatchMod Cluster. To benefit from the parallel running of these research projects, two documents were developed that should support future projects in delivering applicable results: 'Decision Support Systems for water resources management: current state and guidelines for tool development,'[27] and 'Report on end-user involvement in research projects. A telephone survey-based assessment of twelve research projects'.[28] The EUROHARP toolbox for nutrient modelling tools[29] was extended to be more attractive for end-users, HarmoniQuA's Modelling Support Tool was upgraded to a generalized, multi-user Process Support Tool [ProST,[30]], and a web portal to establish a 'community on useful standards in water management – Hydronomy'[31] – was implemented. In addition, Harmoni-CA contributed to establish new links between research groups. The result of this was seen in the consortia that were established for the 6th EC's Research Framework Programme (FP6) research projects.

To further improve the uptake of the results a folder with flyers has been developed.[32] These flyers have been developed in close collaboration with scientists and operational managers (policy implementers). This is reflected through a dedicated section on each flyer that tells what different target groups can expect from the product at hand. Table 2.4.1 provides an overview of the outputs of Harmoni-CA.

Table 2.4.1 Flyers of Harmoni-CA products. Digital versions of these and other flyers can be found on the Harmoni-CA website.[32]

Reference	*Title*
Flyers on synthesis reports and guidance documents	
G 01	Guideline: Model-supported Implementation of the Water Framework Directive – A Water Managers Guide
G 02	Making the Human Dimensions of Water Management More Transparent: Human Perception and Valuation in Model Supported Water Management
G 03	Improving End-user Involvement in Research – Synthesis and Advice Report Based on Interviews
G 04	Handbook on Improving Participation in Water Management: Learning Together to Manage Together
G 05	Key Issues and Tools to Give Support on Economics in the Water Framework Directive
G 06	Harmoni-CA report: Decision Support Systems for Water Resources Management: Current State and Guidelines for Tool Development
G 07	Harmoni-CA Report: Review of Model-Based Tools with Regard to the Interaction of Water Management and Agriculture
G 08	Catchment Modelling Guidance: Good Practice in Joint Use of Monitoring & Modelling
G 09	Synthesis Report: Data Availability and Accessibility – State of the art on existing data required for modelling for research purposes and for the implementation of the Water Framework Directive
G 10	Catchment Modelling Guidance: Uncertainty Analysis
G 11	HarmoniRiB Guidances Supporting Uncertainty in Data and Models for River Basin Management
G 12	Catchment Modelling Guidance: Sensitivity Analysis
G 13	Catchment Modelling Guidance: Model calibration
G 14	Quality Assurance for Model-based Water Management: Better Modelling Practices
Flyers on specific tools	
B 01	Finding Relevant Science and Policy Documents and Tools Made Easy – WISE-RTD: Water Information System for Europe – Research, Technology and Development
B 04	Modelling of Nutrient Losses from Diffuse Sources: The Extended EUROHARP Toolbox
B 12	Developing and Using Water Related Glossaries: Glossary Tool (Glost)
B 14	Supporting Collaborative Multidisciplinary Projects: Process Support Tool (Prost)
B 16	HYDRONOMY: Build the Standards-Community by Sharing your Experiences
B 18	Toolbox Developed in the Harmoni-CA Project: Statistical Toolbox

2.4.3 Analyses and Recommendations from Harmoni-CA's Efforts

The popular mission of Harmoni-CA project's is displayed in the Introduction (Section 2.4.1). The previous sections have already shown that many successful products were developed and significant steps made towards realising the mission. In this section we summarize the main results, draw conclusions and provide recommendations on dissemination and uptake of research results, focussing on results relating to modelling and ICT-tools. Hence this section aims to bring together the main conclusions of the very diverse work carried out over the Harmoni-CA project period. Many of the conclusions are based on the outcomes of specific workshops. In addition, the Harmoni-CA team has derived some more general analyses and conclusions. We distinguish the following activities:

1. Workshops and conferences
2. Development of synthesis reports
3. Development of WISE-RTD and dissemination through leaflets
4. Researching the science–policy interface with respect to model use/research model uptake

We discuss each of these activity lines briefly, providing the key discussions and associated conclusions. For detailed information the interested reader is referred to the reports and software.

2.4.3.1 Workshops and Conferences

One important challenge in organizing the meetings was to identify and attract the 'demand side' (policy maker and operational water manager). Not surprisingly, this was easier if the topic at hand related closely to their daily work, de facto if the topic was closely related to the stage of implementation of the Water Framework Directive, since most topics were targeting activities required to implement the WFD. We found that water managers are certainly interested in incorporating scientific best practice.

In reality, however, it was first of all challenging to identify and attract those responsible for the uptake of research results. Who are these water managers? What is the role of consultancy in the uptake of new insights and tools? The third Harmoni-CA conference focussed on bringing the consultants into the discussion. It was emphasized that they play an important intermediate role to bridge the gap between research and implementation. Still, the double role of keeping an eye on research and on the current needs of water management is often not clearly allocated. But who is pivotal can be identified through local and national (personal) networking. Direct personal relationships have proven to be important instruments to identify the right persons for science–policy meetings about the WFD implementation (on a European level). We found three other barriers hindering the attendance of water managers at European level meetings: (i) lack of time resources to attend meetings, (ii) lack of financial resources to travel and (iii) feeling uncomfortable in communicating in English. Nevertheless, we managed to attract a satisfactory number of water managers, though most meetings were attended in majority by researchers.

The large-scale Harmoni-CA conferences were particularity challenging due to their size and objectives. At the final conference a three-day format was chosen that proved very successful. Starting from a high-level policy day, *via* a research overview day, the three-day event ended in a demonstration and hands-on training day. Such a setting allowed more practice oriented people to get insight and connections in the policy area, and more policy oriented people to be informed on practical developments. Unsurprisingly, in such a setting, the first two days are more interesting to research and policy 'generalists', whereas day two and three are more focussed towards practice. Hence, the second day was pivotal, since the 'overview' day was attractive to all participants, and facilitated interaction between science and policy, generalists and specialists, and different European nationalities. Participants and representatives of the European Commission have voiced their interest in continuing the conferences, applying the same format on a yearly or two-yearly basis.

In conclusion on 'workshops and conferences', due to Harmoni-CA many links were made between the different target groups. This is an essential step to narrow the science–policy gap, to achieve common understanding and, in the end, to create research outputs that meet the demands of application. We believe that as a result of the many activities awareness has been raised that water management can benefit more from research and development and that there are results that can be used, possibly after some (minor) modifications. Hence, the societal added value of research investments should be on the rise. The scope paper[12] on the science–policy interface endorsed by the European Water Directors is one indication of the increased awareness of the opportunities to establish closer links between different societies.

Because science–policy interfacing depends strongly on face-to-face contacts workshops and conferences remain important instruments to achieving interaction between science and policy. On a European level such meetings are important to exchange across borders, but we believe that in a European setting a more regional approach is required as well, to specific regional issues and to reduce barriers such as, for example, language. As a result of the experiences in Harmoni-CA the SPI-Water project[15] has already taken a more regional approach.

2.4.3.2 Development of Synthesis Reports

Harmoni-CA initiated a large number of synthesis reports and guidance documents focussing on narrowing the science–policy gap. We can distinguish the following groups: (i) reports supporting modelling activities and development, *e.g*. quality assurance,[19] sensitivity analysis,[18] or, decision support development;[27] (ii) reports supporting the collaboration in specific different scientific/policy fields such as the reports on modelling & monitoring,[22] water management and agriculture,[25] economic tools in the WFD;[24] (iii) reports on the science–policy interface and end-user involvement.[21,28] Most of the reports have been commissioned to small task groups consisting of both scientists and end-users. The reports include the outcomes of discussions held at the various workshops and conferences. Many of the reports in category (i) and (ii) will be part of an IWA (International Water Association) publication series, greatly improving the visibility of European research on a world-wide level and supporting improved science–policy communication throughout the world. All reports provide a common basis between the different groups: different scientific disciplines, water managers and end-users of research outputs. In the reports, available knowledge is reviewed and added value is provided by picking up recent insights. The reports are essential to improve the quality of communication between the different groups, and make research available to societal use. The reports successfully organize the knowledge, which is otherwise inaccessible because it is mainly published in scientific journals and poorly accounts for people outside academia. Hence, this activity provides a major contribution to improve model-based water management.

The 'Report on end-user involvement in research projects' is built upon a telephone survey-based assessment of 12 research projects.[28] It differs from the other documents since it focuses on the successfulness of such involvement in European research. The analysis shows that the uptake of research outputs by the end-users is frequently taking place, not necessarily by applying the software tools themselves but by incorporating (aspects) of methodologies and techniques. The report provides many recommendations on project preparation, execution and post-project to improve the usability and hence the uptake of research results.

Though the development of synthesis and guidance reports has been successful, their impact on the science–policy interface could be further enhanced by translating the information into other languages. National policy (implementation) is developed in the national language, and English is not often used. Hence, documents in English create an extra barrier. Though we did not ourselves systematically quantify the need for translation, the need for it was frequently voiced. The HarmoniCOP project sets an example: the translation of its handbook[33] into seven languages has greatly increased the interest of end-users in taking up the results.

In conclusion, developing synthesis and guidance documents has provided a process for science–policy interaction and valuable outputs mediating between science and policy. For many other topics in the implementation of the WFD such synthesis should prove useful.

2.4.3.3 Development of WISE-RTD & Dissemination through Leaflets

The WISE-RTD portal, described in Chapter 4.2 and in ref. 13, is a significant step forward to match the needs of policy (WFD) implementation to scientific insights and experiences. It is also a major step towards structured identification of the knowledge gaps for policy implementation. WISE-RTD organizes knowledge according to policy needs; hence it is an essential step towards bridging the gap between development and use. The process leading to the development involved and still involves (in the SPI-Water project) different target groups. Currently, the system is further developed and populated with projects and their results. The system is an essential mechanism to disseminate research to water policy and operational management. However, a system like WISE-RTD can only be successful if maintenance is ensured and content is extended and updated: Practitioners will not use the system if they cannot retrieve up-to-date and accurate information. Researchers will only be motivated to upload information to any system if it is utilized and if the feeling is avoided that the same information is required in a variety of different systems. Hence, WISE-RTD can only support the science–policy interface if it receives prolonged support. Creating links to national information repositories and to other European-wide systems – such as the Community Research and Development Information Service (CORDIS),[34] the European Groundwater Information System EUGRIS[35] (Chapter 4.3) and those developed in water related research coordination projects such IWRM-net (ref. 36, and elsewhere in this book), Circle,[37] CRUE,[38] SNOWMAN,[39] *etc.* – is important to achieve sufficient coverage of the available information. It takes time to develop a fully matured WISE-RTD that truly helps in narrowing the science–policy gap. This means that new initiatives should be carefully aligned to WISE-RTD. A promising development is that the NeWater project decided to use and extend the WISE-RTD to accommodate NeWater's results.

Whereas WISE-RTD is an electronic system backed up by persons, the effort to create product leaflets of research makes the results accessible by providing brief overviews with dedicated sections to bridge the gap between research and practice. Since it is a tangible (non-electronic) effort, the leaflets reach people who are less attracted by electronic systems.

2.4.3.4 Researching the Science–Policy Interface with Respect to Model use/Research Model Uptake

A series of workshops has been devoted to understanding the apparent difficulties between research and application. This has resulted in a publication on understanding the differences between the thinking and driving forces of model developers/scientists and end-users/water managers.[40] Notably, in many European water management organisations the use of models either in-house or indirectly *via* consultancy and research is firmly established; especially in fields of flood warning, sewers, hydrology and to some extent water related spatial planning, tools such as DELFT-FEWS,[41] Sobek,[42] Isis,[43] MIKE-SHE,[44] SWAT,[45] Modflow,[46] and many others are highly appreciated and used. The work carried out in Harmoni-CA focussed on the uptake of models originating from research projects and not with the firmly established models mentioned previously.

It goes without saying that even if research manages to engage in meaningful discussions with users (*e.g.* through dedicated workshops), manages to 'harmonize knowledge' (*e.g.* the synthesis reports), and provides mechanisms to find the results (*e.g.* WISE-RTD, leaflets) the success of a modelling tools depends on its usefulness and acceptability to the anticipated end-user. One key finding of Borowski and Hare[40] was the lack of incentive for researchers to improve their understanding of the water management process. It is of course fundamental for a successful science–policy interaction that researchers understand operational water management. Since researchers are judged by fellow researchers, investing in understanding water management is less rewarding. A second main challenge in research projects is the difficulty to adapt to a real-life management process. Both a project and a real life problem have their own time-constraints. To improve the uptake and appropriateness of research thought should be given to how research programming and research execution can be flexible enough to accommodate interactive real-life water management processes. Instead of water managers, consultants could also play a more prominent role. In many countries consultants are the carriers of scientific knowledge to a broader water management group than research alone can achieve. Knowing the 'market', consultants can play a more pivotal role in making software operational, maintaining its quality, developing documentation and course material, and transferring the software to operational use.

Finally, instead of focussing on the uptake of research models (*i.e.* implementing a model on a local level), focus could shift to the uptake of results obtained using these models elsewhere. Water managers expressed their interest in being able to transfer knowledge generated by models in different settings to their own situation. To improve the uptake of results and hence improve the science–policy interface, learning by analogy or case based reasoning can be an additional instrument.

2.4.4 Conclusions and Discussion

Harmoni-CA has contributed to the exchange between policy makers and water managers with scientists and consultants, developing key knowledge and tools to support water management. It has become clear that both a better dissemination of research results and a more targeted development of research are needed, *e.g.* by involving end-users, including consultants, more closely or by responding more directly to the 'demand' side. This is consistent with the conclusions derived in a recent study on scientific evidence for policy-making undertaken by Directorate-General for Research Directorate Unit Research in the economic, social sciences and humanities.[47]

Our point of departure was European funded research, in which several European Member States are involved. Projects' case studies involve end-users and are distributed throughout Europe. To achieve a better science–policy interface during projects' execution, more attention needs to be given to the end-users. This not only concerns improving dissemination of results and working multi-lingual but also requires improving researchers' understanding of (local) water management processes. With the evaluation criteria used in most calls for research proposals there is little incentive for researchers to prioritise efforts to accommodate local water management needs in the project activities. More or better incentives are required to attract scientists to engage whole-heartedly in local water management. Furthermore, to achieve demand-oriented research, thought should be given to how the required flexibility can be incorporated into projects.

Especially in the case of model and tool development, involvement of consultants should lead to higher uptake of research outputs: Consultants are more likely to document, manage and in particular maintain project results – maintenance and support is a key requirement to obtain the trust of water managers and hence their willingness to apply software tools.

To further improve the science–policy interface beyond the individual project's level, broader science-meets-policy meetings are required. Face-to-face meetings are required to gain insights and reach better understanding. We recommend a regional approach to circumvent barriers such as lack of time, lack of budget and language, but also to allow regional topics to be more specifically addressed. Nevertheless, there is certainly a role for European-wide conferences and activities where science meets policy – at such meetings cross-boundary exchange and cooperation is stimulated and involvement of all levels of European policy and management provides added value to all.

The sheer amount of research results produced needs to be organized and harmonized, keeping the water manager explicitly in mind. The development of WISE-RTD is a major step to ease finding results from the perspective of the Water Framework Directive Implementation. Where many results are available, development of (multilingual) synthesis documents that provide the state of play or harmonize methods should be pursued.

All of this work has been based on the assumption that the science–policy interface and uptake of research results is subject for improvement. Generalisations such as 'researchers do not understand policy' and 'policy fails to see what is available' have been heard during Harmoni-CA's lifetime. But, is this justified? Is the uptake of research poor, is the science–policy interface not functioning satisfactorily? As a final consideration, based on, amongst others, the interviews with end-users of European Research projects[28] we believe that though improvements in the science–policy interface are required and research uptake can be enhanced, one should not forget that today, too, research results do effect water management and are taken onboard. Hence, when are we satisfied with the level of uptake of results? To answer this question one needs to consider first of all what the essence of research is, that some research will produce outputs that are not applicable, that research will falsify promising methods, or research will conclude that results are promising but more research and demonstration is required before products are operational. By its very nature research just adds to the knowledge base, which includes both successes and failures. If we define 'successes' as research results that are picked up by operational water management, what is the success–failure ratio that is satisfactory? In particular in modelling and simulation, how do we assess success? Do we count those research models that are kept alive? How do we measure the impact of scientific publications of a project on other models, developed by non-project partners? These questions confirm the insights from Harmoni-CA that it is important to manage expectations: What can and what may research expect from policy? What can and what may policy expect from science? What can all expect from improved science–policy interfacing?

Acknowledgements

The present work was carried out within the Concerted Action Harmoni-CA funded under EC's 5th Framework Research Programme (Contract EVK1-CT2001-00192). The authors would like to thank the Harmoni-CA team for their contributions in the development of this chapter.

References

1. European Commission, Directive 2000/60/EC of the European Parliament and of the Council of 23 October 2000 establishing a framework for Community action in the field of water policy, *Official J. Eur. Communities* L327/1, 2000.
2. European Commission, Catalogue of Contracts, Sustainable Management and Quality of Water Key Action 1 Energy, Environment and Sustainable Development, http://circa.europa.eu/Public/irc/rtd/eesdwatkeact/library?l=/projects_information/abstractssfp5evk1pdf/_EN_1.0_&a=d, 202p (2002).
3. Directorate-General for Research Sustainable Development (ed), *Global Change and Ecosystems Global Change and Ecosystems*, Catalogue of projects funded during the Sixth Framework Programme, http://www.euresearch.ch/fileadmin/documents/PdfDocuments/media-2007/AllFP6ENV-projects.pdf, 2006, 444 pp.
4. W. J. de Lange, J. P. A. Luiten and M. Blind, Description of Work of Harmoni-CA (Harmonised Modelling Tools for Integrated Basin Management), Contract EVK1-2001-00192, Version 3.0. http://www.harmoni-ca.info/About_Harmoni-CA/DoW-HarmoniCA-v3.0.pdf, 2002, 46 pp.
5. Official Harmoni-CA website: www.harmoni-ca.info.
6. H. Etzkowitz and L. Leydesdorff, The dynamics of innovation: from National Systems and "Mode 2" to a Triple Helix of university–industry–government relations, *Research Policy*, 2000, **29**, 109–123.
7. Official website of the 5th Framework programme on Energy, Environment and Sustainable Development http://cordis.europa.eu/eesd/.
8. P. Winsemius, Gast in eigen huis, beschouwingen over milieumanagement ('Guest in one's own house, thoughts on environmental management', in Dutch). Samson H.D. Tjeenk Willink, Alphen aan den Rijn, The Netherlands, 1986, 227 pp.

9. Anonymous, Common Implementation Strategy for the Water Framework Directive (2000/60/Ec) Strategic Document as Agreed by The Water Directors under Swedish Presidency 2 May 2001. http://ec.europa.eu/environment/water/water-framework/objectives/pdf/strategy.pdf, 2001, 79 pp.
10. WISE-entry to Pilot River Basins: testing the implementation of the WFD, http://ec.europa.eu/environment/water/water-framework/prbs.htm.
11. Central entry point to the CatchMod projects: http://www.harmoni-ca.info/Catchment_Modelling_projects/index.php.
12. G. E. Arnold and Drafting group, July 2004, Scope paper: research supporting the WFD implementation, available on request.
13. Official WISE-RTD (Water Information System Europe, Research and Technology Development) website: http://www.wise-rtd.info/.
14. Official WISE (Water Information System Europe) website: http://water.europa.eu/.
15. Official SPI-Water project website: http://www.spi-water.eu/.
16. H. Madsen, A. van Griensven and A. L. Højberg, Model calibration and validation in model-based water management, International Water Association, 2009.
17. J. C. Refsgaard, J. P. van der Sluijs, A. L. Højberg and P. A. Vanrolleghem, Uncertainty analysis in model-based water management, International Water Association, 2009.
18. V. Vandenberghe, M. Ratto, F. Pappenberger, Review of sensitivity analysis methods International Water Association, 2009.
19. H. Scholten and J. C. Refsgaard, Quality Assurance in model-based water management: Better Modelling Practices, International Water Association, 2007.
20. Harmoni-CA Statistical Toolbox: http://www.harmoni-ca.info/StatisticalToolbox/.
21. F.F. Hattermann and Z.W. Kundzewicz, Model-supported Implementation of the Water Framework Directive – A Water Manager's Guide, Harmoni-CA deliverable D03-10B, http://www.wise-rtd.info/html_docs/planning_guidance/planning_guidance.html, International Water Association, 2007.
22. A. L. Højberg, J. C. Refsgaard and L. Flindt Jørgensen, Good practise in joint use of monitoring and modelling, Harmoni-CA Deliverable no. D4-8, http://www.wise-rtd.info/html_docs/MMGuidance_final.html, International Water Association, 2007.
23. J. C. Refsgaard, L. Flindt Jørgensen and A. L. Højberg, Data availability accessibility – State of the art on existing data required for modelling for research purposes and for the implementation of the WFD Synthesis report, Harmoni-CA Deliverable no. D4-7, http://www.wise-rtd.info/html_docs/DAA_report_final/DAA_report_final.html, 2007.
24. R. Brouwer, D. Barton and F. Oosterhuis, Economic methods, models and instruments for the Water Framework Directive, Harmoni-CA Deliverable no. D 5.1b and 5.4b, http://www.wisertd.info/html_docs/gesamt_D5_1B_4b_oekom_bericht/Harmonised_Modelling_Tools_fo_integrated_river_basins_management.html, International Water Association, 2007.

25. G. M. Bazzani, Model-based tools regarding the interaction of water management and agriculture, Harmoni-CA Deliverable no. D 5.1c and 5.4c, http://www.wise-rtd.info/html_docs/Harmonised_Modelling_Tool_final_for_web_final.html, International Water Association, 2007.
26. P. Bots, G. Gooch, B. McIntosh and C. Pahl-Wostl, Harmoni-CA guiding document on integration of the human dimension in model-supported water management, Osnabrück, University of Osnabrueck, International Water Association, 2007.
27. C. Giupponi, J. Mysiak, Y. Depietri and M. Tamaro, Decision Support Systems for water resources management: current state and guidelines for tool development, Harmoni-CA deliverable no. D6-19, http://www.wise-rtd.info/html_docs/dss/Decision_Support_Systems_for_water_resources_management.html, International Water Association, 2007.
28. S. Rotter and K. Rasche, Report on end-user involvement in research projects. A telephone survey-based assessment of twelve research projects, Seecon Deutschland GmbH, Harmoni-CA deliverable no. D6-23, http://www.harmoni-ca.info/Registered_Users/Document_Store/docs-doc-open-download.php?doc_id=896 International Water Association, 2007.
29. Towards harmonised procedures for quantification of catchment scale nutrient losses from European catchments (Euroharp) toolbox: http://www.euroharp.org/toolbox/index.php.
30. Process Support Tool (ProST) website: http://harmoniqua.wau.nl/ProST/.
31. Build the standards-community by sharing your experiences (Hydronomy, product of Harmoni-CA) website http://www.portolis.nl/hydronomy/.
32. M. W. Blind, European Catchment Modelling Research, Folder with loose-leaf product flyers, digital versions on http://www.harmoni-ca.info/flyers/ (2006).
33. D. Ridder, E. Mostert and H.A. Wolters (eds), 'Learning together to manage together – Improving participation in water management'. *Handbook of the HarmoniCOP Project*, ISBN 3-00-016970-9, 2005, 99 pp.
34. Community Research and Development Information Service (CORDIS) website http://cordis.europa.eu/fp7/projects_en.html.
35. Official EUGRIS (portal for soil and water management in Europe) website: http://www.eugris.info/.
36. Towards a European-wide exchange Network for integrating research efforts on Integrated Water Resources Management (IWRM-net) website: http://www.iwrm-net.eu/.
37. Climate Impact Research Coordination for a Larger Europe (Circle) website: http://www.circle-era.net/.
38. Coordination de la Recherche sur la gestion des inondations financie dans l'Union Europene (Coordination of research financed in the European Union on Flood risk management), CRUE Era-net project: http://www.crue-eranet.net/.
39. Sustainable management of soil and groundwater under the pressure of soil pollution and soil contamination (S.N.O.W.M.A.N.), website: http://www.snowman-era.net/index.php.

40. I. Borowski and M. Hare, Exploring the gap between water managers and researchers: Difficulties of model-based tools to support practical water management, J. *Water Resources Manage.*, 2007, **21**(no. 7).
41. Official DELFT-FEWS website: http://www.wldelft.nl/soft/fews/int/index.html.
42. Official Sobek website: http://delftsoftware.wldelft.nl/index.php?option=com_content&task=view&id=110.
43. Official Isis website: http://www.wallingfordsoftware.com/products/isis/.
44. Official Mike-SHE website: http://www.dhigroup.com/Software/Water-Resources/MIKESHE.aspx.
45. Official SWAT website: http://www.brc.tamus.edu/swat/.
46. Official Modflow website: http://water.usgs.gov/nrp/gwsoftware/modflow2000/modflow2000.html.
47. L. Anastopoulou, *Scientific Evidence for Policy-making*, Directorate-General for Research Directorate L – Science, economy and society, Unit L.2 – Research in the economic, social sciences and humanities – Prospective, ISBN 978-92-79-06973-4, 2008, 32pp.

CHAPTER 2.5

The OpenMI – Science Responding to Policy, Industry and Events

ROGER V. MOORE

Centre for Ecology and Hydrology, Crowmarsh Gifford, Wallingford, Oxon, OX10 6HU, United Kingdom

2.5.1 Introduction

2.5.1.1 Background

The well intended and imaginative policy of developing biofuels and its unforeseen consequences for world food production dramatically underline the interconnectedness of today's world. To reduce the chances of repeating such mistakes, we must not only increase our understanding of the individual processes involved but also their *interactions*. We can then apply that understanding to predicting the wider impacts of events and the policies to mitigate them.

To change the present situation will require vision and a change of mindset. We will need to create a global collaboration between science, industry and the policy-making agencies. But this will be a totally different type of collaboration to anything we have achieved to date. To make collaboration possible, we have to take down many of the walls created to preserve competitive advantage; in the age of the internet, there are far more opportunities to be gained by sharing information than by protecting it. Collaboration requires information exchange, so the development of a minimum set of standards to facilitate that

Water System Science and Policy Interfacing
Edited by Philippe Quevauviller

Published by the Royal Society of Chemistry, www.rsc.org

exchange will be needed. These standards must provide either a common language or a translation service that will enable not just people but computer systems, in this context simulation models, to interact.

A relatively new technology that will be key to the process is integrated modelling – the joining together of models of individual processes to understand and predict how they will interact under different conditions. This will be vital to finding better answers to the complex challenges that face agencies such as the EC or EPA. However, alone, integrated modelling does not solve the agencies' problems; neither could it yet be described as a mature and fully operational science. There are many outstanding problems, which the universities and others must still solve. An effective pipeline has to be constructed for the delivery of the solutions to policy-makers and their agents. As time is probably not on our side, in parallel we must also find ways of driving up the rate of innovation. Simultaneously, we must drive down the barriers to entry; we must make it easy for people to participate. Fortunately, emerging new technologies and social changes have already shown in other areas that these aspirations are not an empty dream. This chapter will discuss how, with drive and imagination, the opportunities can be seized and exploited.

2.5.1.2 A Starting Point

Most regulatory and policy making bodies face an increasingly complex array of questions, for example:

- *Science issues such as:* Will climate change affect ecosystem services in upland habitats?
- *Policy issues such as:* What will be the impact of climate change on flood damage costs?
- *Planning issues such as:* What will be the impact of the medical plan for treating a 'flu pandemic on sewage treatment works and rivers?
- *Operational issues such as:* Why do fish kills regularly occur here?

All the example issues have two points in common. They all involve understanding and predicting multiple interacting processes and they all span many disciplines – environmental, social and economic. At present, we have models of individual processes that encapsulate our knowledge and enable us to make predictions but we have few models of multiple processes. It is rarely feasible to build a single purpose written model incorporating all the processes to be represented. Such models tend to be inflexible, are seldom able to bring the best modelling to all areas and are not often reusable afterwards. They also make poor use of the huge investment in existing models. A much better approach is to make current and new models *linkable*. This is the first reason why linking standards such as the Open Modelling Interface (OpenMI) and others have been developed.

To return to the question, 'Why do fish kills regularly occur here?', Figure 2.5.1 shows the processes that might explain the kills and Figure 2.5.2

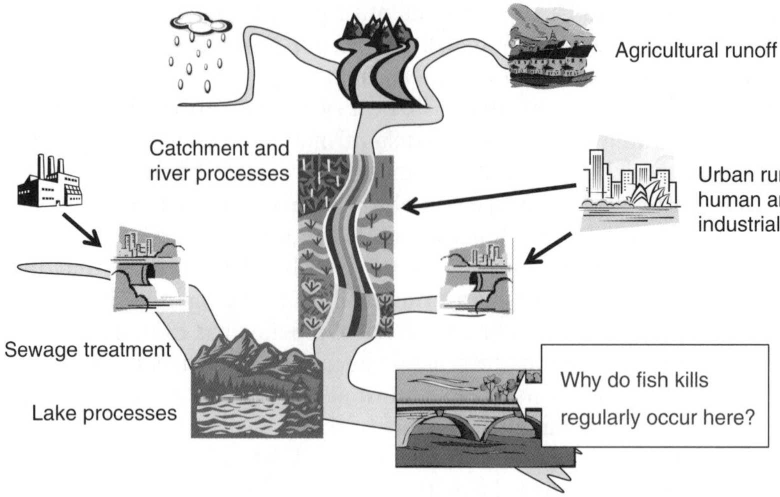

Figure 2.5.1 Processes whose combined effect might explain the fish kills.

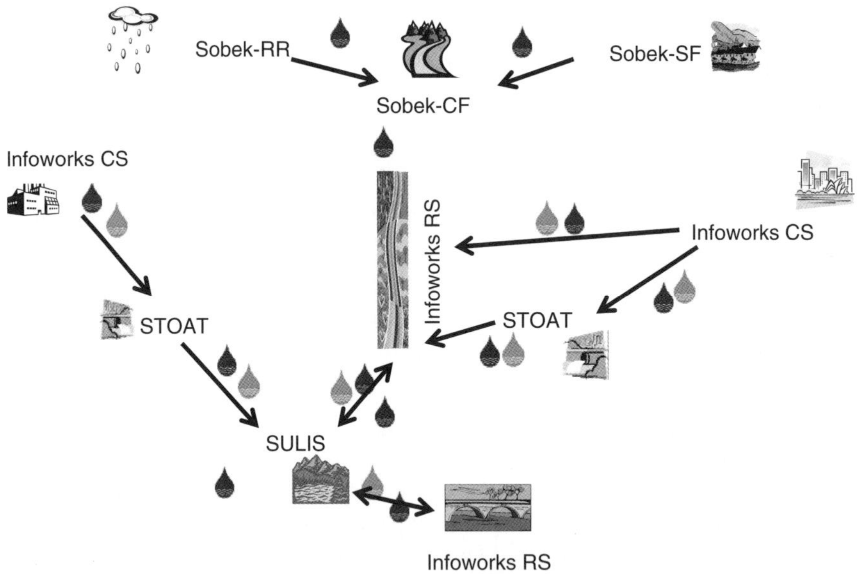

Figure 2.5.2 Models representing the processes and the links required between them.

shows how those processes can be represented by models. However, the real question is, 'Is the collective effect of those processes, after they have interacted, sufficiently toxic to explain the kills?' Figure 2.5.2 also illustrates the links required to answer the question. To date two main options have existed for model linking, serial linking and hard wiring. Serial linking involves running the models sequentially, the output of one becoming the input of the next.

While simple, the disadvantage of this approach is that it does not capture the time step by time step interaction. Hardwiring, on the other hand, involves merging model codes. While this can be very effective for research purposes, it rarely produces operational reusable models.

The OpenMI is being developed to overcome these short comings. It does so by providing a generic interface for linking, which allows the linking process to be reduced to a drag and drop operation. The exercise is challenging because the models may:

- Represent processes from the same or different domains
- Be based on different concepts
- Use any or no spatial representation
- Use different spatial and temporal resolutions
- Run at different time steps
- Use different units and semantics
- Be sourced from different suppliers
- Be written in different languages
- Be running on different platforms
- And so on.

Later, it will be useful to develop this list of challenges further under three headings:

1. The IT aspects of model linking – how do you pass data between independently developed modelling applications and between models and modelling tools?
2. The science aspects of model linking.
3. Creating the working conditions in which integrated modelling is possible, *i.e.* convincing senior managers of the added value of integrated modelling, encouraging mass collaboration and sharing of knowledge amongst modellers, developing a new group with the skills and understanding to implement integrated modelling.

Making models easily linkable is only one aspect of moving integrated modelling out of the research domain into the operational world. The example issues above require knowledge from many domains. The past tendency of organisations was to look inwards for that expertise, to look outwards was often perceived as a failure. Looking to the future, there is now a very important question that all organisations must ask themselves. The question is:

In future, where will most creativity, most knowledge lie – inside or outside your organisation?

Before answering, consider the present and future impact of the internet because it changes everything. It facilitates communication. It alters social customs and working practices. It provides access to knowledge. It links computing power. It is global. And, finally, with respect to knowledge based activity, it is lowering the barriers to entry – all you need is a PC and an internet connection.

The answer to the question is outside and this has profound consequences for our current way of working but also opens up huge opportunities, especially for integrated modelling. How then do we harness the external knowledge and creativity, and can this situation be exploited to drive up the rate of innovation? Companies such as Hewlett Packard, IBM, and numerous drugs companies have reversed past practice and selectively made open and shared knowledge that previously they would have protected. Why this works can be illustrated by the experience of a well-known toy maker. It produced a new version of a construction toy containing a microprocessor and some linked motors. No sooner was this toy on the market than hackers broke into it, improved the software and published what they had done on the internet. The first reaction of the company was to sue. However, it then had a better idea. It published the entire code of the microprocessor, told their customers how to access and alter it and provided them with a web page to say what they had done. In effect they turned their customers into their research and development department. This has allowed them to tap the creativity of the world population as opposed to that of just their own organisation. The result for the companies following this route has often been a huge increase in the rate of innovation. Greatly facilitating the process has been the emergence of loose self-organising collaborations, whose existence is made possible by the internet and social networking. The open source community is the best known example. Lastly, several standards, especially interface standards, have emerged to facilitate collaboration. The OpenMI is such as standard and it exists to enable models from anywhere and any domain to be linked, so that we can explore and predict the outcome of the example issues above.

Up to now the barriers to entering integrated modelling have been very high. The OpenMI brings them down and other events are taking place that will bring them down even further. Notice how, by publishing its API, Google has enabled others to produce a huge number of innovative products around Google Maps. The OpenMI could do the same for integrated modelling and hence greatly facilitate the implementation of integrated management. Hopefully, that will bring very much better thought out answers to the challenging questions ahead. For that to happen, using the OpenMI has to be brought within the range of as many people as possible.

The chapter will therefore start with a description of the OpenMI, its purpose and how it works. This should begin to show the opportunity ahead. However, the OpenMI's authors knew from the start that the world would not just need to be convinced of the value of the OpenMI but of the value of integrated modelling as well. The chapter concludes with the strategy for the future promotion, deployment and development of the OpenMI and how the user community will be supported. Outreach has three threads. The first aims to show the added value of integrated modelling to senior water managers. The second seeks partnerships with universities to solve the science questions that arise when models are linked. A third is intended to build integrated modelling capacity by including integrated modelling in academic courses. The aim will be to put that capability into as many hands as possible.

2.5.2 What is the OpenMI?

2.5.2.1 Introduction

The OpenMI standard defines an interface that allows time-dependent models to exchange data at run-time. When the standard is implemented, existing models can be run simultaneously and share information at each time step.

2.5.2.1.1 Original Drivers behind the OpenMI

The Water Framework Directive (WFD) calls for integrated water management to be put into practice and identifies whole catchment modelling as a key part of integrated management. As has been explained, at present, the only realistic mechanism for whole catchment modelling is integrated modelling.

Until few years ago, no generic operational linking mechanism had been developed. However, technological advances in computing, the impetus provided by the Water Framework Directive, co-funding from the European Commission and support from the leaders of earlier attempts enabled the FP5 project HarmonIT to develop the Open Modelling Interface Standard (the OpenMI). The OpenMI Interface is a standard interface that enables OpenMI compliant components to exchange data as they run. Supporting, but not part of the standard, is a software development kit (SDK). The SDK consists of a set of software tools. They facilitate making new and existing model codes OpenMI compliant and they offer facilities to combine OpenMI compliant components into integrated modelling systems and then run them.

A linkage mechanism, such as the OpenMI, is the key to moving single domain modelling to integrated modelling and integrated modelling from a research exercise to an operational task. It will allow for integrated water management to be put into effect and, hence, the objectives of the WFD to be achieved. However, the OpenMI architecture does not limit its applicability to the water domain only. It can be applied in many more domains, although its base will remain the environmental domain where temporal and spatial variability are key issues in understanding and managing systems.

Adopting the OpenMI requires model developers to make a commitment. Most organizations cannot afford to make that commitment until the OpenMI is widely available in several implementations and is properly supported – in other words it becomes a well-maintained standard. Therefore, the OpenMI Association has been created to promote the OpenMI and achieve critical mass in terms of compliant models and users, to support the user community and sustain the OpenMI into the future.

The following sections provide a brief overview of the OpenMI and the OpenMI Association.

2.5.2.1.2 Aims and Objectives

The aim of the OpenMI is to provide a mechanism by which physical and socio-economic process models can be linked to each other, to other data sources and to various tools at run-time, hence enabling process interactions to be better modelled. Specific objectives are that the mechanism's design should:

- be applicable to new and existing models;
- be applicable to most, if not all, time-based simulation techniques;
- handle feedback;
- impose as few restrictions as possible on the modeller's freedom;
- require the minimum of change to the program code of existing applications;
- keep the cost, skill and time required to migrate an existing model to a minimum so that these factors are not a deterrent to the OpenMI's use;
- be easy to use;
- not unreasonably degrade performance.

2.5.2.1.3 Why should Organizations Adopt the OpenMI?

The discussion above has explained that the need for the OpenMI was created by the introduction of the WFD. What benefits does it bring to the designated authorities, basin managers, regulators, consultants, modellers and model developers responsible for implementing the WFD? Some of the arguments for adopting the OpenMI put forward by organizations that have already adopted or are considering adopting the OpenMI are:

- the simplification of the model-linking process, leading to an improved ability to model process interactions;
- the ability to use appropriate model combinations and to swap between different models of the same process, assisting sensitivity analyses and benchmarking;
- the protection and enhancement of existing investment in model development (*i.e.* it is not necessary to rewrite model codes completely for them to become linkable);
- a reduction in development time and hence cost for decision support systems;
- an increased choice for model users, in that they will be able to 'mix and match' models from different sources;
- increased opportunities for model developers in that individual models become more saleable because they can be linked to established systems, enhancing the value of both;

- the increased opportunities for the creation of Small and Medium Enterprises (SME), especially from the academic sector;
- the increased opportunities to contribute to the implementation and evolution of EU policies;
- the opportunity for model developers to concentrate on their core business (*e.g.* computational cores) because they will be able to buy in OpenMI-compliant tools such as GUIs and post-processing tools where these are peripheral to the business;
- the availability of the OpenMI Environment tools for migrating and linking models and monitoring linked model runs (which are available free under an Open Source licence and would otherwise have to be written by the developer);
- the small cost of conversion compared with the cost of writing a whole catchment model from scratch or redeveloping existing models;
- the ability for model users to run third-party computational cores in their own environments;
- no need to understand other organizations' I/O procedures;
- the ability to change a model's code without affecting the linking process or interface.

Interestingly, many of these arguments are as much about the practical implementation aspects of integrated modelling as about the science.

2.5.2.1.4 Development History of the OpenMI

HarmonIT Project (2002–2006). The first version of the OpenMI was developed by a team drawn from 14 organizations and seven countries co-funded through the European Commission's Fifth Framework programme under contract number EVK1-CT-2002-00090 (the HarmonIT project). Led by the UK Natural Environment Research Council's Centre for Ecology and Hydrology (CEH), the team consisted of the Institute for Inland Water Management and Waste Water Treatment (RIZA), DHI Water and Environment (DHI), WL Delft Hydraulics (WL|Delft), HR Wallingford Group (HRWG), Universitat Dortmund, Instituto di Ricerca Sulle Acque (IRSA), the National Technical University of Athens (NTUA), WRc plc (WRC), DHI Hydroinform a.s., Povodi Labe s.p. (PL), Hydroprojekt a.s., Alterra B.V. and the Centre National du Machinisme Agricole, du Genie Rural, des Eaux et des Fôrets (CEMAGREF).

Design and development was conducted in an incremental way, taking use cases as a basis for iterative development. The development was primarily undertaken by the three major commercial model developers, DHI Water and Environment, WL Delft Hydraulics and HR Wallingford. The role of the other organizations was to manage the project, to support the design and development and to test the standard and SDK rigorously.

To ensure that the work met the standards required by the Commission and the scientific and user communities, a panel of experts consisting of leading scientists from around the world reviewed all key documents and advised the Steering Committee. The project's quality assurance plan established procedures for the critical areas of work and covered document and code version control.

OpenMI-Life Project (2006–2010). To turn the OpenMI from a research outcome into a sustained standard for operational practice, a second project has been initiated under the policy area 'Sustainable management of ground water and surface water management' of the European Commission's LIFE Environment programme (Contract no: LIFE06 ENV/UK/000409). The technical work (maintenance and improvement) concerning the OpenMI is being led by the same team and in a similar way to that followed in the HarmonIT project. However, the team is growing as users outside the project begin to contribute to the support and development process. To demonstrate and evaluate the OpenMI and integrated modelling under operational conditions, new members from the Competent Authorities have joined the project. Led by the UK Natural Environment Research Council's Centre for Ecology and Hydrology (CEH), the team consists of DHI, WL Delft Hydraulics (now part of Deltares), Wallingford Software Ltd. (WSL), the National Technical University of Athens (NTUA), RIKZ (now part of Deltares), Aquafin, Vlaamse Milieu Maatschappij (aka VMM), Flanders Hydraulics (FH), Université de Liège (ULg) and University of Thessali (UTH).

The objective of the project is to transform the Open Modelling Interface and Environment from a research output into an operational supported standard; activities include:

- Building capacity
- Demonstrating and evaluating under operational conditions:
  - The use of the OpenMI in integrated water management/modelling in:
    - i. The Scheldt (Belgium) – four cases
    - ii. The Pinios (Greece) – three cases
  - The support organisation [now in existence and called the OpenMI Association (www.openmi.org)]
- Disseminating information.

2.5.2.2 Overview of the OpenMI

This section describes the main user requirements for model linking, the functions currently provided by the OpenMI interface and the scenarios against which the standard has been tested.

2.5.2.2.1 Original Requirements for the OpenMI

To be useful, the OpenMI must be able to link any models whose interactions need to be simulated in meeting the requirements of the Water Framework Directive. Against this background, the key requirements of the OpenMI are perceived to be:

- linking models from different domains (hydraulics, hydrology, ecology, water quality, economics, *etc.*) and environments (atmospheric, fresh-water, marine, terrestrial, urban, rural, *etc.*);
- linking models based on different modelling concepts (deterministic, stochastic, *etc.*);
- linking models of different dimensionality (0, 1, 2, 3D);
- linking models working at different scales (*e.g.* a regional climate model to a catchment runoff model);
- linking models operating at different temporal resolutions (*e.g.* hourly to monthly or even annual);
- linking models operating with different spatial representations (*e.g.* networks, grids, polygons);
- handling feedback;
- handling iteration;
- the ability to save its state and revert to a previous state;
- linking models using different projections, units and categorizations;
- linking models to other data sources (*e.g.* databases, user interfaces, instruments);
- linking new and existing (legacy) models with the minimum of re-engineering and without requiring unreasonably high level IT skills;
- low barrier to entry;
- not impairing performance, especially of large models;
- being based on proven and available technologies (and, in particular, the architecture must be component-based and multi-layered);
- linking models running on different platforms (*e.g.* Windows, Unix and Linux);
- being 'open' (the interface specification should be placed in the public domain);
- allowing components to be developed using at least the following programming languages: C/C+ +, C#, Fortran, Delphi/Pascal, Java and Visual Basic.

The remainder of this section shows how these requirements have been met.

2.5.2.2.2 Use Cases

To check that the requirements were correctly expressed and to ease the development of an architecture for the OpenMI, a range of scenarios or 'use cases' were identified. Some of the initial cases drawn the hydraulic domain are shown below. Since then, the use cases have been drawn from a much wider range of domains:

- connect two 1D hydrodynamic river models;
- connect a 1D hydrodynamic model with a water quality transport model;
- connect a 1D river model with a 3D groundwater model;
- connect a 1D hydrodynamic river model to vegetation and habitat models;
- connect a 3D coastal model to a 1D river model;
- connect a 2D polygon-based root zone model to a 3D regular grid groundwater model;
- calibrate a rainfall runoff model linked to a hydrodynamic sewerage model;
- model the propagation of uncertainty through a chain of models;
- use different units of measurement for the data to be exchanged between models;
- connect to an agent-based model.

This list has been further extended during the OpenMI-Life project, where it is being evaluated by the competent authorities in Belgium, The Netherlands and Greece against operational water management problems. In the USA, it is being extended into a much wider computing environment.

2.5.2.2.3 Terminology

Several terms are used when describing the OpenMI standard.

As shown in Figure 2.5.3, the term *model application* encompasses all parts of the modelling system software that is installed on a computer: for example Mike11, PHABSIM and InfoWorks-RS.

Typically, such systems consist of a *user interface* and an *engine*. Usually, the engine is a generic representation of a process and this is where the calculations for simulating or modelling that process take place. The user supplies

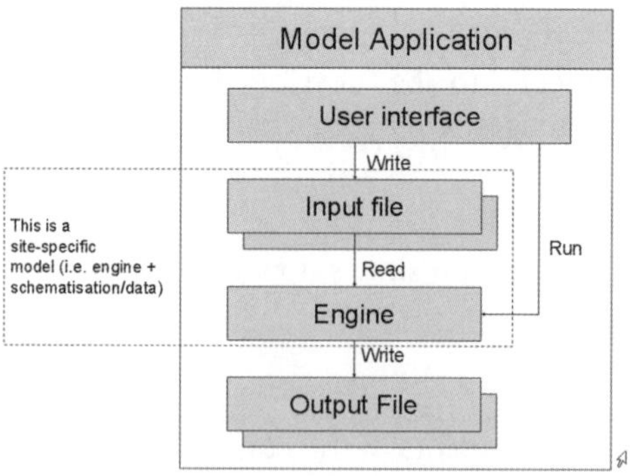

Figure 2.5.3 General structure of a model application.

information through the user interface and this is converted into the input data for the engine.

The data describes a specific scenario in which the process is to be simulated: for example the flow in the river Rhine during a time of extreme rainfall. The user runs the engine by selecting an option or pressing a button on the user interface. The engine reads the input, performs the calculations and outputs the results to files or displays.

When an *engine* has read its input it becomes a *model*. For example, an engine may represent the generic process of water flowing in an open channel. When it has read in the data describing the channel network of the Rhine, along with any boundary conditions and rainfall data, it becomes a model of the Rhine in the scenario to be simulated.

If the code for an engine can be instantiated separately and has a well-defined *interface* through which it can *accept* and *provide* data, then it is an *engine component*. (The engine's *interface* is the part of the code that handles the transfer of data to and from the engine; it should not be confused with the *user interface*, which is the part of the application that the user sees.) The key to enabling models to exchange data lies in standardizing the design of the engine interface. When an engine component implements such a standard interface it becomes a *linkable component*. An engine that implements the OpenMI interface is called *OpenMI compliant*.

2.5.2.2.4 OpenMI Standard Interface

The OpenMI defines a standard interface that has three functions:

1. *Model definition*: to allow other linkable components to find out what items this model *can* exchange in terms of quantities simulated and the locations at which the quantities are simulated.
2. *Configuration*: to define what *will* be exchanged when two models have been linked for a specific purpose.
3. *Run-time operation*: to enable the model to accept or provide data at run time.

Figure 2.5.4 shows two model applications whose engines have been made OpenMI-compliant. Their overall structure remains unchanged but each engine is now a component with an OpenMI interface and one component can now get values from another.

Figure 2.5.5 illustrates some of the information held in the model definition about the quantities that two models can either accept or provide. The arrow represents a link between the two models and indicates that, in this particular case, runoff produced by the Rainfall Runoff Model will be used to represent lateral inflow in the River Model. There is no requirement to harmonize the terminology; the linking process creates the appropriate cross-reference table.

Figure 2.5.6 shows the geographical matching of elements in a river model to those in a groundwater model. The river model is a vector model and each

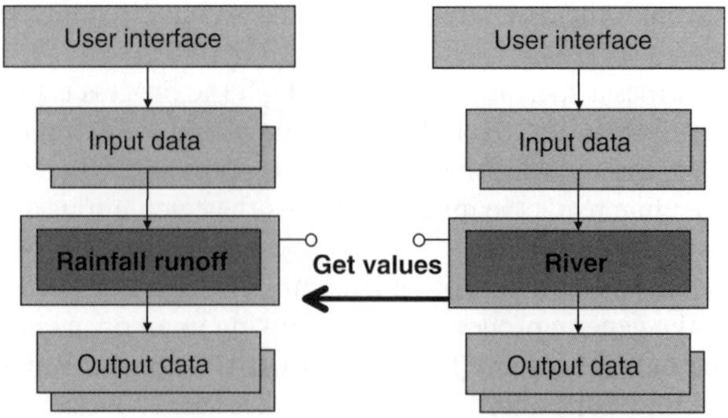

Figure 2.5.4 Two applications after migration to the OpenMI standard.

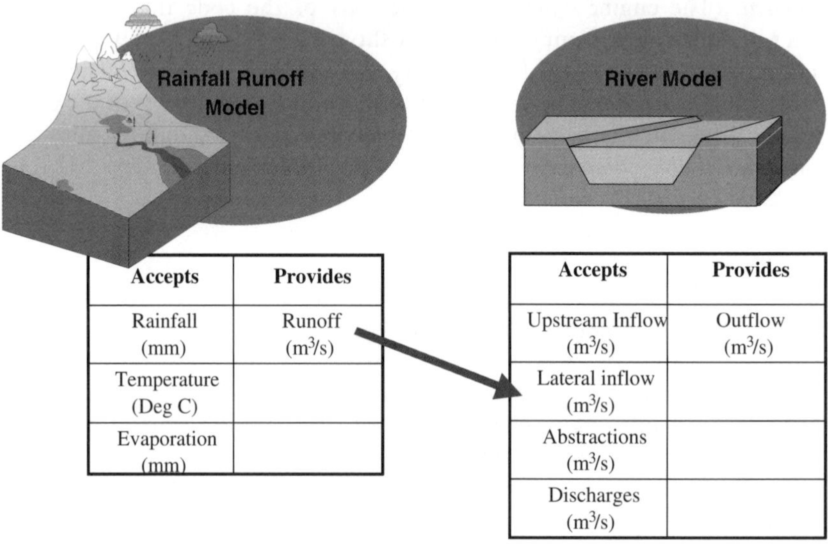

Accepts	Provides
Rainfall (mm)	Runoff (m^3/s)
Temperature (Deg C)	
Evaporation (mm)	

Accepts	Provides
Upstream Inflow (m^3/s)	Outflow (m^3/s)
Lateral inflow (m^3/s)	
Abstractions (m^3/s)	
Discharges (m^3/s)	

Figure 2.5.5 Showing and linking quantities.

element represents a single stretch; the groundwater model is grid-based, each node being an element. Therefore, to link the two models, each element in the river model will usually be linked to several elements in the groundwater model. In any non-trivial situation, this will require the matching of thousands of elements and therefore the process is automated.

2.5.2.2.5 An Interface-based Open Standard

The OpenMI provides an intelligent mechanism whereby models running simultaneously can exchange data time step by time step. It thus enables process interaction to be represented more accurately than is possible by sequential

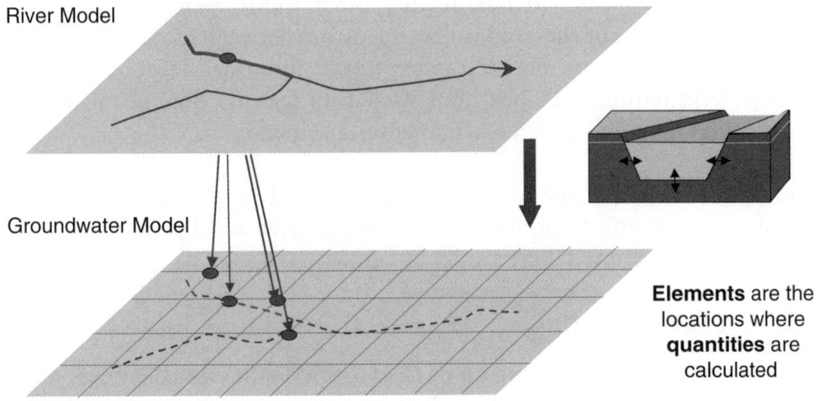

Figure 2.5.6 Linking element sets.

linkage. It is important to explain that the OpenMI is *neither* a common data-model specification *nor* is it an integrated modelling system. It is a standard.

The OpenMI is 'interface-based':

- its 'standardized' part is defined as a software interface specification;
- this interface acts as a 'contract' between software components;
- the interface specification is not limited to specific technology platforms or implementations;
- the interface implementation may be limited by the technology supported in a specific release;
- by adopting the implemented interface a component becomes an OpenMI-compliant component.

The OpenMI is 'open':

- its specification is publicly available *via* the Internet (www.OpenMI.org);
- its source code is open and available under Lesser GPL licence conditions;
- it enables linkages between different kinds of models, different disciplines and different domains;
- it offers a complete metadata structure to describe the numerical data that can be exchanged in terms of semantics, units, dimensions, spatial and temporal representation and data operations;
- it provides a means to define exactly what is linked, how and when;
- its default implementation and software utilities are available under an open source software license.

The OpenMI is a 'standard':

- it standardizes the way data transfer is specified and executed;

- it allows any model to talk to any other model (*e.g.* from a different developer) without the need for co-operation between model developers or close communication between integrators and model developers;
- its generic nature does not limit itself to a specific domain in the water discipline or even in the environmental discipline.

The OpenMI cannot guarantee that the representation of the process in the component or the link to another component is scientifically valid. That is the responsibility of the modeller, model integrator and user, as it should be.

2.5.2.3 Current Status

At the time of writing, Version 1.4 of the OpenMI standard has been released for .NET and Java. It is available under non-restrictive open source licence and can be downloaded at www.openmi.org. Support is provided through the OpenMI Association, an international, not-for-profit, open, democratic organisation established under Dutch law. Version 2.0 is due for release late 2009.

2.5.3 OpenMI Association's Strategy

2.5.3.1 Background

For regulatory organisations, organisations charged with safety critical tasks and commercial companies, the adoption of a standard on which much of their work will become dependent is a significant decision. To make that decision they must have confidence that the standard's keepers know where they are going, have a credible strategy for getting there and the resources to implement the strategy. The OpenMI Association has therefore made the development and maintenance of its forward strategy a priority. It is now used to guide all its activities.

The association realised early on that there was very much more to making integrated modelling a tool fit for operational use than just solving the IT issue of passing data between models. It was clear that as soon as it became easy to link models then a host of scientific and cultural issues would emerge. Some of the science issues would be generic and some would be domain specific. Many would arise because existing models were being pushed beyond their design limits. A collaboration would need to be established with the research community to ensure that as these problems emerged they were solved.

A parallel demonstration exercise would be needed to show policy makers and managers that integrated modelling brought real benefits. In anticipation of the exercise being successful and managements adopting integrated modelling, then there would need to be a body of modellers in place by the time that moment arrived; modellers who understood and could apply integrated modelling effectively. To create this body, integrated modelling would have to be incorporated into the university syllabus.

As the OpenMI standard matured and major organisations began to consider seriously its adoption, further considerations came to the fore. In summary, the user community, that is policy makers, managers, consultants, researchers and developers, needed to be convinced that:

- Integrated modelling added value
- There would be a body of practitioners who could implement it
- The OpenMI was reliable
- Development would continue
- Users could influence development
- The OpenMI users would be supported
- The science issues would be resolved
- The backing was credible and sustainable
- Users couldn't be held to ransom.

Through the OpenMI-Life project and with European Commission support, the OpenMI Association has now been formed to ensure the long-term sustainability, support and development of the OpenMI standard. Its first task has been to develop the forward plan setting out its vision of the future, the role of the Association in achieving the vision and a strategy for getting there. These are set out below.

2.5.3.2 OpenMI Association's Objectives According to its Charter

The OpenMI Association is an association ('vereniging') established under Dutch law. It is open to international membership for both organisations and individuals. It is governed by a Charter, which sets out in very high level terms its objectives and how it will achieve them, as follows:

The promotion of the development, use, management and maintenance of the Open Modelling Interface (the OpenMI), a standard for the exchange of data between computer software in environmental management.
The Association seeks to achieve this goal, among other things, by:
Exchanging information with regard to the standard specified in the objective, by word, writing and by means of electronic devices such as a website, both within the Association and with other organisations that pursue a similar objective
Stimulating the maintenance and development of the OpenMI Standard and its supporting software and managing their release
Stimulating the provision of information and promoting discussion on the OpenMI in Europe and across the world by the organising of and participating in events
All that which may be further conductive to the objective

The Association's strategy translates these aspirations into a vision of the role of modelling within the management of land and water, a mission statement for the Association and a strategy for achieving the mission.

2.5.3.3 Vision

The essentials of the Association's vision have been set out in the introduction. The only area that is not fully addressed in the current vision is how openness, sharing and collaboration are likely to be key to driving up the rate of innovation and making integrated modelling an operational tool. In summary the vision is:

- Governments and their agents will face increasingly complex issues
- Many of these will be global in extent and span many disciplines
- It will become increasingly important to predict the wider impacts of the policies adopted to manage the issues
- Integrated modelling will become accepted by authorities as leading to better decisions
- The mode of working will become increasingly collaborative; sharing, openness and standards will be key components of the new collaborations, especially with respect to research results.

2.5.3.4 Mission

The attainment of the vision will require the combined efforts of developers, modellers and users. Within this context, the role, and hence the mission, that the OpenMI Association has set itself is to:

- promote integrated modelling as a means of achieving better management decisions, so that resources are more fully exploited and impacts are better understood;
- develop and support the OpenMI Standard, so that it becomes the first choice for model integration.

The OpenMI Association has defined for itself the following boundaries:

- technical scope:
  - the exchange of data between independent models and modelling components either at run-time or when the models are run sequentially;
  - facilitating the embedding of linked models and components in decision support systems (DSSs);
- application domain: environmental management with the priority on land and water systems;
- intended user groups of the OpenMI and its applications: model developers, the modelling community and the competent authorities;
- geographic focus of dissemination activities: although the OpenMI Standard can be applied world-wide, the priority regions for the OpenMI Association in the first years will be Europe and the United States of America.

2.5.3.5 Implementation Strategy

To achieve its mission, the OpenMI Association will focus on the following key actions:

- creating a culture that eases the adoption and acceptance of integrated modelling as a concept and the use of the OpenMI as the recognised option for integrated modelling;
- ensuring that the OpenMI remains relevant, easy to use, of high quality and available under acceptable conditions;
- supporting the community of OpenMI users and providing a compliancy service for developers who have adopted the OpenMI Standard in their products;
- disseminating necessary and useful information;
- enabling the user community to influence the development of the OpenMI and its Association;
- being the legal body that holds the ownership of the OpenMI;
- securing the necessary resources to fulfil its mission and implement its strategy.

2.5.3.5.1 Creating a Culture for Integrated Modelling

Changing attitudes to integrated management and modelling will require the provision of information and the creation of new skills at both high and low levels in the user, modelling and developer communities.

The top-down approach will be to ensure that key decision-makers (the competent authorities) are aware of the added value of integrated modelling and the role of the OpenMI Standard. Their needs will be identified so that they can be provided with the information that will enable them to make well-informed decisions about the use of integrated modelling and the OpenMI.

In parallel, the OpenMI Association will identify the most widely used model codes and decision support systems and encourage their developers to make them OpenMI compliant (if not yet so). To do so, the Association will endeavour to drive down the barriers to participation in integrated modelling.

The bottom-up approach will be to create a new body of scientists, engineers and IT professionals with integrated modelling and OpenMI skills. This will be achieved by liaising with respected higher educational institutions and encouraging the inclusion of material on integrated modelling and the OpenMI in appropriate undergraduate, MSc and PhD courses, workshops and seminars. The Association will co-operate with the educational institutions to create course material. The development of training services by the Association's members will be similarly encouraged.

2.5.3.5.2 Supporting the OpenMI user Community

Through its Technical Committee, the OpenMI Association will provide and supervise a forum for discussion and questions (http://sourceforge.net/projects/

openmi/); this forum is intended primarily for developers. Additionally, the Association will provide and maintain a website and a 'wiki' site (wiki.openmi.org). The web site will be the entry point for the Association and the OpenMI. It will contain final copies of all key documents, news and announcements and will be designed with end users and those new to the OpenMI in mind. The site will link to the wiki, which will be the home page of the Technical Committee. The wiki will have a more informal style and will provide access to detailed technical information and reports of user experiences with the OpenMI.

The Technical Committee is not resourced to maintain a staffed help desk and cannot guarantee that all questions on the discussion forums will be answered by them. However, the Association will foster the creation of a mutually supportive OpenMI Community in the open source spirit.

The Association will always recommend end users to go first to their software supplier for support. The OpenMI Association will encourage software suppliers to build up OpenMI skills among their support staff, so that they can help their users with OpenMI-related queries. Hence, from an end user's (modeller's) point of view, the preferred channels of communication are the same as those that exist in today's single-domain modelling. All users seeking support will be welcome to explore the information available through the website and the wiki. All users, but particularly developers, should feel welcome to report bugs and make proposals for change to the OpenMI.

The OpenMI Association will supervise the co-ordination and organisation of training sessions for model developers to help them make their modelling software OpenMI-compliant. The Association will not provide the training itself but will collaborate with dedicated third-party trainers, who can advertise their services on the Association website. Registration for training sessions will be co-ordinated through the website. Links to published training materials and tutorials will also be provided.

End users will be asked to contact their own software providers for specific OpenMI-related training. Generic examples and general information on the functionality and use of the OpenMI User Interface, as offered by the Technical Committee, will be provided by the Technical Committee as soon as resources permit.

2.5.3.5.3 Dissemination of Information

The main medium for disseminating information will be the OpenMI Association website (www.openmi.org), through which all formal information about the OpenMI Standard and the OpenMI Association will be made available. This website will also provide information on events, training and where to find OpenMI-compliant software.

Members will be allowed to advertise their products and services on the website. More informal information, primarily focused at developers, will be held at the OpenMI Association's wiki, which will be accessible *via* the website. In addition, the Association will provide a forum on Sourceforge for debate, feedback and the informal exchange of information and experience relating to the OpenMI, model integration and integrated modelling.

The OpenMI Association will convene specific sessions on the OpenMI at conferences. It will also publish papers and articles in the scientific and popular press and maintain a newsletter.

Basic material for creating OpenMI-related presentations will be made available through the website.

2.5.3.5.4 Availability of the Technology

The Association will impose no restrictions on the use of the OpenMI for research or commercial purposes, and will not impose any royalty charges or licence fees.

The OpenMI Association will maintain, develop and release the OpenMI Standard and provide access to it through its website. The Standard will be released under the Lesser General Public Licence (LGPL) (http://www.gnu.org/licenses/lgpl.html). The Association will provide all necessary documentation relating to the Standard.

To maintain the high quality and reliability of the Standard, the OpenMI Association will adopt quality assurance procedures and quality control standards. It will establish a group to monitor the driving forces, user needs and new technologies requiring further development. In this way, the OpenMI Association will seek to ensure the relevance of the OpenMI Standard. New releases will be timed to achieve a balance between the need for stability and the need to move with the times.

The OpenMI Association will ensure that there is always at least one documented Software Development Kit (SDK) and Graphical User Interface (GUI) available under suitable open-source licence conditions. It will also allow and encourage third parties to develop, deliver and maintain other versions of the SDK and GUI, or any other tools that will simplify the migration (making modelling components compliant), linking and running of integrated models.

2.5.3.5.5 Compliancy

The OpenMI Association will develop a procedure for establishing that components comply with a particular release, and will allow developers to label such components as 'OpenMI version *n.m* Compliant'. The Association will hold a register of OpenMI compliant components. The Association will not make any statement of warranty about OpenMI-compliant components. The Association will work towards establishing an auditing service.

2.5.3.5.6 Association as a Legal Entity

The Association will be an international, non-proprietary and not-for-profit organisation. New members will always be welcome, particularly those willing and able to contribute actively.

Figure 2.5.7 Logo of the OpenMI Association, which is a registered trade mark.

The OpenMI Association, as a legal entity, will safeguard the copyrights and intellectual property rights (IPR) relating to the OpenMI Standard and any related products created by the OpenMI Association. The OpenMI logo and the name 'OpenMI' have been registered as a trademark – see Figure 2.5.7; the domain names www.OpenMI.org and www.OpenMI.com have also been registered.

2.5.3.5.7 Secure Resources

The OpenMI Association will seek the resources to support its activities by membership fees, contributions in kind, donations, project funding or any other resource that is approved by the General Assembly.

To keep the required resources to a minimum, the OpenMI Association will actively encourage services to be carried out by third parties, thus creating business development opportunities for the OpenMI community.

2.5.4 Where Next?

2.5.4.1 Standards Take Time

The development and roll out of new technologies and practices, such as the OpenMI and integrated modelling, are inevitably long term and costly processes. They require sustained commitment over many years. The reasons are simple but not widely appreciated. A standard such as the OpenMI has to be written and tested to a very high level. While the initial idea can often be conceived and demonstrated in a short period, producing a robust standard that will stand up to operational use often takes many years. For the OpenMI, it took four years to deliver the first working version. However, that first version was not surrounded by the all the support and back up necessary for it to become acceptable to the user community. The user community of competent authorities, consultants and developers also need time to accommodate change. Before a regulator can introduce a new practice such as integrated modelling into its operational work, there will rightly be a long period during which the benefit of the new practice is evaluated. Only rarely will the evaluation be a priority process. For model developers, the decision to incorporate the OpenMI into their products can also take time. The OpenMI is controversial because it changes the market from one for major systems to a market for modelling components. The former lock customers in for long periods, thereby providing a long-term licence income. The OpenMI allows customers to pick and choose items from different suppliers. In the past, it was by no means clear that the OpenMI was a good idea commercially. The counter argument, which has gradually prevailed, is that if systems become linkable, then other models can be joined to them, enhancing the value of both the original system and the models added. Openness not protectionism is likely to be the key to making integrated modelling a field where the maximum number of players can benefit.

The decision making process is often extended by the funding process. Once an idea has been formed, it is necessary to seek funds. The OpenMI was fortunate in that the policy need for integrated modelling coincided with an appropriate research call and with a change in the funding policy of the European Commission. The size of projects jumped from €500000 to €6M. Without this increase, it would not have been possible to produce the professional level standard achieved so far. Also important to success was the presence in the project of commercial partners, who needed the model linking capability but individually did not have the capacity to produce it. They brought energy and drive to the team together with the business sense to start procuring the next round of funding as soon as the research phase began. This was crucial to having follow-on funds in place by the end of the research phase. There is, however, a major problem in European research and development; there is no European funding mechanism for taking research projects to operational status, especially where the main beneficiary is the public sector. The public sector does not see this as part of its responsibility.

How is the OpenMI implementing its strategy and meeting these challenges?

2.5.4.2 Implementing the Strategy

All aspects of the strategy are now being implemented. Some are simple one-off actions that have been completed and about which there is little to say, for example, creating the Association as a legal entity. The only, but important, lesson here is that had the task been handed to a lawyer earlier it would have been a very much quicker process.

The sections that follow highlight aspects of the implementation process that have been challenging or valuable experiences.

2.5.4.2.1 Creating a Culture for Integrated Modelling

There are three aspects to the Association's work in this area: demonstration to senior managers of the value of integrated modelling to integrated management, the inclusion of integrated modelling in university courses and solving the science questions raised by the new ability to link models.

To start the process of demonstrating the value of integrated modelling, seven operational problems have been identified that require an integrated approach to their management. A typical example is achieving operational efficiencies through the joint management of the rivers and sewers around the city of Leuven in times of flood. This involves linking an InfoworksCS sewer flow model to an InfoworksRS river flow model of the River Dijle in the drainage area of the town of Leuven. Although, there are strong similarities between rivers and sewers, the problem is useful because it requires the participants to harmonise the different ways in which the capacities of sewers and rivers are assessed by sewer and river modellers. Until this is done, questions such as, 'What is the effect of sewer flooding on river flooding?' cannot be answered. There are several hundred connections between the rivers and sewers, so the problem is a non-trivial test of model linkage as well. A more complex problem in terms of linking processes from different domains is presented by a requirement to investigate the water quality upstream of the Pinios river junction with its Enippeas tributary. This involves linking of two National Technical University of Athens models, a hydraulic model (RISH-1D) and a water quality model (R-Qual) with a commercial rainfall-runoff Model, MIKE-11. This problem is also an excellent test of the OpenMI standard and its supporting information. The models have all been written independently.

The other five demonstrations have been similarly chosen to explore the benefit of integrated modelling in different contexts. An essential point about these demonstrations is that staff members of the competent authorities are undertaking all the work of preparing and carrying out the demonstrations.

It is not only the managers who have to be convinced. Scientists need to be convinced too. An example of a scientific demonstration is the WADES project. Here integrated modelling is being used to evaluate the impact of climate change on ecosystem services. This is an issue of great scientific concern at present. The exercise should reveal the strengths and limitations of integrated modelling as a tool for understanding and predicting process interactions.

As the demonstrations bring models of different processes together, so a range of IT and scientific issues are emerging that require resolution. An example IT issue might be the need to link .NET models to models in the High Performance Computing (HPC) environment. The science issues cover a huge range, from numerical instability, *via* validation and optimization, to discipline specific topics. Here the Association is building links to universities and encouraging them to offer these problems to PhD students.

To encourage the flow of ideas back to the integrated modellers, several external organisations are creating modelling platforms, such as OpenWeb or CHyMp. These are essentially open clubs. The club provides linkable models and tools. Members can use these as they wish but are encouraged to contribute new components or improve existing components.

To help create the new body of integrated modellers that will be required as integrated modelling takes off, the Association is encouraging universities to include integrated modelling in university courses.

2.5.4.2.2 Supporting the OpenMI user Community

The OpenMI Association has been established and is now actively pursuing its mission of promoting integrated modelling, pushing forward the development of the OpenMI and supporting its users. The Association has a very simple structure (Figure 2.5.8).

The work of the Association is done by the Technical and Dissemination working groups (officially they are called Committees, but the title is misleading; their function is to do the work). The Technical group is responsible for developing the OpenMI, supporting its users and advising the Executive Committee. It has finite resources. Therefore, it is drawing on the creativity and experience of the user community to improve the OpenMI and to support the users. The aim is to create a mutually satisfying collaboration. At the same time, the Dissemination group is promoting the OpenMI. It does so by running sessions in conferences, publicity, facilitating training and maintaining the website, a wiki and a discussion forum. Building a community around the OpenMI is an important part of its function.

These two working groups receive their remit from and report to the Executive Committee, the main decision making body. Its primary function is to ensure that the OpenMI continues to meet the needs of its users. All key decisions relating to the OpenMI are made by it. The Committee also oversees the day-to-day running of the Association. The members elect the Executive Committee at the Annual General Meeting and it is answerable to them.

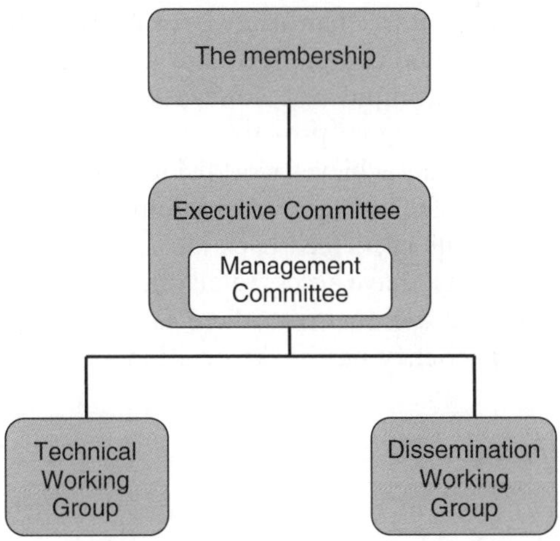

Figure 2.5.8 Structure of the OpenMI Association.

In turn, the Executive Committee elects four office holders, the Chairman, Vice Chairman, Treasurer and Secretary, who form a Management Committee.

2.5.4.2.3 Availability of the Technology

A very important lesson from the OpenMI is the crucial need to lower the barriers to entering the integrated modelling world. For the OpenMI, the initial barriers have been:

- making it easy to find
- making it easy to download
- making it easy to use.

The first two are largely a matter of having good internet skills. Making it easy to use is far harder. It will require a long and probably painful evolutionary process. In this respect, it is useful to look back over the development of word processors. The first versions were far harder to use and infinitely slower and less reliable than the state of the art typewriters of the day. Few people would wish to go back to a typewriter now. Integrated modelling has to go through the same process. Already there are encouraging signs of progress. An independent company is replacing the original model configuration editor. This was designed only to test the OpenMI and was never intended for operational use. The new editor was specifically conceived to make the life of Fortran modellers easier. Other initiatives are in hand.

The original decision to make the OpenMI open source is a good example of the right decision for the wrong reasons. There was a fear that no one would use the OpenMI unless it was open. This fear resulted from a complete misunderstanding both of the open source paradigm, how and why it works and of the potential user community. The reason that the OpenMI is open today is that the Association gains more than it gives away. By making it open, users can download and use it at no charge. If it does not meet their needs, they can improve it. Although not compulsory, convention expects that the improvements are feedback to the original provider, thus creating a virtuous circle. If the OpenMI is successful and achieves worldwide usage, then it will be able to access a pool of creativity far in excess of anything it could fund itself, even if only a tiny proportion of its users feed back ideas. An important advantage of openness to users is that they have access to all information about the OpenMI. Should the Association cease to exist, there is no threat to their business because there is no dependency on the Association.

2.5.4.2.4 Compliancy

It is valuable to developers to be able to advertise that their models are compliant. It is equally valuable for users to know which models are linkable. Therefore, gradually, the Association is introducing a compliancy service, though it may encourage others to provide the service, if it becomes burdensome. The opportunity for an outside provider is to offer a more rigorous test than will be possible for the Association.

2.5.4.2.5 Secure Resources

This is by far the most challenging task facing a standards organisation, especially at the outset, when the standard is in advance of the market. The initial research and development has been achieved with funds provided in equal measure by the industrial partners and the European Commission. It is unlikely that the Commission will continue to provide funding because the work has passed out of the development phase. It has no mechanism to fund the next stage. Therefore, the continuation will have to depend on the industrial partners until more partners can be brought in. Their participation will be dependent on their seeing a market – this why the Association has placed great emphasis on achieving wide usage of the OpenMI in the US by public bodies. That effort is now beginning to bear fruit; the major players are beginning to include the OpenMI in their systems.

CHAPTER 2.6

Stakeholder Responsive Research as an Approach to Support Adaptive Integrated Water Management: Examples from the NeWater Project

CLAUDIA PAHL-WOSTL, BRITTA KASTENS, ILKE BOROWSKI AND DARYA HIRSCH

University of Osnabrueck, Institute of Environmental Systems Research, Barbarastrasse 12, 49076 Osnabrueck, Germany

2.6.1 Introduction

In the past decade, a major change in the rhetoric surrounding water resources management has become evident. The debate is now dominated by an increased awareness of integrated management approaches, taking into account environmental, economic and social considerations, and by the search for strategies that go beyond technical end-of-pipe solutions.

Water crises have often been caused rather by problems of governance and inefficient and ineffective management than by resource or technology problems. Accordingly, more and more voices have advocated the need for a radical change, for a paradigm shift in water management. The arguments put

Water System Science and Policy Interfacing
Edited by Philippe Quevauviller

Published by the Royal Society of Chemistry, www.rsc.org

forward differ in detail and emphasis but not in the essential elements of the nature of the needed paradigm shift, which are:[1]

- move towards participatory management and collaborative decision making,
- increased integration of issues and sectors,
- management of problem sources not effects,
- decentralized and more flexible management approaches,
- more attention to management of human behaviour by 'soft' measures,
- include environment explicitly in management goals,
- open and shared information sources (including linking science and decision making),
- incorporating iterative learning cycles.

The paradigm shift in water management may be interpreted as a sign of an increased awareness of complexity and a fundamental change in understanding what management implies that is not only limited to the field of natural resources and water.[2] More recently the debate on improvements to water management has been increasingly dominated by the need for adaptation to climate change. This debate has shifted the importance of addressing increasing uncertainties to the centre stage. Experience from the past is no reliable guide for the future, and water management is confronted from all fronts with unprecedented situations.

However, current water management regimes have evolved over decades and provide in general no enabling conditions for implementing and sustaining innovative management approaches. Major structural changes are often required. One cannot expect that design and implementation of integrated and adaptive management regimes will be based on a full understanding of the interaction between the various elements forming a management regime. Essential elements are, for example, governance style, information management, risk management or technical infrastructure. Some regime properties are emergent and path-dependent, and will unfold during the implementation process. Hence the whole process of change, the transition towards integrated and adaptive management regimes has to be regarded as a kind of adaptive management process as well.

The challenges of improving the scientific foundations of adaptive and integrated water management (AIWM) and of understanding the transition towards adaptive and integrated management regimes have been tackled in the European project NeWater (New methods for adaptive Water management under uncertainty).

AIWM acknowledges the different kinds of uncertainty: that there is (and will always be) lack of knowledge on how the different parts of the water system interact, how the system works and how it will change with time. Therefore, a central part of this management approach is not only to look for new information, *e.g.* with tools to support adaptive monitoring as well as tools to better assess future development. It is equally central to improve the capacity of the actors to process this information and draw meaningful conclusions from it:

Adaptive management is learning to manage by managing to learn.

Unlike other management approaches, AIWM asks at its core for an assessment cycle that builds on the participation of all relevant actors during the management process. To become fully effective, the management process needs to encourage change in a way that is open and understandable to all actors. Joint decision making with the aim of sustainable water resources management should not be torpedoed by different capacities of stakeholders to protect their interests and/or to acknowledge the necessity of a whole-system approach. By fostering social learning collective action builds trust and social networks, both factors that are considered to be essential for the capacity of social–ecological systems to adapt to and shape change.[3,4]

Among environmental scientists, the concept of adaptive management is generally associated with the tradition developed in the field of ecosystem management. The idea of adaptive management has been discussed in ecosystem management for quite some time.[5–8] It is based on the insight that the ability to predict future key drivers influencing an ecosystem, as well as system behaviour and responses, is inherently limited. Hence management must be adaptive and include the ability to change management practices based on new insights. What has been perceived as the most effective form of adaptive management employs management programs that are designed to experimentally compare selected policies or practices, by evaluating alternative hypotheses about the system being managed (*e.g.* refs. 9–11). This implies that hypotheses can be generated and that the outcomes of experiments allow us to distinguish the comparative advantages of different hypotheses. Hypotheses refer to working assumptions that allow one to structure arguments and the policy debate. Modelling has been used for careful elaboration of the system structure and quantitative relations implied in hypotheses such that participants in the process can explore the dynamic implications of each hypothesis under different scenarios. Policy within the scientific context is seen more as an expression of how to test the hypotheses or questions generated during assessment. This experimental framework relieves decision makers of the responsibility to deliver policies as 'solutions,' but it challenges them to join the experiment and test assumptions, not defend them – an essential condition for learning.

In the context of NeWater (and this chapter), adaptive management is defined as a systematic process for improving management policies and practices by systemic learning from the outcomes of implemented management strategies and by taking into account changes in external factors in a pro-active manner. By doing so it explicitly recognizes uncertainty and complexity. Systematic process refers to a structured approach to addressing the need for learning rather than a mere muddling through. Systemic learning emphasizes the need for a holistic approach.

In the following, we present experiences on how NeWater research processes contributed to systemic learning processes by following a stakeholder-responsive research approach.

2.6.2 Scientific Background

Water management is a political process and science can only be effective when it actively takes part in it. However, it is important that science remains stakeholder-responsive and does not become stakeholder-driven and hence at risk of supporting vested interests. It has been emphasized for years that innovative approaches are required to shape the science–policy interface.

Adaptive management as developed in the 'Holling tradition' has been associated with an expert mode relationship between science and policy. Consequently, some scholars introduced the notion of adaptive co-management[12,13] to link learning by experimental approaches as promoted by the adaptive management tradition in ecosystem science and stronger sharing of responsibilities and co-decision making as promoted by collaborative management approaches. However, adaptive co-management could also be characterized as a sub-category of a more broadly defined adaptive management concept rather than as an equivalent new term. Therefore, here this distinction is not made but the term adaptive management is defined more broadly to encompass a wider range of approaches.

According the model of ref. 14 the interaction of different streams shapes the dynamics if a topic is effectively taken up in the policy process: problem (agendas), policy (alternative solutions) and politics (actors, strategies).

These streams may each show windows of opportunity at different times without overlap and hence a problem may not be tackled and scientific knowledge may not be used until the problem has reached such a level of urgency and damage that action is mandatory. However, policies for adaptive and integrated water management should operate in an anticipatory fashion. Hence it is important for science to take an active role and connect these streams.

The approach pursued in NeWater can be characterised as 'action research'. As outlined by Argyris *et al.*[15] action research produces knowledge that can be used to formulate policies and to produce action, while at the same time contributes to a theory of action. It involves carefully designed experiments – interventions into social processes to solve problems. At the same time carefully designed analyses, monitoring and evaluation allow the development of a better understanding of social processes. In proposing an action science, Argyris *et al.*[15] articulated the features of a science that can generate knowledge that is useful, valid, descriptive of the world, and informative of how the world might be changed.

This concept of action research blends well into the concept of social learning that has been identified to be of crucial importance to support change. Social learning in river basin management refers to developing and sustaining the capacity of different authorities, experts, interest groups and the public to manage their river basins effectively. Collective action and the resolution of conflicts require that people recognize their interdependence and their differences and learn to deal with them constructively. In action research, science is an active participant in such social learning processes, often taking the role of the facilitator. Science must establish for itself the role of the honest broker but not that of the issue advocate in danger of supporting vested interests.

2.6.3 Science–Policy Interface in NeWater

NeWater research has been developed – from the beginning– to meet the needs of water policy makers and practitioners. The policy–science interface was strengthened by continuous and active involvement of stakeholders such as water managers in the research processes in the different basins. At a less basin specific level they were informed and consulted during activities such as the policy–science day that took place during the first Conference on Adaptive and Integrated Water Management in Basel, Switzerland in November 2007. Further, NeWater results are presented in a multitude of ways, including 12 Synthesis Products to raise broader awareness for the concept of AIWM and the project's results.

In the following, we briefly present our main motivation and incentives for such intensive stakeholder participation and describe our experiences with it.

2.6.3.1 Stakeholder involvement in NeWater

With the rise of public participation, it becomes more and more important that water management research processes are also opening up to include non-scientists, such as water managers, planning authorities and other stakeholders. To better deal with the rising complexity of water management issues, the aim is to give them a role already during the shaping of research agendas, to generate and communicate various kinds of knowledge and to foster social learning.[16,17] Transformation processes towards adaptive management often start in so-called 'niches' at lowest administrative or geographical levels. These niches often have the advantage of deviating from the status quo and manifest themselves as new ideas, initiatives or innovative techniques.[18,19] Together with local and regional stakeholders, the NeWater project made use of such a niche by initiating participatory processes in the different river basins. Stakeholder participation in NeWater was used as a means to promote new ways of bridging science, policy and implementation of adaptive management. New methods of adaptive management were developed and tested in case studies in Europe (Elbe, Guadiana, Rhine and Tisza), in Central Asia (Amudarya) and in Africa (Nile and Orange). The involvement started at an early stage, when setting priorities in the research agenda, and focused mainly on the operational level of the basins. Emphasis was put on a stakeholder-responsive approach to guarantee that the methods and concepts developed meet the stakeholders' demands and to take into account the concerns and expertise in the basins. Further, stakeholder participation was considered as crucial in NeWater for providing a test bed for the plausibility and applicability of the new methods and approaches under different environmental and societal conditions. In the following, the processes of stakeholder participation in two NeWater case studies, one European and one non-European example, will be described in more detail. The content of Section 2.6.3.1.1 is mainly based on two publications, *i.e.* refs. 20 and 21.

2.6.3.1.1 Meeting the Challenges of the WFD: Stakeholder Participation in the Wupper Basin

The European Water Framework Directive (WFD) is the first material European law that explicitly combines environmental policy goals with public participation. It particularly demands the member states to encourage active stakeholder involvement during the Directive's implementation (Art. 14 WFD). In the NeWater project, researchers scientifically supported the implementation of a stakeholder process in the Wupper catchment. The aim was to provide a stakeholder analysis, the methodological background for participatory approaches and to give support in planning implementation and evaluation of the stakeholder process in the context of the WFD. The Wupper is a sub-catchment of the Rhine basin located in North-Rhine Westphalia (NRW), Germany. Research involving stakeholders in the Wupper basin was mainly done on two levels: the whole Wupper basin and the Dhünn basin, which is a sub-basin of the Wupper. The participatory process linked to the WFD implementation was co-organized and scientifically supported on sub-basin level (Dhünn), which also belongs to the pilot projects for implementing the WFD in NRW.

The NeWater process built on the results and the structure of previous informal round tables that were initiated by the regional water board, the Wupperverband. The novelty was the aim to involve the different parties and integrate different approaches of the various sectors located in the area in the run-up to the official participatory process of the WFD, but without being bonded to formal necessities. This allowed the creation of a 'win win co operation between scientists, consultants and practitioners' by enabling innovative methods and participatory tools of active stakeholder involvement to be used.

In more detail, the goals of the process were:

- to develop solutions for the Untere Dhünn and its tributaries to achieve the WFD's goals together with all potential actors effected by these solutions;
- to jointly discuss possible measures and compromises in the case of conflicts in the use of water;
- to jointly formulate concrete results in a consensus-document that, besides the measures, also indicates the issues of conflict;

- to encourage networks and long-term co-operations and to improve the public awareness for water management issues;
- to support learning in methods for public participation to make later official participatory processes easier.

The Dhünn process was composed of actors from various sectors, such as water management, nature and landscape conservation, agriculture, forestry, regional planning and industry. Both researchers and a group of moderators of the NeWater project accompanied the process. Besides three stakeholder workshops that took place in 2007 and 2008, expert interviews and questionnaires supported the research. Further, iterative evaluation procedures were carried out to allow continuous process improvement.

During the first workshop the different perspectives of the various stakeholders were discussed and the moderators collected open questions that were ranked according to their importance by each stakeholder. In total, 61 questions were noted and each participant was asked to answer those questions that were possible to be answered by this person. The answers were distributed among all participants before the second workshop and were used as the discussion basis for the next meeting. After a couple presentations by experts, *e.g.* from regional water boards and landscape planning consultants, during the second workshop the answers were further discussed within breakout groups and the results of these groups were fed into the plenary discussions again. On the basis of breakout group and plenary results, a final consensus-document was created and distributed among all participants, who were asked to comment on the document. The third and final workshop was used to discuss open or conflictive issues of the document and to jointly agree upon the document's content. Moreover, further activities were discussed during this final meeting.[22]

The consensus-document includes, along with a description of the process and its methods, the jointly developed thematic results and conclusions regarding both the process itself and the different content-wise aspects of water management discussed. Thematic issues were related to temperature and discharge, structural quality of the Dhünn and its tributaries, continuous water flow and further development goals for river banks. The process' results fed into the official participatory processes of the WFD implementation in NRW.

The process at the Dhünn took place outside formal decision-making structures, which allowed a higher flexibility in process structure. In the literature, such structures are considered as one of the preconditions for transformation processes,[23] since they allow the emergence of informal networks facilitating learning processes by information and knowledge flows.[24] Concrete outcomes in the Dhünn process were less linked to formal necessity, thus allowing a higher flexibility in the process structure and its outcomes. The process opened the way for participatory experiments outside the boundaries of existing patterns, *e.g.* by involving new 'methods of recognizing and responding to emerging issues by organizing such an informal participatory platform'.[25]

The joint conclusion of the process in the consensus-document was that the discussion of the open questions regarding the WFD and possible measures for its goal achievement allowed all actors to get to know the various perspectives of the different stakeholders in the region. The participants also emphasised the importance of being more informed by a user-friendly form of documents and presentations that helped to better understand both the demands of European water management and the ecological and socio-economic processes in the Dhünn catchment. This also helped to balance the different stages of knowledge the stakeholders had in the beginning. Further, the participants complimented the open process atmosphere and the possibilities of bringing in their own views. The regular evaluation of the meetings allowed them to contribute to the iterative improvement of the process.

Stakeholders saw the limited number of workshops as drawback – three meetings were considered too few for a very detailed development of measures under consideration of the various uncertainties in water management. The participants highlighted that it was nevertheless possible to collect various suggestions for measures and at least to decide which measures are worth following up for future planning. Many stakeholders underlined that they will use the new and improved networks with other participants to further follow the developed approaches for water management to be jointly prepared for the official process of measurement planning during the WFD's implementation at the local scale.

2.6.3.1.2 Experiences from the Amudarya Case Study

Stakeholders of the Amudarya basin had been involved in the design and implementation of the research from the beginning of the project. The process in the Lower Amudarya river basin was planned to generate feedback to the water managers on the ongoing work in NeWater and to jointly develop an understanding of the features of the current regime and possible interventions for improvement and changes in water management.[26,27] The aim of this approach was to elicit a wide range of stakeholder perspectives and preferences and to promote horizontal and vertical cooperation among stakeholders and experts.

The participative approach serves as an example of how to support the development of new water management policies by identifying potentials and gaps in the current regime through an integrative approach that takes the various views of stakeholder into account.[27] The series of workshops also intends to improve the communication between different levels of water management, water management authorities and water users.

The thematic scope of the participative exercise was three predefined topics – improving water and soil quality, improving the provision of ecosystem services, and improving coping with extreme events. The timescale was on the short term, *i.e.* within the next 5–10 years, but also intended to go beyond the planning towards developing visions of what a future water management could look like.[27,28]

Eight main workshops attended by 30–40 stakeholders each and ten smaller ones focusing on specific topics were carried out on a regular basis from 2005 to 2008. The main workshops brought together people from different management levels, different interest groups, different sectors and backgrounds.

Cognitive mapping and group model building were used during the workshops since they had proven successful to engage people, encourage them to express their opinions and contribute their knowledge in a rather informal and non-hierarchical way, avoiding some of the common problems of workshop settings in the former Soviet Union republics (*e.g.* lack of open discussions, very hierarchical contributions and involvement).

The NeWater Team invited relevant stakeholders either from one type of organization, which are subordinated to each other (*e.g.* the Ministry of Agriculture and Water Management, its province and district departments), or from those organization that could eventually work together but currently do not make use of cooperation (*e.g.* NGO, water management organization, nature protection organization, water users, fishermen), and who are relevant for a certain thematic issue.

In the small workshops the focus was on individual themes such as transboundary information exchange, social aspects of water management or environmental flows. Here European and local project partners analyzed the stakeholders, their roles, views and adaptation options at different scales of the river basin from the transboundary to the local level in the lower reach of the Amudarya river. Moreover, additional information (database, documents, references, former studies and surveys) was collected on topics – such as governance, institutions, participation, transboundary water management, new approaches of information management, transition to adaptive water management regimes (climate change and barriers for changes), water quality, quantity, ecosystems, incorporating environmental flows into integrated water resources management, effect of water management on livelihoods and local population – that were then addressed in stakeholder meetings.

The workshops resulted in suggestions for measures that will inform ongoing research activities, but also should be transported to the management and policy making levels. For instance, measures to cope with drought and enhance the preparedness for extreme events suggested by the participants for Uzbekistan included:

- improvement of water management laws;
- observation of water quotas within Uzbekistan and among the Central Asian countries;
- increase of the authority of water practitioners to separate their decision-making mandate from the agrarian practitioners due to development of joint water policy that enclosed all water related issues;
- introduction of water saving measures (economic, legal, technical), increase in water use efficiency, change in cropping patterns;
- capacity building of water practitioners and water users;

- decrease water discharge in the middle reach of Amudarya to guarantee better quality irrigation water.

Most of these measures are concrete suggestions on what the contacted stakeholders can do to improve water management based on this exercise and other activities of the project.

None of the presented exercises were directly linked to a policy process. Nevertheless, we think that some of the results may inform ongoing policy processes. In the Amudarya river basin, representatives of local administrations became interested in the activities and their results as a means to learn about the views and needs of their stakeholders and about measures that are acceptable to them.

One of the major achievements in the Amudarya Case study was the involvement of stakeholders at different levels in workshops and the successful application of participative methods in the Uzbek political and cultural context. Many of the participants appreciated the possibility to voice their ideas and concerns and the holistic approach that the methods used made possible. However, we have not succeeded in actually demonstrating how stakeholder views can be taken into account in decision making. Given the current political situation in the river basin this will be a long process; however, some attempts have been started at the local level. It is those processes we target with much of our research and stakeholder activities. We create awareness for uncertainties and show the need to take them into account and develop strategies to cope with them. Also the need to take the views and needs of diverse users into account and find ways to incorporate them into management decisions was spread due to NeWater work in the Amudarya Case study.

2.6.3.2 CAIWA Science-to-Policy Day

In November 2007, the NeWater project initiated the first international conference on adaptive and integrative water management (CAIWA). The CAIWA, which took place in Basel, Switzerland, included a dedicated 'Science-to-Policy Day'. The aim was that scientists, policy makers and practitioners from different disciplines and levels analyse progress, explore new research directions and eventually highlight policy implications of scientific findings on adaptive and integrated water resources management in the light of climate change. The insights of the policy day are summarized in a NeWater policy brief.[29]

The policy day was composed of a start-up plenary session, seven panel discussions with either issue or regional focus and a closing plenary to synthesise and wrap-up the main results. Panel discussions took place on the Rhine and the Guadiana river basins, international cooperation between IWRM & EU, with the example of the Orange river basin, water quality and ecological status, water scarcity and droughts, flooding and governance and institutional aspects. From guiding questions regarding the most harmful practices in river basins, the panel discussions emphasized the need to increase adaptiveness and, greatest of all, needs in co-operation between research and policy-making.

One main conclusion for policy makers and practitioners relates to collaboration and management. It was emphasised that policy-makers and practitioners must increase their cooperation efforts to balance the differences in understanding present opportunities and to increase knowledge. This is also the case for differences in comprehension of 'good water management' and 'good responses to climate change', where economic, social and cultural backgrounds are crucial for making sense of the different meanings of these terms. Further, it was considered as very important that knowledge gaps should not be used as an excuse for delaying action against climate change.

Communication and participation, *e.g.* in the form of the involvement of society in the debate on finding solutions to climate change and its impacts, was emphasized as a key factor. To this end, mechanisms are needed that allow linkages between the technical discourse and the political dialogue at multiple administrative levels and between multiple sectors. Participatory methods are considered as supportive here.

To better deal with uncertainty the overall motto should be 'learning to live with uncertainty and being comfortable with it'.[29] Long- and short-term planning has to be combined by policy and practice. Moreover, recent experiences with extreme events show that, still, probability-based and reactive risk actions seem to be given priority against integrated management styles, where anticipated impacts are included, even though they do not (yet) have statistical evidence.

Regarding the linkages between researchers and policy-makers, research can be more effective when research duplication is avoided, *e.g.* by a better coordination between countries. In particular, research on climate change adaptation in developing countries has to be increased. Here, new means of channelling donor money need to be identified, including a shift of budget responsibility to these countries.

A further issue discussed referred to new implementation strategies, as wide deficits have been experienced in the implementation of environmental policies and particularly also already in the implementation of the WFD.[30–32] The political willingness to ambitiously follow the comprehensive implementation of the water management policies has thus to be ensured over all stages and levels of implementation. In this context, it also has to be acknowledged by all actors involved in the implementation procedure that water management is a cross-sectoral issue and that so are adaptation actions towards climate change. The WFD's statement that water is more than a commodity (preamble No. 1 WFD) was jointly underlined. A pivotal step to foster the policy–science interface during the implementation of water management policies would be the support of demonstration activities that consider all aspects of adaptive management, the socio-ecological environment, the technical infrastructure and the management style itself. The latter includes organisational adaptability and the analysis of the existing administrative structures with regard to their adaptive system capacities. The pilot projects conducted during the implementation of the WFD can serve as an example on which such demonstration activities can build.

The key messages for researchers first of all referred to the form of information and knowledge as requested by policy makers and practitioners. One main message formulated was that emphasis in research should be put on supporting the understanding of organisational cultures, amongst others, to enable practitioners to effectively deal with the inevitable interlinks of sustainability and climate change adaptation. Supportive means by research could be the simulation of behavioural change, the engagement in demonstration projects and, particularly, also on how to involve stakeholders and the public when dealing with uncertainty.

Moreover, collaborative research and communication needs to become more specific in terms of solutions provided and actions recommended. Currently, the policy-oriented results delivered by research are often too abstract for policy-makers to be translated into concrete actions.

One means to achieve this could be to include policy-makers and practitioners more intensively in the design of applied research. Another means is to take better notice of which data and information is really relevant to policy makers and practitioners. Moreover, co-operation with private sector research should be increased to achieve stronger synergies. Finally, researchers should be aware that science might take a major role in monitoring progress in implementing IWRM.

2.6.3.3 Synthesizing insights for Water Policy Makers and Water Managers

Taking the experiences throughout the project, NeWater realized that, especially towards the end of project time, it is important to raise again the awareness for the project's results and to devote efforts to tailoring products to target groups and current needs of policy processes. For this reason, the central insights from the NeWater project have been synthesised into 12 Synthesis Products (SPs), some of which explicitly address water policy makers and water managers. For example, to ensure that the current 'burning' policy questions are addressed by the project one of the 12 synthesis products in NeWater 'merged' the present 'burning' EU water policy questions linked to adaptive water resources management and climate change with key insights from the NeWater project. The questions were identified during interviews with seven EC policy makers. They confirmed two central areas of interest on the policy side for AIWM. The first refers to the need for more information (assessing status of, trends in and impacts on water environment), the second concerns the need to both reflect on and improve the approach to water management (water governance) as central elements for approaching the different aspects of uncertainty and complexity. In general, researchers face the expectation from water management of providing information and concepts in these areas. In the context of NeWater, this has been responded to by different synthesis products, which are introduced below.

2.6.3.3.1 Role of Science in providing Information on Water Environment: Assessing Status, Trends and Impacts

Scientific knowledge on the interaction between different parts of the water system, on pressures and impacts and on the effectiveness of measures is a key issue for water management and decision-making. For instance, strong demands are expressed by water policy makers for knowledge on the impact of climate change and related uncertainties, including a good understanding of the hydrological and ecological context related to vulnerability, uncertainty, *etc.*. NeWater's focus on the issue of climate change reflects, in the case studies, the perception of changes and the lack of knowledge on how to assess the development of climate change. Scenarios have been used here as one tool for support. Special attention to climate change is given in three synthesis products (SP). A climate and water adaptation book (SP4) provides insights into different adaptation strategies for climate change. SP5 presents an evaluation of water resources scenarios for the case study regions, taking up most recent results of climate development simulations. Finally, a special effort was undertaken to cross compare the adaptation strategies across regions in terms of the differences in awareness and strategies with regard to climate change. Furthermore, much effort has been made to describe the approach for analysing dynamic vulnerability and adaptive capacity of both the ecological system and the social system (SP 7).

2.6.3.3.2 Science providing Concepts and Support for Management: Water Governance – Dealing with Uncertainty, Integration, Public Participation

In the face of a critical public, water managers are tempted to focus on the generation of sufficient information to make the 'right' decisions. The complexity of the social–ecological water systems demands also to reflect on the way these decisions are reached. Two themes are central in this context: the approach to uncertainties inherent in the socio-ecological system or due to lack of (or inaccessible) information on the system. NeWater tackles this in Uncertainty Guidance (SP2), providing insights into how the different types of uncertainty can be approached in adaptive management processes. Another crucial element is the establishment of monitoring systems that allow gathering information on the water system to reflect development and changes, including the impacts of measures on the water system. However, establishing monitoring systems is often (perceived as) resource intensive: devices are needed to collect, to store and to analyse data. Often, (similar) data is required at different scales for different purposes, *e.g.* allowing the monitoring of the state of a water system for the whole basin or in contrast to a water body. Centralized (*e.g.* national) data collection often provides only limited support to decentralized management approaches. Monitoring systems are, for example, set up to support over-regional reporting needs, thereby losing touch with local water managers and thus not providing them with much support. In response, NeWater's focus on monitoring was on developing a GIS-based advanced monitoring and information system (AMIS) that can integrate and structure local and expert knowledge. In the Amudarya case study the developed monitoring methodology addresses the topics of soil salinity assessment and monitoring, and of wetland ecosystem monitoring. In both cases local knowledge was used to amplify the existing environmental database.

2.6.4 Conclusions

NeWater has invited water managers and policy makers to engage with research processes on various occasions, both to enrich the practical relevance of research and to provide a 'safe space' that allows for (experimental) changes in the approach to the water system. The project makes this experience-based and thus rather intense and individual policy–science interface accessible to further water managers who want to learn from it through a series of training and education materials. For example, the *Guidebook on Adaptive Water Management* explains the benefits, outcomes and lessons learned from adaptive water management. Training materials build the capacity of those who 'train' the water managers, while the online curriculum provides academic instructors with a curriculum for the purpose of teaching adaptive river basin management. In addition to the project's website (www.newater.info) the results are also distributed through the WISE-RTD (Chapter 4.2) and the Global Water Partnership to multiply the different access points.

Communication across the science–policy interface cannot be limited to glossy brochures, but needs shared practices to find new ways towards transition in water management. NeWater has interpreted its role at the policy–science interface as twofold: first, stimulating capacity building and learning among those water mangers directly involved in participatory management processes. Second, NeWater has provided information on the socio-ecological water system. Feedback of the involved stakeholders shows that, especially, the first approach is more efficient in the adaptation of perspectives and learning. In particular, research on AIWM needs to take this up strongly to enable better the necessary processes of transition.

Collective action with stakeholders and support to bridge the gap between science and policy in research projects such as NeWater can only be seen as a beginning. The time horizons of research projects are relatively short compared to the long-term implementation of European policies. This applies all the more where uncertainties in the environmental system are addressed and institutional adaptation has only started to be developed, as is the case with climate change issues.

The strength and future perspective of projects like NeWater lies, *inter alia*, in the (further) development of various methods and tools to support processes of learning and scenario planning to take into account uncertainties. Particularly active research in this field offers great potential for further applications to support, for example, development and implementation of adaptation strategies for climate change, where barriers often reside in communication problems.

References

1. C. Pahl-Wostl, The importance of social learning in restoring the multifunctionality of rivers and floodplains, *Ecol. Soc.*, 2006, **11**(1), http://www.ecologyandsociety.org/vol11/issl/art10.
2. C. Pahl-Wostl, Transitions towards adaptive management of water facing climate and global change, *Water Resources Manage.*, 2007, **21**(1), 49–62.
3. C. Folke, J. Colding and F. Berkes, Building resilience and adaptive capacity in social-ecological systems, in *Navigating Social-Ecological Systems*, ed. F. Berkes, J. Colding and C. Folke, Cambridge University Press, Cambridge, 2003, pp. 352–387.
4. C. Folke, et al., Adaptive governance of social-ecological systems, *Annu. Rev. Environ. Resources*, 2005, **30**, 441–473.
5. C. S. Holling, (Ed.), *Adaptive Environmental Assessment and Management*, John Wiley & Sons, Ltd, Chichester, 1978.
6. C. Walters, Adaptive management of renewable resources, in *Biological Resource Management*, ed. W.M. Getz, MacMillan, New York, 1986, p. 374.
7. C. Pahl-Wostl, *The Dynamic Nature of Ecosystems: Chaos and Order Entwined*, John Wiley & Sons, Ltd, Chichester, 2005, p. 288.
8. K. N. Lee, Appraising adaptive management, *Conservation Ecol.*, 1999, **3**(2), 3.
9. L. H. Gunderson, *Barriers and Bridges to the Renewal of Ecosystems and Institutions*, Columbia University Press, New York, 1995, **XIV**, 593 S.
10. B. D. Richter, et al., Ecologically sustainable water management: managing river flows for ecological integrity, *Ecol. Appl.*, 2003, **13**(1), 206–224.
11. C. F. Kiker, J. W. Milon and A. W. Hodges, Adaptive learning for science-based policy: the Everglades restoration, *Ecol. Economic*, 2001, **37**, 403–416.
12. P. Olsson, C. Folke and F. Berkes, Adaptive comanagement for building resilience in social–ecological systems, *Environ. Manage.*, 2004, **34**(1), 75–90.
13. R. Plummer and D. Armitage, A resilience-based framework for evaluating adaptive co-management: Linking ecology, economy and society, *Ecol. Economics*, 2007, **61**, 62–74.
14. J. W. Kingdon, *Agendas, Alternatives and Public Policies*, HarperCollins, New York, 1995.

15. C. Argyris, R. Putnam. and D. McLain Smith, *Action Science.*, Jossey-Bass, San Francisco, 1985.
16. J. Newig *et al.*, The role of formalisation, participation and context in the success of public involvement mechanisms in resource management, *Systemic Practice Action Res.*, 2009, 21(6), 423–441.
17. S. O. Funtowicz and J. R. Ravetz, Science for the post-normal age, *Futures*, 1993, **25**(7), 739–755.
18. R. Kemp, J. Schot and R. Hoogma, Regime shifts to sustainability through processes of niche formation: The approach of strategic niche management, *Technol. Anal. Strategic Manage.*, 1998, **10**, 175–196.
19. R. van der Brugge and J. Rotmans, Towards transition management of European water resources, *Water Resources Manage.*, 2006, **21**(1), 249–267.
20. Seecon Deutschland GmbH, Institute of Environmental Systems Research/ University of Osnabrück, and Wupperverband, Participative planning of measures in water management. Results of the Workshop 'Untere Dhünn' (in German). Available online at http://www.wupperverband.de (08.01. 2009) (2008).
21. S. Moellenkamp *et al.*, Linking adaptive water management and stakeholder participation. Insights into niche-finding, collaborative design and outcomes from a case study in the Rhine basin, *Ecol. Soc.*, in review.
22. K. Speil *et al.*, Systematische Gestaltung eines partizipationsprozesses: Ziele, methoden und herausforderungen am bespiel der Workshops 'Untere Dhünn'. UVP-Report, in *Umweltprüfung und wasserwirtschaftliche Planung*, 2008.
23. L.H. Gunderson et al., Water RATs (resilience, adaptability, and transformability) in lake and wetland social-ecological systems, *Ecol. Soc.*, 2006, **11**(1), 16. Available online: http://www.ecologyandsociety.org/vol11/iss1/ art16/ (2006).
24. P. Olsson *et al.*, Shooting the rapids: navigating transitions to adaptive governance of social-ecological systems, *Ecol. Soc.* 2006, **18**. http:// www.ecologyandsociety.org/vol11/issl/art18.
25. S. L. M. Moellenkamp and E. Ebenhöh, Institutional elements for adaptive water management regimes. Comparing two regional water management regimes in the Rhine basin, in Adaptive and Integrated Water Management. Coping with Complexity and Uncertainty, ed. C. Pahl-Wostl, P. Kabat and J. Möltgen, 2008, pp. 147–166.
26. M. Schlueter (ed.), Research Action Plan Amudarya, Report of the NeWater project-New Approaches to Adaptive Water Management under Uncertainty, Deliverable D 3.6.1, 3.6.2, D 3.6.3, NeWater Report, Leipzig, 2006.
27. D. Hirsch *et al.*, Public participation in water management in Uzbekistan: How the peculiarities of the region are taken into account in participative processes within an international research project, *Ecol. Soc.*, in review.
28. D. Haase *et al.*, Enhancing stakeholder participation in river basin management using mental mapping and causality models in the Tisza, Orange and Amudarya basins, *Ecol. Soc.*, in review.

29. D. Ridder and C. Pahl-Wostl, Key messages of the CAIWA conference 2007, in *NeWater Policy Brief No. 7*, 2007.
30. C. Knill and A. Lenschow, (Eds.), *Implementing EU Environmental Policy: New Directions and Old Problems*, Manchester University Press, Manchester, 2000.
31. B. Kastens and J. Newig, The water framework directive and agricultural nitrate pollution: will great expectations in Brussels be dashed in Lower Saxony?, *Eur. Environ.*, 2007, **17**, 231–246.
32. C. Demmke, Towards effective environmental regulation: innovative approaches in implementing and enforcing European environmental law and policy, in *Capacity Building for Integration. European Environmental Policy: The Administrative Challenge for the Member States*, ed. C. Demmke and M. Unfried, European Institute for Public Administration (EIPA), Maastricht, 2001, pp. 81–311.

Section 3: Links to Water National or Regional Research, Policies and Management

CHAPTER 3.1

Building a National Strategy on Science–Policy Interface in Support of Water Policies – The Case of France

MARIE-PERRINE DUROT AND PATRICK FLAMMARION

ONEMA, the French National Agency for Water and Aquatic Environments, Hall C, 5 square Felix Nadar, 94300 Vincennes, France

3.1.1 Introduction

Stakeholders in the water sector, whether they are national authorities, regional and local governments, companies, or research institutes, share common long-term visions and goals, but there is a huge need for coordination among them. These actors interact at various geographical and administrative scales and management of these multiple-scale interactions is essential. For example, the Water Framework Directive (adopted in 2000) has been adopted at European level, while its implementation depends upon local and regional aspects. Conversely, certain regional issues are very general in nature and may require relays at national and European levels. A good coordination mechanism between scales and actors is a pre-condition for the water sector to be able to respond to major driving forces, such as climate change, globalisation, demographical increase, quantity and quality water stress. In light of the above, there are also great expectations concerning research and development transfer

Water System Science and Policy Interfacing
Edited by Philippe Quevauviller

Published by the Royal Society of Chemistry, www.rsc.org

systems between scientists and policy-makers and implementers at every level, *i.e.* local, catchment basin, national and European. The links between research, the formulation of public policies and their implementation in the field are key factors in reaching the ambitious goals set by France and Europe for sustainable water management.[1]

3.1.2 Plans for Research and Development Support of Public Water Policies in France

The Water framework directive (WFD), adopted on 23 October 2000, aims to establish overall consistency between the tools and methods set up to protect water resources in different zones or according to uses, and to take action against the release of dangerous substances. The WFD strongly calls towards an integrated water resources management (IWRM) approach at the level of each river-basin districts.[2] To this end, river basin management plans (RBMPs) have been prepared in each district and provide a framework for the implementation of the Directive.

The WFD imposes an *obligation* to produce results in terms of the good chemical and ecological status of water resources by 2015. The return to good condition and protection of aquatic environments are essential steps to ensure the continued use of water, whether for supply of drinking water, economic activities or recreation. Though exceptions to the 2015 deadline are possible (to 2021 or 2027), they must be justified by an overriding public interest and reasons having to do with the 'technical feasibility' or 'disproportionate costs' of the measures required to reach good ecological status.

The river basin management plans must be reviewed every six years, taking into account any new knowledge. Similarly, technical and economic studies must be updated along the same period. This *cyclical* calendar places considerable pressure on all stakeholders in the water sector, proportionate to the importance of preserving the water resources of Europe.

In France, the 2006 law on water and aquatic environments acknowledged this situation and the related challenges facing water stakeholders. It highlighted the need to create the means for *high-level scientific* and *technical assessment* in the field of water and aquatic environments to support implementation of public water policy. The law set up the National agency for water and aquatic environments (Onema, see Box 3.1.1) with the aim to preserve water quality and good ecological status of aquatic systems. To this end, Onema has a role to play in integrating the action of the various stakeholders involved in the water sector, from the researchers in their laboratories to the people in the field. By assigning an explicit task in research and development to Onema, lawmakers wanted to reinforce the role of scientific and technical knowledge in supporting the formulation and implementation of public policy in the water sector.

Box 3.1.1 Onema, Playing the Role of an Interface

Onema is a national agency active in the field of environmental public service. Overseen by the Ministry in charge of ecology, it organises and produces high-level scientific and technical advice to assist in formulating, implementing and evaluating public water policy. Its mission is to contribute to overall and sustainable management of water resources and aquatic ecosystems, with the view to restoring water quality and reaching the good chemical and ecological status set by the European Water framework directive, adopted on 23 October 2000.

Established in April 2007, Onema has launched a three-part strategy:

- *understand*, through its research and development support activities;
- *learn*, by collecting data, monitoring and supporting the water information system;
- *act*, through its technical support in the field of ecosystems management for other stakeholders in the water sector, and *via* its capacity to police water uses.

Onema is a participant in implementing the national strategy for water and freshwater/coastal environments. Onema works closely with Governmental bodies on the European, national and local levels, as well as with other public agencies, notably the River basin agencies. It also develops partnerships with research organisations and maintains contacts with water users such as companies, environmental-protection groups and fishing federations. It has 900 personnel, both at the headquarters (near Paris) and in the field through the local implantation of Onema in nine inter-regional offices and 77 local offices.

The combination of research, development and the monitoring system is the starting point for a win-win relation between researchers and the people in the field. The goal is to prepare the design of future methods and observation systems by serving as the *interface* between information-system management and the scientists who are the most apt to make use of the data, notably *via* modelling. Knowledge is required for action. To reach the objectives of public policy, *i.e.* good ecological status, a halt to losses in biodiversity, sustainable management of water resources, good governance of water services and treatment, we must be capable of measuring the differences between the current situation and the goals, and thus organise the action on the local, national and European levels. Similarly, the *combination* of policing and monitoring responsibilities, calling on the know-how of professionals in the field, is the means to design tools suited to actual conditions and specific aspects of each territory.

By combining these various elements in a national agency that can profit from strong local contacts and key partnerships, France has set up an interface to meet the challenge of a research and development sector needed to assist public decision-making and action in coming to grips with the future issues in the water sector.

The recent 'Grenelle environmental process' in France has confirmed the national ambitious targets for water and aquatic-environment quality, namely good ecological status of water with particular weight placed on the concept of green and blue corridors to interconnect the areas supportive of biodiversity, protection of wetlands, 50% reduction in pesticides by 2018, efforts concerning dangerous chemicals, phosphates and medical residues, identification of the drinking-water resources the most at risk, upgrading of wastewater treatment stations, *etc.*[3] These heightened ambitions have produced a notable acceleration in schedules for people in the water sector and made clear, in terms of research, the need to organise, raise awareness and disseminate more efficiently existing work. The importance of managing and organizing the research and development sector by bringing people together was further underlined to:

- *develop new knowledge* (biodiversity, adaptation to climate change, dangerous substances and mixtures, ecotoxicology, chemical and biological markers, risk assessment, ecological continuity, the specificities of overseas territories, *etc.*);
- *experiment and develop new techniques* (wastewater, treatment of emerging pollutants and polluted sediments, rainwater recovery, use of rain or treated water, reloading of aquifers, *etc.*);
- *develop research bringing together the human and social sciences* (to ensure the suitability of technologies and their management techniques, balanced and appropriate management of water resources and aquatic environments, assessment of the cost effectiveness and benefits of management policies, *etc.*).

In the end, the Grenelle environmental process clearly confirmed the need for *better links* between short- and mid-term operational issues and the major scientific issues situated at the crossroads between several disciplinary and topical approaches. The goal is to reinforce the ties between scientific results and how those results are used and transferred. The priority is to organise relations between the people using the results and the know-how that exists or must be developed in the scientific community.

3.1.3 The Research and Development System and its Management in France

In France, the scientific and technical personnel working on research targeting directly useable results in the field of water and aquatic environments represents approximately 3000 FTE (full-time equivalent jobs in the public sector). These personnel are split among about 80 structures, (public research institutes, universities and so on). Five national organisations represent one half of the total, the remainder being scattered in small research teams (less than 30 people), notably in the universities. In addition, no major regional cluster can really be identified: no region concentrates more than 20% of the workforce and no world class clusters dedicated to water yet exist in France. Teams tend to concentrate around specific platforms or in observation and experimentation zones, *e.g.* in the Environmental-research observatories or in certain major catchments such as the Seine, and the Rhône, or work together on specific research topics. Unfortunately, these efforts to regroup have not reached the critical mass required to influence the major national and European decisions in the field. The need for strategic leadership has often been expressed by researchers, notably those in the smallest structures.

Another important aspect of the national research and development system concerns the programming tools capable of managing and structuring the research community as well as the links between these tools and the expectations expressed by stakeholders. In France, there are several programming tools for the water scientific community. On the national level, the National research agency (ANR) runs a minimum of nine research programs topically oriented but addressing partially the water sector. In addition, there are several research programs set up by the Ministry in charge of ecology and the programs co-funded by the regions. The European funding programs, *e.g.* the Framework program for research and technological development (FP7), see Chapter 1.3 of this book, and the more applied programs such as the financial instrument for the environment (LIFE and LIFE +) constitute other sources of financing and structure for French scientific teams that enable them to take part in European projects and contribute to shared issues. Finally, Onema and the River basin agencies handle another aspect of the water research and development programming system, focusing on the research needs in terms of management and forward-looking analysis on the national level and that of the catchment basins. They orient activities through framework agreements and contracts for research and studies, while maintaining subsidiarity with respect to the other research funding agencies. The funding granted by the River basin agencies and Onema for research and studies is estimated at 13 M€ per year, *i.e.* on a par with research-support credits attributed by ANR to water and aquatic-environment topics (11.5 M€ per year from 2005 to 2008), to which the Ecology ministry adds approximately 1.5 M€ to fund various programs. It is estimated that the annual financing of the European commission to French scientific teams working in the water sector is a little under 10 M€.

3.1.4 Building a Strategy for National Research and Development on Water and Aquatic Environments Targeting Operational Stakeholders

The goal is to enhance knowledge through research and study, to support public and private stakeholders with scientific advice, to develop engineering tools and to make knowledge available in the form of summarised data, technical guides and training courses.

3.1.4.1 Coordination between the National and Catchment-basin Levels

In view of gaining knowledge, as well as protecting and monitoring water and aquatic environments, Onema offers a framework designed to rationalise on the national level the various activities concerning studies, scientific advice and research and development, and to mutualise certain aspects of their implementation. One particular goal is to launch concerted scientific and technical efforts between the River basin agencies (Box 3.1.2) and Onema.

Box 3.1.2 River Basins Management in France

Since the 1964 Water law, France has been divided into six river basins covered by six River basin agencies (Figure 3.1.1). The agencies are in charge of implementing river basin management plan (RBMP/SDAGE) for their basin while promoting balanced and sparing organisation of water resources and aquatic environments, supply of drinking water, high-water regulation and sustainable development of economic activities. The multi-year river basin management plan and program of measures of each agency sets the conditions for its activities as well as the financial resources for action and work in the public interest in the basin.

The governing body of each agency is *the basin committee* made up of stakeholders, public and private, in the water sector (consumers, elected officials, the State). It debates and collectively defines the major policy guidelines for water resources and the protection of aquatic environments in each major river basin. The novelty of this entity lies in its geographic zone based on actual territorial considerations, the river basin, in its specific mission to collectively debate and decide on policy, and in its wide ranging and diverse composition.[4] There are seven basin committees in France (Adour Garonne, Artois Picardie, Corse, Loire Bretagne, Rhin Meuse, Rhône Méditerranée, Seine Normandie) and four others in the overseas territories (Guadeloupe, Martinique, Guyane and Réunion).

Objectives are to jointly develop tools and methods designed to meet the needs of managers and stakeholders, to ensure synergy between the studies and research carried out or supported at the basin and national level, to limit any redundancies among initiatives taken in the various catchment basins to meet needs not met elsewhere and to make efforts to distribute and ensure effective use of the results obtained.

River basin agencies and Onema have built a common strategy to better organise and detect synergies in the research and development work that they

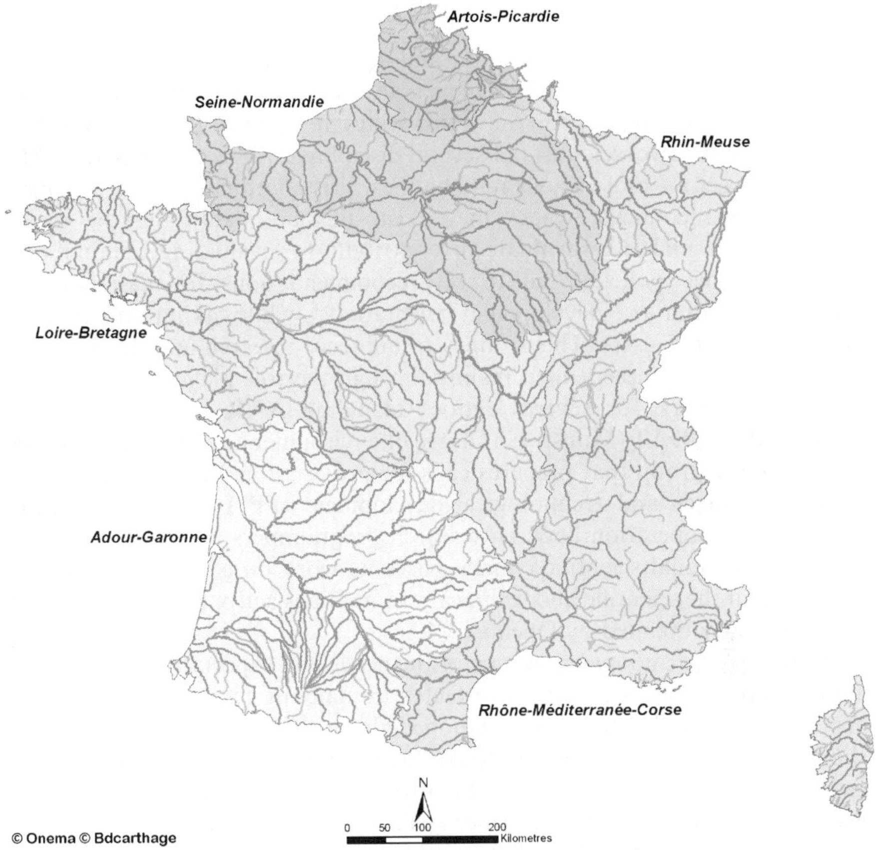

Figure 3.1.1 Territory covered by each of the six River basin agencies.

manage, starting with identification of problems and needs and continuing with the creation of new knowledge, the development of tools to assist in effectively implementing measures to restore and protect aquatic environments, and finally with assessment systems to evaluate the environmental and social-economic effects of the measures.

3.1.4.2 A Research and Development Strategy Interfacing with Stakeholders

On the basis of this partnership between the national and basin agencies, a shared research and development strategy on water and aquatic environments has been set up. It consists of three parts:

1. Forward-looking analysis, strategy and assessment
2. Research programming and funding
3. Transfer and best use of research results.

3.1.4.2.1 Forward-looking Analysis, Strategy and Assessment

A national scientific council on water and aquatic environments was recently established and is supported by Onema. It deals with strategic planning of research and development for the water sector in France. It is composed of twenty-four high-level scientists of several disciplines, representing both public and private sectors, and various nationalities. It is coordinated with the river-basin scientific councils, where they exist. Beyond identifying the major challenges, the goal is to evaluate or contribute to 'evaluating national and European public policies and to set recommendations for national and European research and development orientations', while taking into account the expectations of users. This strategy to influence research and development policy priorities is implemented mainly through the role played by Onema in the planning work for several entities including ANR (French Research Agency), MEEDDAT (Ministry in charge of environment, research grants), the national topics group that prepares calls for projects for the Framework program for research and technological development (FP7) and the European networks for cooperation between research programmers and funders (ERA-NET). The national scientific council bases its work on a national and European map of the scientific and technical communities active in the water sector that is regularly updated by Onema.

Attentiveness to and *understanding* of the needs and expectations of research and development users is a decisive factor in planning research and development for the scientific challenges mentioned above. The goal is to identify needs (short-, mid- and long-term) and make them known to the suitable research institutes on the European, national and transnational levels as well as to its own research and study units. Further goals to that end are *participation* in the French groups of research and development users (*e.g.* the WFD working groups organised by the Ecology ministry, groups of electricity producers, scientific and technical associations) and European groups [groups involved in the WFD Common implementation strategy (CIS), standardisation groups, *etc.*], as well as *consulting* users during preparation of research programming undertaken by Onema, River basin Agencies and the research institutes. Elsewhere, goals include designing new and participative 'bottom up' approaches to identify needs for multi-sectoral and multi-disciplinary research.

3.1.4.2.2 Research Programming and Funding

To address the major issues in the water sector, Onema and River basin agencies have set up multi-year programs. The priority topics are listed below:

1. *The dynamics of water resources in the context of climate change* (resource availability, hydrology of low-water periods, quantitative management, *etc.*)
2. *Physical restoration of hydrosystems* (sediment dynamics, sediment transit, hydromorphology, ecological engineering, *etc.*)

3. *Dynamics of aquatic living communities* (dynamics of species subjected to climate change and anthropogenic pressures, evaluation tools, restoration engineering, adaptation of methods to specific overseas conditions)
4. Chemical contaminants and ecotoxicological risks.
5. *Environmental technologies* and methods to reduce concentrated and diffuse forms of pollution
6. *Social-economic value of water and aquatic environments*, social-economic tools for water management, evaluation tools for public water-management policies.

In this context, joint research and study teams have been set up between Onema and research institutes/universities on a couple of specific topics, *e.g.* on the ecological continuity of rivers or on analysis and modelling of interaction between the physical, chemical and biological compartments of rivers. In addition, Onema signs three-year *framework agreements* with the main public institutes for applied research in the given fields, notably INERIS for chemical substances, BRGM for groundwater, CEMAGREF for surface water, IFREMER for coastal waters, INRA, *etc.* These agreements take into account the work by the River basin agencies with the same institutes in each catchment basin. Identification and planning of projects is carried out jointly by Onema, the River basin agencies and the Ministry in charge of ecology with an effort to bring in the end-users needs from the start of the research programming process.

The research targets two time horizons, the *short term* for operational results required within specified time limits and the *mid term* for work on future or more complex issues.

3.1.4.2.3 Transfer, Best use of Results and Distribution

The results of the research and studies discussed above are presented during various events and symposiums, in technical guides and notes, in tools and instruments for managers, and in training courses. The goal is to make the results readily available in various manners suited to the needs of users. Certain uses of results designed on the national level are *tested* in the basins before being generalised. In addition, the agencies and/or Onema, *via* its local offices, take part in demonstration projects and local pilot tests in the basins (*e.g.* LIFE, INTERREG, *etc.*). These *in situ* tests are essential steps in the development and implementation of agency Programs of Measures. Sufficiently advanced basin projects can then be *implemented* on the national level; conversely, results acquired on the national level can be validated or developed on the basin level. In addition, a special system will be set up by Onema to detect projects resulting from national and European research programs (not funded by the agencies or Onema) and producing valuable results that must be acted upon or made available to water managers. An internet portal will be set up for technical documents in French, with further access to all the available scientific resources. Included will be all the study and research and development documents produced or financed by the agencies and Onema. *Technical training courses* will be created and offered in the framework of mutualisation efforts to facilitate implementation of new public policies (Green and blue corridors, wetlands, *etc.*).

3.1.5 Conclusion. A European Platform to Share Information and Experience

France has setup a new mechanism that spans the entire research field from strategy to making results available to professionals in France and, in part, on the European level. It fosters cooperation between stakeholders at the national level. Ideally, the interface role on the national level should be shared at a European level with the other member states, which have to tackle similar WFD issues, and with the European commission, from which a great deal of water legislation originates. Several initiatives have been undertaken at a European level concerning the interface between science and public policies, but there is not yet a mechanism, or an entity, having the mandate and ability to implement a global and operational approach to the links between research and public water policy. To be effective, such a mechanism must be able to call on national and regional relays, similar to the system set up by Onema in France on the national level or to the system of research & policy centres on the regional level that is presented in Chapter 3.4.

References

1. Ph. Quevauviller, P. Balabanis, C. Fragakis, M. Weydert, M. Oliver, A. Kaschl, G. Arnold, A. Kroll, L. Galbiati, J. M. Zaldivar and G. Bidoglio, *Environ. Sci. Policy*, 2005, **8**, 203.
2. European Commission, Directive 2000/60/EC of the European Parliament and of the Council of 23 October 2000 establishing a framework for Community action in the field of water policy, *Off. J. Eur. Communities*, L 327, 22.12.2000, p. 11.
3. Rapport au Ministre d'Etat, Ministre de l'écologie, du développement et de l'aménagement durable. Comité opérationnel 'eau' du Grenelle de l'environnement, présenté par le député André Flajolet, 12 March 2008.
4. La politique de l'eau (1964–2004): 40 ans d'une gestion décentralisée, La documentation française, April 2004.

CHAPTER 3.2

Links to Water National or Regional Research, Policies and Management – IWRM-Net

NATACHA AMORSI,[a] PETER ALLEN-WILLIAMS,[b] MICHIEL BLIND,[c] DANIELA HOHENWALLNER,[d] IRENE HUBER,[e] XAVIER LAFON,[f] STEPHEN MIDGLEY[g] AND DANIELA PAST[d]

[a] Office International de l'Eau (OIEAU), 15 rue Edouard Chamberland, 87065 Limoges Cedex, France; [b] Environment Agency, Waterside House, Waterside North, Lincoln, LN2 5HA, United Kingdom; [c] Deltares,, PO Box 85467 3508 AL Utrecht, The Netherlands; [d] Umweltbundesamt, 5 Spittelauer Lande, A-1090 Vienna, Austria; [e] PTKA, Hermann-von-Helmholtz-Platz 1, 76344 Eggenstein-Leopoldshafen, Germany; [f] MEEDDAT, 20 avenue de Ségur, 75007 Paris, France; [g] SNIFFER 25 Greenside Place, EH1 3AA, Edinburgh, United Kingdom

3.2.1 Introduction

Implementation of the Water Framework Directive[1] raises many challenges in terms of research, policies and water management and leads stakeholders involved at the different local, regional and national levels to adopt integrated approaches on both governance and science sides. Although Integrated Water Resource Management (IWRM) is not a new concept, the WFD has made it even more accurate and has reinforced the necessity to find operational IWRM applications.

Water System Science and Policy Interfacing
Edited by Philippe Quevauviller

Published by the Royal Society of Chemistry, www.rsc.org

IWRM can be understood as part of the Science Policy Interface (SPI) in the particular domain of water. Both concepts aim to figure out how to integrate bottom-up and top down driven approaches to ensure a transfer of knowledge among stakeholders. Here, different levels are to be considered: first the governance (*i.e.* how to ensure the widest range of stakeholders' interests from different fields is taken into account in order to implement relevant water policies) and, second, the environmental knowledge (including 'soft' and 'hard' sciences in an interdisciplinary approach).

IWRM-Net [ERAC-CT-2005-026025] is an ERA-Net project funded by the European Commission. The ERA-Net scheme set up new tools to support collaboration among research programmes managers tender to implement transnational research activities. This chapter aims to show, based on IWRM-Net experience, how activities related to the implementation of transnational research can be perceived as an example of the science–policy interface. After setting the scene in the Introduction (Section 3.2.1) and Section 3.2.2, describing IWRM-Net, its context and links with the WFD, Section 3.2.3 highlights water specificities of the SPI. Subsequent sections give more details of IWRM-Net activities over the first two years and a half of the project. Section 3.2.4 details IWRM-Net strategies and related SPI actions, and Section 3.2.5 shows our current limits and gives feedbacks on our first lessons learnt.

3.2.2 IWRM-Net Context

The WFD is an ambitious European legislation to achieve a good ecological status for Europe's fresh and transitional water resources by 2015. It strongly leans on the concept of Integrated Water Resources Management (IWRM). The integrative aspects of the WFD concern not only the integration of chemical-physical domains such as ecology and hydromorphology, but also include socio-economic aspects such as cost-recovery and cost-effectiveness. Finally, the WFD requires public consultation. As the Directive is a Framework, its implementation is carried out by Member States and in transboundary basins, transnational collaboration is required. The Common Implementation Strategy (http://ec.europa.eu/environment/water/water-framework/objectives/implementation_en.htm) supports the implementation of the WFD and provides some coordination.

The WFD has resulted in a large number of research initiatives both in individual domains (*e.g.* ecology) and in 'integration of domains'. Much of the research is carried out on a national level, supporting the national implementation of the WFD. The Europe's Lisbon Agenda aims to make the European Union 'the most dynamic and competitive knowledge-based economy in the world capable of sustainable economic growth with more and better jobs and greater social cohesion, and respect for the environment by 2010'. It was set out by the European Council in Lisbon in March 2000, along with the European Research Area. One of its aims was to:

Optimise and open European, national and regional research programmes to support the best research throughout Europe and coordinate these programmes to address major challenges together.

As a result the ERA-net scheme came into place, facilitating the collaboration between research programmes.

Making both ends meet, the IWRM-Net ERA-net project was launched in January 2006 for five years. Its overarching objective is to improve the research collaboration in the field of IWRM among research programme managers at the national and regional levels, the exchange of available knowledge, but also to identify those issues of joint, transnational interests and, if appropriate, tender for research. Though IWRM-Net deals with IWRM in general, its agenda is set by the WFD. For example, the focus of the first call, launched in November 2007, is on applied research that will support the development of the second river basin plans.

In 2006, IWRM-Net started with 17 partners throughout 14 Member States. As stated in the ERA-Net scheme, they all share the same particularity of being research programme managers, some of them working at the national level such as the French and Romanian ministries of environment, others having a more regional focus. As IWRM-Net is open to new partnerships, its network has progressively increased. Currently, 20 partners are actively involved (see http://www.iwrm-net.eu/spip.php?rubrique7). Three organisations have joined the network in order either to participate to IWRM-Net first call or to get involved and prepare the forthcoming common research activities scheduled in 2009.

Through the implementation of common research call(s), IWRM-Net acts as an interface between objectives set by the WFD (top down approach) and stakeholders in charge of implementing the WFD but also in charge of providing the appropriate knowledge to make it possible (bottom up approach). As presented below in Section 3.2.5, the process of research needs identification, of drawing up principles of the call, and of disseminating IWRM-Net state of progress are good examples of what can be the interface between science and policy.

3.2.2.1 ERA-Net Scheme

The ERA-net scheme is one of the instruments initiated by the Commission to develop the European Research Area. According to the official ERA-net scheme webpage (http://cordis.europa.eu/coordination/era-net.htm):

The objective of the ERA-NET scheme is to step up the cooperation and coordination of research activities carried out at national or regional level in the Member States and Associated States through: (1) the networking of research activities conducted at national or regional level, and (2) the mutual opening of national and regional research programmes. The scheme will contribute to making a reality of the European Research Area by improving the coherence and coordination across Europe of such research programmes. The scheme will also enable national systems to take on tasks collectively that they would not have been able to tackle independently.

The ERA-net scheme is a bottom-up approach, meaning that development of ERA-nets is demand driven: If there is sufficient interest in a certain discipline, the European Commission will support the development of an ERA-net project. The scheme complements national and European instruments for research funding and collaboration. ERA-nets typically carry out several of the following activities:

1. exchange of best practices in research programme management;
2. exchange of information about and results of ongoing national research;
3. exchange of information about research programmes;
4. identification of common research needs;
5. implementing joint transnational activities, including joint calls addressing aforementioned research needs.

As the objective is to foster transnational collaboration on research programmes, ERA-nets are run by research funding institutes. In the field of fresh water management, several ERA-nets are ongoing:

- BONUS for the future of the Baltic Sea (Bonus era-net)- http://www.bonusportal.org
- CIRCLE: Climate Impact Research Coordination for a Larger Europe – http://www.circle-era.net
- CRUE: Coordination de la Recherche sur la gestion des inondations financée dans l'Union Européenne – http://www.crue-eranet.net
- SPLASH: EU water initiative era.net – http://www.euwi-era.net
- SNOWMAN: Sustainable management of soil and groundwater under the pressure of soil pollution and soil contamination – http://www.snowman-era.net.

Several others deal with the marine environment. The ERA-net SKEP (Scientific Knowledge for Environmental Protection – http://www.skep-era.net) is described in Chapter 2.1. The SKEP network aims to 'facilitate the improvement of science into policy processes, and to support evidence-led modern regulations'.

3.2.2.2 Water Framework Directive and Integrated Water Resource Management

The Global Water Partnership (GWP) defines Integrated Water Resources Management as follows:[2]

IWRM is a process which promotes the co-ordinated development and management of water, land and related resources, to maximize the resultant economic and social welfare in an equitable manner without compromising the sustainability of vital ecosystems.

In recent years some criticism of IWRM has been voiced, focussing on the lack of scientific evidence, of measurable indicators and dealing with uncertainties in IWRM.[3] Integration is required over several axes, for example:

- integration of freshwater management and coastal zone management;
- integration of land and water management;
- integration of surface water and groundwater management;
- integration of quantity and quality in water resources management;
- integration of upstream and downstream water-related interests;
- cross-sectoral integration in national policy development;
- integration of all stakeholders in the planning and decision process, and so on.

IWRM should in our view be seen as foremost a policy concept. The underlying science consists of all environmental disciplines (hydrology, ecology, morphology and so on), many socio-economic scientific areas (economy, decision making, management, public administration, sociology and so on) and the integration and uncertainties of all these scientific domains. Instead of speaking of IWRM research, one should speak of IWRM related research, consisting of both the specific scientific disciplines and the scientific aspects of integration. Owing to this broadness of scientific domains, the IWRM-Net project focuses on IWRM-research that is relevant to the implementation of the WFD where IWRM is the key concept underlying the WFD. In popular terms the WFD provides a framework to achieve 'good ecological status' of Europe's fresh water resources by 2015, contributing to 'the provision of the sufficient supply of good quality surface water and groundwater as needed for sustainable, balanced and equitable water use'. 'Framework Directive' means that the European Member States (MS) are required to implement the WFD, but have significant liberty on how to achieve the objectives.

3.2.3 The Science–Policy Interface and its Water Specificities

3.2.3.1 Policy Life-cycle and Interface with Science

It is often suggested that the impact of science on policy could be improved. To achieve such an improvement, representatives of both science and policy need to be brought closer together. Hence, working on the 'science–policy interface' means working to improve the quality of policy decisions through the provision of 'scientific' evidence as well as improving the uptake of scientific outputs by policymakers and implementers. Science–policy interfacing is relevant for most policy fields, but one should not forget that not all science should serve policy nor all policy relies on science. Additionally, science–policy is also an oversimplification of the interactions. Industry plays a pivotal role, and many efforts that aim to improve uptake of science are based on improving the 'triple-helix' of collaboration between knowledge institutes, industry and government. Likewise, the general public, stakeholder and non-governmental organisations may play a role in the science policy interface.

The policy life-cycle consists of the following steps: (i) policy preparation, (ii) policy formulation, (iii) policy implementation and (iv) policy evaluation. Science may play an important role in each step. Even prior to policy preparation, science plays a role in agenda setting. Climate change is a good example, where scientists have played a crucial role in early-warning. The Club of Rome is an earlier example, publishing *Limits of Growths* in 1972, putting the environment onto the political agenda. However, policy is often only prepared after serious incidents. For example, the consecutive oil slicks of *Erika* (1999) and *Prestige* (2002) led the European Commission to consider marine security as an important political topic. During policy preparation, the role of science is often to deliver the rationale for the policy. The importance of such interaction is illustrated by current policy developments in biofuels. EU-policy in a 'renewed policy directive' (http://ec.europa.eu/energy/renewables/biofuels/biofuels_en.htm) aims for 10% biofuels in 2020. However, this policy is currently being reconsidered due to scientific insights of the effects this policy might have on worldwide food prices as well as its environmental consequences. Current developments of the Flood Directive (2008/60/EC) provide another example of interaction between science and policy. During policy implementation, the role of science is to provide the insights required to achieve the policy objectives. This includes both sound knowledge to design policy implementation and development of indicators to measure the progress towards such objectives. Such indicators are also required during policy evaluation. It is in these latter two steps, where science and policy must find common ground, *e.g.* indicators need to be scientifically sound, relate to the objectives and be of use to policy. While the foregoing appears fairly straightforward, the science–policy interface is much more complex in (European) reality.

The WFD implementation, with its many ambitious and innovative features and its requirement for a strong scientific basis, produces complex technical challenges that may be very difficult to manage. For example, it is often difficult for water managers to find reliable and acknowledged scientific evidence to design programmes of measures, doubts exist in the definition of good ecological status and their links to good chemical status, *etc.*

The research life-cycle goes through approximately the same set of stages as policy. It needs to identify research topics, develop a research proposal and search for funding; then research is implemented within a project and managed to reach objectives and deliver outputs that will be evaluated. Yet working on the 'science–policy interface' may often simply mean working on improving the uptake of scientific outputs by policy. It is apparent from this analysis that the interface can be activated at any of the stages in research life-cycle or policy life-cycle, and in any direction, science to policy or policy to science. As a consequence, it is necessary to specify the questions that require specific focus. A preliminary step is to clarify SPI water specificities as discussed below.

3.2.3.2 SPI-WFD Water Specificities

For the purpose of this chapter the specific criteria of SPI in Integrated Water Resources Management should be considered in the context of the implementation of the WFD. Without aiming to be comprehensive, this chapter elaborates on several issues:

- the difference in dynamics and expectations;
- the difference in languages;
- the complexity of Integrated Water Management in general and the implementation of the Water Framework Directive, and in particular:
    - — the challenging amount of scientific disciplines to integrate,
    - — the variety of situations encountered within the European Union,
    - — involvement;
- the lack of overview of existing knowledge;
- the difficulty of aligning research and application;
- the challenge of programming research to policy needs;
- the different drivers.

3.2.3.2.1 Difference in Dynamics, Expectations and Language

Though one would like to see otherwise, research results are difficult to plan and buy 'off the shelf'. Research tends to bring about more questions than answers. Policy, and in particular policy implementation, is usually governed by a timeline, *e.g.* the steps of the WFD implementation are set. To voice the common cliché: Policy makers and implementers hence require input today, whereas researchers always need more time. Furthermore, science is often expected to provide straightforward questions, whereas scientists will in general always present results with a level of certainty that can be wide ranging. However, though clear answers are expected, policy makers and implementers always make policy decision under uncertainty. The aforementioned example of biofuels perfectly illustrates a situation in which policy was developed based on the insights and challenges at the time, but has or is being changed based on new (scientific) insights.

To voice another cliché: policy representatives expect researchers to elicit the right research questions from policy documents; whereas scientists expect policy makers to specify policy to a level that research needs are clearly specified.

A key barrier in the science–policy interface is the differences in language and understanding. For example, the WFD speaks of 'water bodies'. This term is new to the scientific domain.

3.2.3.2.2 Complexity of Integrated Water Management in General and the Implementation of the Water Framework Directive

Clearly, from the long list of topics to integrate in IWRM, it is a challenge to integrate the large variety of scientific disciplines. The WFD adds more ambition to these challenges due to its geographical size and variations in governance and culture.

Referring to the policy life-cycle, the project's focus implies that we mainly deal with the stages of policy implementation and evaluation. The SPI in these stages is, however, complex since there are so many policy levels and scientific domains involved:

- Different policy (implementation) levels: the European Commission, National Ministries, International basin authorities, national basin authorities, regional and local authorities are all involved in developing specific policies and deciding on appropriate measures to meet the demands of the WFD.
- Different scientific domains and different spatial scales: besides the various domains mentioned above, it must be recognized that there is no such thing as 'one size fits all' science; for example, the ecological issues in the alpine region are different from those in lowlands, indicating that, on the level of implementation, local knowledge, science and policy must come together. In contrast, whereas ecological sciences may need to deal on a very local scale that is determined by environmental aspects, scientific developments in the field of, for example, water-economy relate to larger spatial, administrative boundaries (*e.g.* national policies on cost-recovery).
- Different institutional and cultural settings: though the WFD has invoked some streamlining of institutional settings by requiring the identification of a competent authority per river basin (Article 3) this does not imply consistency in mandate of these competent authorities and the challenge they face in coordinating the implementation of the WFD. Furthermore, cultural differences between Member States (MS) mean that implementation options in one MS may, due to political/cultural reasons, not be feasible elsewhere.
- The extraordinary number of stakeholders in IWRM/WFD implementation is in itself a challenge. Policy makers and implementers on all levels, scientists, consultants, non-governmental organisations and the public all play a role in IWRM and, consequently, all may influence the science–policy interface.

Hence, where IWRM already implies an extraordinary effort to 'integrate', the IWRM-Net projects' focus on the implementation of the WFD provides some clarity on the scope, but on the other hand adds complexity due to the spatial variety in the environment of the European Union and the variety of administrative, institutional and cultural settings.

3.2.3.2.3 Lack of an Overview of Existing Knowledge

The amount of research carried out within IWRM and the WFD is extensive. Additionally, a huge amount of knowledge is gathered by the implementation. From the policy perspective, finding the right science is a major challenge, not only because it is so extensive but also because it is presented in scientific papers, in (national) grey literature, or in people's heads. Also, it is believed that much research output is not visible on a European level, since it has been funded and developed on national levels.

To 'build' a picture of the national and regional priorities in terms of water related research, IWRM-Net has developed a Knowledge Management tool (KM tool), storing 180 research programmes and projects funded by Member States. Its user-friendly search engine allows drawing heuristic, geographical, calendar and tabular knowledge maps. In the first instance, the KM tool aimed to provide IWRM-Net partners with accurate information on research carried out by Member States in order to make it easier to work out common priorities for their own transnational activities. We are now entering a phase of opening up access to the KM tool to make it available, as well as our database, to exchange information on programme contents and management (see: http://www.iwrm-net.eu/spip.php?rubrique3). The IWRM-Net KM tool approach is complementary to WISE-RTD tool, which aims to organise knowledge in such a way that it is accessible starting from the WFD policy document (described in Chapter 4.2).

3.2.3.2.4 Difficulties in Aligning Research with Applications

As has been stated before, the timing of research outputs is often not aligned to the needs of policy. However, ongoing research could possibly improve the relevance of outputs. To make this happen, research programming and execution may be altered to enhance the relevance of the outputs. This topic is discussed in Chapter 2.4.

3.2.3.2.5 Challenge of Programming Research to Meet Policy Needs

The European Commission funds only about 3% of European research. This means that most research is programmed and funded on the national level.

Previously it has been stated that IWRM-research concerns both the research on individual scientific domains and the integration as a research issue in itself. Owing to the focus on implementation of the WFD, the science developed needs to be 'applied science'. Broadly speaking such research can be funded *via* three mechanisms on a national level:

1. Directly by local/regional (water) authorities: This 'research' is usually denoted 'consultancy services', and has the strongest focus on application.

2. By national/regional science foundations. This research typically emphasizes the scientific innovation, and has significantly less focus on application.
3. By national/regional research programmes run by 'water/environmental ministries': This research can be both of the 'consultancy' and the 'innovation' type.

All three funding mechanisms are relevant for a successful implementation of the WFD.

In the particular case of IWRM-Net, where research can both deal with domain specific science and with science on integration, the Knowledge Management Tool has listed some 180 programmes and projects, as stated previously within this number only a few are specific to water management. Many have water as a subject within an interdisciplinary programme, *e.g.* the RELU programme in the UK looks at Rural Economy and Land Use and funds a total of 29 projects, of which water management has been the subject of 15. This highlights one of the difficulties in cooperation. Within these programmes the criteria for the programme are wide ranging and on the face of it not compatible, but similar projects can be funded.

3.2.3.2.6 The Different Drivers

The differences in drivers of researchers and policy makers are a fundamental problem in improving the science–policy interface. Policy makers are driven by societal needs and politics. Researchers are driven in the first place by curiosity. The fact that researchers are increasingly judged by their scientific publications results in a strong commitment to publish, allowing only limited time to interact with policy.

3.2.3.2.7 Steps Already Taken to Harmonize the WFD Implementation

Several mechanisms, however, are in place to harmonize the implementation of the WFD and improve the science–policy interface. Firstly, the Common Implementation Strategy has developed several non-mandatory guidelines for implementation on a European level.[5] Scientists and policy makers were both involved in the drafting of these documents. Secondly, several reporting requirements are in place that are, or will be, used to monitor progress and assess deviations in approaches. This will lead to transnational exchange of good scientific practice. Thirdly, in the case of trans-boundary river basins, collaboration is required, invoking some transnational harmonization. Of course, several other activities take place to achieve consistency. One important activity is the scientific development co-funded by the European Framework Programmes for Research (FP). To name just a few:

- Training courses on Analytical Quality Control and method validation in support of the Water Framework Directive (TAQC-WFD);

- Development and Testing of Practical Guidelines for the Assessment of Environmental and Resource Costs and Benefits in the WFD (AquaMoney);
- Institutional Benchmarking for the Implementation of the Water Framework Directive and Integrated Water Resources Management in South Eastern Europe (WFD-SEE);
- Screening method for Water data Information in support of the implementation of the Water Framework Directive (SWIFT-WFD);
- Science–policy interfacing in support of the water framework directive implementation (SPI-WATER).

One of the key challenges in EC-funded research is that the research usually targets large parts of the European Union. As a consequence results are often difficult to apply since it is difficult to produce a 'European' output and simultaneously provide results required for local use ('one size fits all'). However, in several EU funded projects representatives of the Common Implementation Strategy were directly involved to improve the usefulness of outputs.

3.2.4 IWRM-Net Strategies and Actions Related to Science–Policy Interface

As stated before, the science–policy interface deals with different levels of complexity. A first level of complexity could be described as intra-complexity inside each sphere: science with its interdisciplinary challenges and policy with its governance issues aiming to reconcile divergent systems of values, needs and interests. Another level of complexity could be described as inter-complexity between each sphere (the link between science and policy) and concerns the interface itself. Inter-complexity refers, for example, to the production of knowledge, the access and transfer of information among stakeholders evolving in a sort of fragmented context.

We have already stated that IWRM has to tackle the same kind of challenges by combining environmental knowledge based on an interdisciplinary approach with a growing concern for governance able to take into account public consultation to implement accepted and relevant water policy. In this broad picture, research and ability to identify the 'good' societal and scientific issues play a major role. Taking into account the WFD and subsequent issues in terms of policy and research, IWRM-Net has set a strategy to elaborate scientific specification for transnational collaboration.

3.2.4.1 Ambition of IWRM.Net

The 20 partners of the IWRM-Net(http://www.iwrm-net.eu), from 14 European member states, involved in the project agreed on a shared Vision of what should be IWRM-Net by 2010:

1. the source for knowledge about IWRM-research being undertaken in Europe at Member-States level, with a focus on the WFD;

2. a forum for future perspectives and co-ordination of research needs and programmes on related issues in different countries, including accession states and EU neighbours;
3. the link between research and water policy makers and managers to bridge the communication gap;
4. a facilitator for bringing together researchers and funders from different countries to work on joint research programmes;
5. a forum for exchanging best practices on administrating research programmes across Europe.

Like any vision, IWRM-Net's is very ambitious. IWRM-Net strategy is based on a bottom-up and top-down approach. Stakeholders (research programmes managers, researchers, practitioners, non-governmental organisation) are closely associated with each stage of IWRM-Net. On the other hand, the institutional context gives the main orientations of the strategic activities that will be implemented by the consortium. IWRM-Net hence mainly focuses on some critical aspects of SPI.

3.2.4.2 SPI Focus for IWRM-Net: Topics, Critical Aspects and Actors

Partners can be considered as the primary actors in the SPI within IWRM-Net. They are developing a role to coordinate across the programmes within Member States the sharing of information. The initial process is between the partners themselves and then the second stage is to ensure the research and information is transmitted to the important actors in the scientific and policy spheres across Europe.

The critical aspects of SPI for IWRM-Net hence mainly focus on:

- the difference in processes and expectations of partners – by providing a transparent platform for communication;
- involvement and engagement of a wide range of stakeholders – by providing an easy method for involvement in the network;
- the lack of overview of existing knowledge – by providing links towards knowledge repositories and engaging in discussions to combine these knowledge repositories;
- the challenge to programme research to policy needs – by providing a research programmes analysis.

As can be recognized from the vision, many aspects relate to bringing together people. IWRM-Net wants mainly to be a facilitator, delivering, for example, a forum to water management and research, such that collaboration between the different target groups improves.

A key issue within IWRM-Net is that the project is 'not alone'. Especially with respect to the overview of available science, there are numerous tools, websites and projects that attempt(ed) to deliver access to water policy science. Hence a key strategic issue is that IWRM-Net focuses on liaising with similar initiatives, and prevents repeating what has or is done elsewhere.

3.2.4.3 From Vision to Actions, IWRM-Net's Activities

To reach the vision many tangible activities need to be carried out. Table 3.2.1 shows what is to be undertaken and developed, and to which items in the vision the activities contribute.

The next section goes into more detail about the specific activities of IWRM-net to consider the SPI.

3.2.4.4 Research Needs Identification

The process implemented by IWRM-Net partners to identify research needs in order to set transnational research calls combines both a top down and a bottom up approach. The former deals with the objectives set by the WFD, as reminded above. The latter relies on gathering water stakeholders' needs. This section specifically focuses on the bottom up approach.

3.2.4.4.1 IWRM-Net Launch Event – An International Workshop for Water Policy and Research Managers

IWRM-Net launched its external publicity efforts by sending out invitations to a high profile international workshop for water policy and research managers held in the heart of London:

- Invitations to the workshop were sent out to a wide variety of water research managers and policy makers in IWRM-Net countries, non-IWRM-Net countries in the EU and also Accession countries. They served the dual purposes of informing the recipients of IWRM-Net's existence, as well as encouraging them to attend or send delegates to interact with their peers in a workshop surveying EU research needs within the water sector.
- Participation in the workshop and accompanying social event supported the development of working relationships for a mixture of nearly 60 research managers, policy makers and implementers and researchers from many different countries, including those already in IWRM-Net, other EU countries and some of the union's eastern neighbours.
- Presentations from a carefully selected representative group of speakers illustrated the scope and variety of research needed to support policy implementation.
- Individual workshop sessions of mixed participants sparked discussions that further refined and prioritised overall EU water research needs.

Table 3.2.1 Activities developed within IWRM-Net to reach the vision.

Activity	*Paragraph*	*Vision 1 source for knowledge*	*2 forum for co-ordination*	*3 link research and water policy*	*4 joint research programmes*	*5 exchanging best practices*
Website		The website itself is the source of knowledge	FORUM in the community section	Provides links to policy documents and policy websites	Provides a common platform for management supporting the secretariat	Sharing of information *via* the website
Database		The tool for managing the information and the repository of information that links all aspects of the vision				
Research needs foresight	3.4.2	Collation of needs into strategic document	Links to website and use of forum section	Workshops bring people together to develop needs	Network supports collaboration	Via the network information is shared using different methods & building relationships
Short-term research needs identification	3.4.1	Collation of needs into strategic document	Links to website and the use of the Forum section	Workshops bring people together	By collaborating on the development of needs it improves common understanding	Via the network information is shared using different methods & building relationships
Research calls	3.5	Coordinated MS calls means they can be accessed by a wider range of people	The main purpose is to bring together the national programmes	The variety of programmes allows a range of target audiences for the research	Not applicable	All eranets consider the management process within their work programmes to share experiences
Conferences, workshops and other dissemination activities		Awareness of IWRM-Net and its vision	Developing personal networks between staff from different M. S. to support exchanges	Content of events designed to attract research managers and policy makers	Events both publicise and allow opportunity for development of joint call content	Events include both formal and informal opportunities for exchange of good practice

- A professional co-ordination team managed both the form and content of the event and drafted the final report and conclusions on European water research needs.

Detailed information on the form, content and participants of the workshop can be found in the Workshop Report– see: http://www.iwrm-net.eu/IMG/pdf/IWRMNet_London_wshop_rep_final-2-2.pdf.

One direct outcome of the workshop was issues to set the content of the first transnational call. The results of the workshop can also be considered as the starting point of IWRM-Net reflection on research needs classification. Both aspects have been further developed through activities undertaken to identify both short- and long-term research needs.

3.2.4.4.2 Identification of Short-term Research Needs

As part of the process for developing calls in the era-net, SNIFFER has led the identification and development of short term research needs. This process is based on the process used by SNIFFER to focus the UK programme management for water. It was then adapted to the necessities for collaboration and cooperation between national and regional programmes as required by IWRM-Net.

The normal process for identification of research needs has been to bring interested stakeholders together to discuss the needs. For the first call this was a two-day meeting in Edinburgh, where proposed subjects were presented to the participants who then debated the specific questions within the themes. Following agreement on the subjects the participants then voted to prioritise the subjects.

Invitees to the workshop were carefully selected to bring a balance between researchers, research funders and those who would use the knowledge generated (both policy makers and the operational managers who create organisational policy for implementing regulations and laws).

Recording this information has proved very important. The method by which you record the discussions on research needs impacts on the way in which you can present this information to those who need to make decisions based in it. Policy makers and operational managers have different requirements for the information generated and so it often needs to be presented differently. Short precise headline information is required for policy makers who read many documents on many different subjects and often cannot afford the time to develop in-depth knowledge on a particular subject. More detailed information is required for other participants.

What has been found is that the level of detail changes from policy makers to scientists. As one would expect, those with a high level of technical knowledge require more detail for them to classify their knowledge. This can be highlighted by the difference in complexity between the final presented research needs and the questionnaire that the workshop delegates and sounding board completed. Again the process of transforming the detailed questions into research needs requires facilitation from an individual. After the technical details are simplified they are then given to the policy makers, in this case the Senior Water Managers and Water Directors who provide the overview on direction of the CIS process. Without the work of the project board then the levels of communication would remain too far apart to allow valuable interaction.

Within the research need workshops organized by SNIFFER we started the process of transforming the questions raised by policy makers and water managers into scientific questions. Table 3.2.2 is an example of the questions raised within the workshop. These questions would then be summarized into headline themes to provide a simple voting mechanism and thus direction to the partners of IWRM-Net as to where funding priorities lie.

The facilitation process during workshops is necessary and requires an understanding of the issues being discussed. Again it is about finding a balance between the differing members of the audience. It is very easy for a technical expert be interested in the finer details of the subject as their level knowledge

Table 3.2.2 Example of research needs identified at Sibiu Workshop, October 08.

Management of non-native species

- Under which circumstances should non-native species be considered as substitute for native species (*e.g.* biodiversity, biomass) and thus be integrated into WFD assessment tools?
- How do invasive non-native species affect native biocoenosis and thus prevent from achieving a good ecological status?

Climate change

- How will climate change affect hydrology (high/low flow amplitude, frequency, seasonal/annual variability)?
- How will rising sea levels, altered flow regimes and sediment transport affect coastal areas (deposition or erosion, management) and ecosystems (changing salinity)?

Erratic Flows

- How can seasonal changes in erratic flow be managed (*e.g.* storage capacity, aquifer recharge)?

Management of river basin/flood plain

- How can priorities be defined in river basin/flood management (*e.g.* priority habitats, priority economic drivers)?
- How can impacts on ecosystems from economical drivers (*e.g.* improved navigation routes) be avoided or limited?
- What are the benefits of re-establishing functional aquatic ecosystems, and how can they be evaluated financially (*e.g.* tourism and nature benefits for communities)?

can allow them to understand this level of detail and use a technical language. Those within the meeting who do not have a technical knowledge often require a summary of the subject and confirmation of understanding. Bringing discussions to a good conclusion in the time available may require a facilitator to curtail speakers who go into too much detail (this also requires an understanding of what is too much detail) and then summarise the debate in easy to understand language, which is also recorded for external parties. In Edinburgh for the first call and subsequent 2nd call meetings this has proved very successful in bringing a different audience together for a successful and productive meeting. Table 3.2.3 lists the headline thematics.

This prioritisation is then incorporated into the knowledge classification headers created my merging the work of the CIS and also OIEau in the pre-project phase. More information on the CIS is given below.

The Common Implementation Strategy was set up by the European Union to allow the engagement of scientists and policy makers in the implementation of the Water Framework Directive. While not directly associated with the dissemination and communication of science it created a framework for communication between the scientist and water managers so that issues of knowledge could be developed and many of the issues raised were then put into research to allow knowledge gaps to be filled. One aspect of this was in Information Exchange and Research Needs, which is an activity under Working Group B (Integrated River Basin Management) under a section of work that was entitled 'Moving to the next stage in Common Implementation Strategy for the Water Framework Directive – Progress and work programme for 2005 and 2006. This was agreed by the Water Directors during their meeting in Amsterdam (December 2004) (Table 3.2.4). The work programme set up a questionnaire to find the research needs for the implementation of the WFD in 2005–2009, with a project team of four experts from around Europe and then a sounding board of 13 people. Delegates were then invited to a workshop to validate the work. An expert group was set up to undertake this task and came up with the following classification (Table 3.2.4).

Table 3.2.3 Example of debates summarized using votes.

Headline research themes – Sibiu meeting	*Votes*
Integrated pollution management (incl) • Hazardous pollution • Arsenic in groundwater	9
Management of river basins and floodplains	8
Hydro-power	6
Reference conditions	6
Good ecological status	6
Integration of socio-economic and environmental evaluation	5
Groundwater management	5

Table 3.2.4 Common implementation strategy, Working Group B classification.

Water resources and demand management
Groundwater management
Knowledge on physical processes
Knowledge on ecological processes
Monitoring
Pressure impact relations
Data management
Measure assessment
WFD policy questions
Policy assessment
Socio-economy

The management of these groups is very much a soft science skill that needs developing and recognising in this debate on the science–policy interface. The use of classification systems for knowledge should be recognised as a tool for management of information, and as knowledge is dynamic then the means to manage and classify this should also be dynamic – a flaw in the classification systems created.

3.2.4.4.3 Long-term Need: Foresight for Research Needs Identification

In the field of research and innovation policies, 'foresight' helps to plan research and innovation efforts corresponding to future needs of society.[7]

WFD long-term implementation will require review and update of RBMPs (River Basin Management Plans) in 2015, 2021 and 2027. Rather than react to research needs driven by policy demand in 2015, 2021 and 2027, IWRM-Net aims to identify the research needs to support the reviews and updates of RBMPs, *i.e.* aims to be pro-active.[8]

A workshop has been organised using common practices and methodologies, and inputs from other foresight networks related to water.[7]

As stated for the Launch Event and for the identification of short-term research needs, intense preparatory work has proven to be very useful. Three topics were decisive:

- careful choice of the participants with a mix of competences, positions and geographic origins;
- topics to discuss and background components to share with participants;
- facilitation process during the workshop, joining a water expert and a moderator with good experience in foresight.

More specifically, identification of future research needs requires exploring collectively what could be potential threats, opportunities and likely future developments that are at the margins of current thinking and planning. Indeed, research outputs will be useful and used only if expected by water managers. This means that the strategic research needs should be legitimated by a common commitment from scientists and water managers.

Key findings of the working sessions covered: (i) main driving forces of future change, (ii) the potential changes, (iii) how they might influence water management, especially WFD implementation, and (iv) the most important research needs to respond to these potential changes.

A strong, common emphasis from water managers and scientists was made on a few research topics to be developed in the near future:[8]

- What is the resilience of the ecosystems when faced with extreme perturbations?
- How to change social values and practices to improve the legitimacy of measures to conserve water and the perception of fairness and accountability?
- How will the social values of water and the people's behaviour and practices respond to higher prices of water use (full recovery of costs)?
- How to make the change from supply driven to demand driven water management and balance uses with ecosystem needs, but also balance between different uses?
- Does the current governance of the WFD allow it to be adaptive considering the future potential changes?

This workshop reached its goals because we managed to overcome differences, between science and policy, in languages, temporalities and focuses, to achieve a common understanding of the main driving research needs. This was made possible thanks to the freedom of every participant to consider future projections.

The main conclusions of the workshop reveal a common interest for research on adaptation of societies, governance and ecosystems in a world whose evolution is seen as more unpredictable and unstable.

3.2.4.5 Joint Calls

All the work done to gather and classify research needs dealing with IWRM represents the raw materials of transnational calls.

Joint call activities have become an important pillar of ERA-Net performance because, in general, it is a means to practice joint European research and development activities. In addition, joint calls allow for tackling of specific themes of European interest that are not yet covered by either national research programmes or the European Framework Programmes (FP).

For IWRM-Net such transnational funding initiatives are especially attractive because they enable a direct reaction to short-term demands that are put forward by WFD implementers, policy makers, scientists and other stakeholders and thus provide an effective instrument that:

- enhances a scientifically sound implementation of the Water Framework Directive;
- meets practical requirements;
- supports policy and decision makers dealing with integrated water resource management.

This is also reflected by the selection criteria for common joint activities that have been applied to the selection of successful transnational project consortia:

- added value for the WFD implementation/IWRM policy and potential for solving problems;
- relevance/extent to which the proposed project addresses the objectives of the Call;
- Scientific quality of the project approach and innovation (clear progress beyond the current state-of-the-art);
- Inter- and trans-disciplinarity;
- demonstration/case studies;
- dissemination/expected exploitability of results;
- transnational linkage and benefit of cooperation;
- quality of consortium management and expertise of the project coordinator and the project partners;
- prospects for success with regard to the work and financial plan/time schedule.

To allow for evaluation from a scientific but also practical point of view, IWRM-Net has included scientific experts and practical WFD implementers in its international evaluation panel.

Apart from the above-mentioned criteria scientific excellence has a major impact on the final funding decision. Owing to limited national funds the strict assertion of scientific excellence sometimes, however, is compromised and illustrates the limited potential of a 'virtual common pot'-model that is commonly used by ERA-Nets because European national funding regulations are still quite heterogeneous.

New models and a more flexible or common framework of European regulations are needed to allow a development towards a homogenous European Research Area.

Once transnational project consortia come into action, project monitoring will provide another field of interaction between science and policy. Progress of the funded projects is subject to national evaluation. Nevertheless, the scientific evaluation panel responsible for the evaluation and thus selection of the project proposals should be involved. For practical reasons, and since no budget for the payment of external experts is allocated in the ERA-Net contract, IWRM-Net will focus on national evaluations but will open up the projects and their progress for the expert panel during a final presentation.

A major objective of IWRM-Net is the transnational knowledge transfer of common results amongst the participating countries and beyond. The dissemination of project results should generate multiplier effects within Europe. Therefore, detailed dissemination plans are a prerequisite of successful applications. Next to the proactive promotion of knowledge and results arising from the work undertaken by the project consortia, public access to deliverables will be guaranteed by the IWRM-Net website (www.iwrm-net.eu), WISE-RTD and an IWRM-Net conference in 2010.

3.2.4.6 Developing the Network – Liaison and Communication Activities

The project itself is the creation of a network that brings together scientists and policy makers. Within the work programme there is a strategy for liaison to coordinate how and with whom the project coordinates. For all meetings and events the project considers who would be the target audience and how we improve communication with these people. For each event there are specific outcomes that require certain individuals and would also be of interest to others. By coordinating this within the network we aim to become efficient at engaging with the right people at the right time without duplicating efforts of other networks and organisations such as WSSTP, EURAQUA and the CIS process.

3.2.5 Our Current Limits and First Lessons Learnt

IWRM-Net is still an ongoing project. Four research projects, funded by IWRM-Net, begun at the end of 2008, from which we already have some feedback experience and evaluation. From our concrete examples of practices, products and processes, we have learned to improve our actions.

3.2.5.1 Be Desirable to Target Actors

With so many activities in the field of IWRM many of the individuals and organisations that IWRM-Net wishes to target are busy with many pressures on their time. When targeting certain groups there needs to be an efficiency of communication and the products and outputs delivered should be specific. The example of the workshops for research need identification highlighted the requirement to ensure we had clear subjects and a clear process for reaching the prioritisation of research questions; more information is available below.

3.2.5.2 Content of IWRM-Net Calls – Identifying Research Priorities

The process of collective identification of research needs between water managers and scientists proved to be very useful yet difficult to implement and to make long lasting. Collective identification of research needs is efficient in bridging the gap between science and policy because it helps to:

- Reduce the potential misunderstanding related to expectations from research results, both with regard to the timeframe and to the applicability. The collective identification of research needs implies arguments about feasibility of the research, potential results reachable, limitation of applicability and conditions for knowledge transfer.
- Offer an alternative source of knowledge to water managers that often choose to finance technical surveys and short expertise when they lack some scientific evidence to support their policies. Rather than competing for funding, the different sources of knowledge (research project, technical survey or short expertise) are complementary one to the other for the water manager.
- Give rise to an expectation from water managers for the results of research that they contributed to the launch of. SPI too often focuses on improving the uptake of scientific outputs by policy. Here, we go further by improving the uptake of research questions by policy, therefore preparing the suitability of further scientific outputs.

Collective identification of research needs is also a good means to design a customised call for research projects dedicated to improve public policies. Therefore, it can prove to be a useful tool to develop new, probably flexible, calls in coordination with some water managers directly interested and ready to fund.

IWRM-Net provides a systemic approach to water issues and therefore covers most of the research questions, including technology, ecology, socio-economic and governance issues. One option to further attract and support policy makers might be to broaden our focus beyond the WFD to other IWRM-related research questions. The already identified topics would make up the basis for a European Water Research Agenda that also implements the research questions developed in the ERA-Net CRUE (floods), SNOWMAN (groundwater & soil), other ERA-Nets covering water-relevant research, scientists and other stakeholders.

Based on their national interest and intention, funders and other policy makers could sign up for specific fields or topics of this European Water Research Agenda. Once interested parties have committed themselves to a specific field or topic, they could form a so-called action group (name derived from the model of ERA-Net TRANSPORT: www.transport-era.net/research-calls.html) and work together on the specific research field/topic.

There are different levels of collaboration that might be chosen dependent on interest:

- information exchange, joint conferences and likewise;
- joint calls and actively search for co-funders;
- joint programming as follow-up activities.

The above-outlined model would make European Water Research, and water policy requirements much more transparent for all stakeholders, irrespective of ERA-Net membership or likewise involvement. By this, operational water managers, researchers, water policy implementation agencies, and policy makers from Europe and beyond its borders will open up and may share water related research information and operational practices more freely. This will help in facilitating the transfer of WFD and IWRM-Net experiences within EU and towards non-EU countries and paves the way for a true European Research Area.

3.2.5.3 Implementation of the First IWRM Transnational Call

The process of sharing information on programme management practices is an important step in a developing a common understanding. From the agreement made in April 2007 for the scientific specification there followed subsequent meetings on the process by which the joint call was managed. It became clear that the process is slow, that it is sometimes difficult to reach agreement and understanding and that there were still issues encountered at the project selection stage one year later. Issues about national priorities and requirements were not fully resolved and a question remains about the objectives of the research. A clear understanding of the type of research undertaken in the projects is an important issue that needs to be clarified early in the process, and from this a clear strategy for how the process is managed can be set out. The role of the secretariat in coordinating the process should not be underestimated.

3.2.5.4 First Lessons Learnt and Corroboration of our Aims

The key issue is communication. First, communication within the project and between partners is vital to develop a clear understanding of the objectives and outcomes from the collaborative research programme. This includes knowledge of partners' internal management processes to identify possible obstacles. Then, the identification of research needs should be clearly delineated so that the right actors can be engaged at the right time to make maximum use of people's time and to avoid false expectations. Finally, the funding and management of projects can be clearly set out with time allowed for the interface between science and policy. One of the most important lessons the project has learnt is the costs and time involved with this activity and planning a programme accordingly.

3.2.6 Conclusion

The chapter has focused on the experience of the IWRM-Net ERA-Net project to highlight links to national or regional water research, policies and management. After having set the scene by explaining the IWRM-Net context, recalling how the Water Framework Directive influences water policies and related research activities, we showed how the different phases aiming to implement transnational research programmes can be perceived as science–policy interface illustrations.

ERA-Net project activities are thought out from the beginning to allow partners to find ways to implement common research activities. In fact, this goal also represents the most visible action of ERA-Net projects network. Nevertheless, IWRM-Net cannot be reduced to this single activity. As stated in this chapter, IWRM-Net carries on its own vision, explaining where the 20 partners to reach by the end of the project and what they intend to achieve to do so. Behind each vision item lays the overarching objective of allowing transfer of knowledge between different kinds of stakeholders sharing different interests, systems of values and involved at different levels of action from the local ground to the national and European arenas. Integrative Resource Management and Science Policy Interface tackle the same challenge of crossing bottom up and top down approaches. The complexity arises due to the different levels to be taken into account on both sides (science and policy). The WFD gives the regulatory context (impossible to modify except at the occasion of regular reviews at the end of each river basin management planning cycles) and sets up strong objectives. It gives the main incentives to support Member States to carry on reaching good ecological status by 2015 without providing common tools. On the operational side, Member States have to figure out appropriate water policies according to their national and regional characteristics and constraints as stakeholders need to find appropriate ways to identify and address their research needs. The communication gap between science and policy requires an interface to allow efficient transfer of knowledge. This is where IWRM-Net steps in. The interface between science and policy promoted by IWRM-Net relies on two main processes:

1. Give access to the current water related research state of the art. IWRM-Net has developed a Knowledge Management tool giving access to 180 research programmes funded at the national or regional levels. A user-friendly search engine enables the user to produce heuristic, geographical, tabular and calendar knowledge maps. A forum is now available on research needs identified during IWRM-Net regional workshops. The objective is to get stakeholders participated to comment, prioritise and/or suggest research needs to support IWRM-Net in identifying its transnational programmes specifications and set its research agenda.
2. Suggestion: Meeting of stakeholders. Allow stakeholders from different fields and at different levels to meet. Being able to address the research needs on which IWRM-Net partners would agree to launch common

research activities requires, on one side, tackling the needs of the funding organisations and, on the other side, finding compromises of interests among the water managers and policy makers confronted with what researchers are able to deliver.

Considering the WFD context, understood as the top-down legislation obligation, IWRM-Net has started, over two and half years, to progressively build the interface between science and policy in the specific area of transnational research activities implementation. Based on its agreed vision, IWRM-Net is open to new research programmes managers and supplies technical and network opportunities to 'water stakeholders' to find convergence and fill the communication gap. Networks such as IWRM-Net, by improving their capacity building in terms of activities required to implement transnational research activities, provide 'new' skills often missing to support the creation of an interface between science and policy. During the process, information is made available *via* technical tools, but the success of such initiatives requires time and ability to find its place among the on-going other water related initiatives.

Useful Websites

http://ec.europa.eu/environment/water/water-framework/objectives/implementation_en.htm
http://www.iwrm-net.eu
http://cordis.europa.eu/coordination/era-net.htm
http://www.bonusportal.org
http://www.circle-era.net
http://www.crue-eranet.net
http://www.euwi-era.net
http://www.snowman-era.net
http://www.skep-era.net
http://ec.europa.eu/energy/renewables/biofuels/biofuels_en.htm

References

1. European Commission (2000), Directive 2000/60/Ec of The European Parliament and of The Council of 23 October 2000 establishing a framework for Community action in the field of water policy, *Off. J. European Communities* L327/1 (22-12-2000).
2. GWP, Technical Advisory Committee (TAC), *TAC Background Papers No.4: Integrated Water Management*, ISBN: 91-630-9229-8, 2000, 67 pp.
3. C. P. Pahl-Wostl and J. Sendzimir, The relationship between IRWM and adaptive management, Discussion input for NeWater international platforms, NeWater Working Paper 3, D5.1.1., http://www.newater.info/ 15p. (2005).

4. Winsemius, Gast in eigen huis, beschouwingen over milieumanagement (Dutch). [Guest at home, reflections on environmental management]. Samson H.D. Tjeenk Willink, Alphen aan den Rijn, 1986.
5. CIS 2004-2007, Several guidelines supporting the implementation of the WFD: http://circa.europa.eu/Public/irc/env/wfd/library?l=/framework_directive/guidance_documents&vm=detailed&sb=Title.
6. D. H. Meadows D. L. Meadows, J. Randers and W. W. Behrens, *Limits to Growth*, Universe Books, ISBN 0-87663-165-0, 1972, 205 pp.
7. S. Treyer, Methodological panorama mapping of research related water foresights, *IWRM-NetDeliverable*, 2007, 31.
8. L. Citeau and X. Lafon, 2015 and beyond: emerging issues on water for policy and research – an expert perspective, *IWRM-NetDeliverable*, 2008, 34.

CHAPTER 3.3

Linking Water Science to Policy: A Canadian Experience

KARL A. SCHAEFER,[a] ALEX T. BIELAK[a] AND LEAH E. BRANNEN[b]

[a] Environment Canada, Science and Technology Branch, 867 Lakeshore Road, P.O. Box 5050, Burlington, Ontario L7R 4A6, Canada; [b] Environment Canada, Science and Technology Branch, National Hydrology Research Centre, 11 Innovation Blvd. Saskatoon, Saskatchewan S7N 3H5, Canada

3.3.1 Introduction

Science to inform decisions is the main driver in government-led research. While basic research (discovery, identification of emerging issues) remains an important activity, most people with interests in this field agree that publicly funded research should be effectively applied in the decision-making process. Clearly, the water resources sector is no exception. Water science should help shape federal and provincial policies, management of programs, setting of standards and guidelines, implementation of regulations, and a range of other strategic initiatives (all of which are more generally referred to in this chapter as policy- or decision-making instruments). Water science also plays an important role in informing decisions at the municipal level, within community-based organizations and in many industry sectors.

The well-documented divide between science and policy stems from the fundamental differences between the culture of researchers and that of the decision-making community.[1–5] Although policy makers, program/operational managers and other practitioners are often distinct communities with distinct

Water System Science and Policy Interfacing
Edited by Philippe Quevauviller

Published by the Royal Society of Chemistry, www.rsc.org

science needs, in this chapter we use these terms interchangeably, referring collectively to those individuals who need and use science in their work to develop and/or manage policy, programs, strategies, guidelines, *etc.*

Researchers and policy makers vary considerably in their work timescales (long-term hypothesis testing *versus* short-term policy horizons), in the language they use and in the incentive structures available to them (*e.g.*, publish or perish in the case of researchers). There is also a divide in how they deal with uncertainty and standards for evidence. Within organizations, the two groups are typically siloed, and often physically separated in different locations. In terms of perception, science is often seen as pure, objective and definitive, whereas policy is viewed as messy, subjective and approximate, particularly so in the environmental field. With such a distinct divide between cultures, it sometimes seems astonishing that the two communities can come together as often as they do.

If we focus on the environmental field in the Canadian context, potential remedies for bridging the divide are also well documented. Suggestions abound for improved training to better understand respective work environments and expectations, for development of science and policy teams to work jointly on given projects, for more opportunity for learning sabbaticals and co-location, and for improved recognition/rewards systems. Much of the existing documentation has been conceptual, however, and offers relatively little in terms of practical experience.[6–11] In practice, few such suggestions (*e.g.*, sabbaticals in the 'other' community) are actually implemented in anything but pilot fashion, if at all. In Canada, discussions on the need for the development of science–policy integration mechanisms continue with little resulting tangible action evident.

The science–policy divide has consequences for the decision-making process. This process has been typified by a linear progression of science investigation, followed by a dissemination phase driven by communication specialists, and then full hand-over of the issue to the policy arena. Many argue though that this paradigm is increasingly outdated and should be replaced by a model where science is socially distributed, demand-driven, trans-disciplinary and subject to multiple accountabilities.[12–15] In short, science should evolve from a one-way linear process to a multi-party, iterative dialogue – a more optimal approach to foster strengthened science–policy links.

With this evolution in the science process comes a need for evolution in science communication.[16] In government settings particularly, communications groups have typically flourished by pursuing a practice of broadcasting or 'pushing' consistent, overarching messaging about their scientific advances or corporate agendas, largely to the public and/or media. Unfortunately, this practice has often resulted in a 'closing down' of the science communications process, effectively burying uncertainty and staving off debate, contributing to a loss of trust in government science.[17]

As Bielak *et al.* argue:[18]

Traditional, big 'C' broadcast models of 'pushing' science to undefined audiences are losing ground to more nuanced approaches that recognize that different players prefer to receive science information in different ways . . . Communicating science, therefore, has expanded to include knowledge translation in which science information is packaged to the preferences, channels and timescales of particular audiences, and knowledge brokering in which intermediaries (knowledge brokers) link the producers and users of knowledge to strengthen generation, dissemination, and eventual use of that knowledge. Effective science communication now includes the full spectrum of approaches from broadcast to iterative dialogue.

This approach provides for greater opportunity for improved and sustained linkages between science providers and science users.

The literature on science–policy linking has consistently yielded two fundamentally important observations: (i) a need to customize and target science knowledge to the user audience to improve uptake and utility and (ii) a need to develop mechanisms or opportunities for sustained interaction between these two groups not only to ensure a regular 'push' of science knowledge, but also to allow science users to inform/influence the research agenda ('policy pull').

In the remaining sections, we provide a Canadian context for the science–policy interface, review experience with a series of water resource science–policy workshops, including impact on decision making, and identify further knowledge translation and brokering activities currently in progress, highlighting additional needs.

3.3.1.1 Canadian Context for Science Policy Linkages

In recent years, many national governments have struggled to better connect science with policy, driven largely by a need for greater transparency and accountability in decision making and delivery of services and programs to the public.[19,20]

Canada has had its share of incidents in which, for various reasons, the role of robust scientific evidence in developing key policy issues has been questioned. Crises such as the Atlantic cod fishery collapse[21] and tainted blood scandal[22] led to renewed government-initiated dialogue on how science informs policy.

At the federal level, we now have a growing body of knowledge dedicated to better understanding the science–policy dynamic and to fostering ways of improving production, dissemination and uptake of policy-relevant research. Three reports stand out. The Council of Science and Technology Advisors ('SAGE') report, *Science Advice for Government Effectiveness*,[8] outlined principles for providing effective science advice. The Government of Canada responded by developing the *Framework for Science and Technology Advice*,[23] and the *Creating Common Purpose* report[6] explored ways to improve the use of science in developing federal policy. The science–policy interface continues to attract attention from many groups, including the Professional Institute of the Public Service of Canada's major science–policy symposium in 2007[24] and another planned in 2010, and a recent federal S&T community workshop on this issue in March 2009. Water-related conferences now regularly feature, or have recently featured, the science–policy link: for example, the Central Canadian Symposium on Water Quality Research, the A.D. Latornell Conservation Symposium, and meetings of the Canadian Water and Wastewater Association and the Policy Research Initiative.

Strengthening the science–policy interface continues to concern many in the federal government and those who interact with it. Other government departments and research councils and networks have shown considerable interest in the concepts outlined by Bielak *et al.*[18] Feedback sought after a series of solicited presentations indicates there has been uptake of these ideas, and various groups are currently more actively considering how to address the science–policy interface.

Environment Canada is both a significant environmental science performer and the responsible federal authority for policy and regulation development, programme delivery, and enforcement in a range of environmental areas. Consequently, the interface between science and policy is critically important to the department, as emphasized in Environment Canada's *Science Plan*: 'Recognizing that transmitting new scientific knowledge to decision makers is a key role of government science, the [Department's] S&T Branch will promote more effective communication between scientists and decision makers'.[25]

In Canada, the science–policy issue has been more comprehensively addressed by the health science community. Both the Canadian Health Services Research Foundation (CHSRF) and the Canadian Institutes for Health Research (CIHR) give considerable weight to knowledge translation, '. . . bridging the gap between research and healthcare management and policy' (www.chsrf.ca). Also, the National Collaborating Centres for Public Health (NCCPH) have been established with the explicit mandate of translating existing and new evidence produced by academics and research in the public health field. These initiatives have helped spawn novel approaches such as a health evidence site (http://health-evidence.ca) where the intent is to distill and disseminate the best available evidence to inform and improve public health policy and practice. In the health sciences, academic expertise in the knowledge transfer and exchange field is growing rapidly.

Within the water sector, expertise and experience with science–policy interfacing are less developed, though beginning to grow. As well as the emergence of science–policy themes at water-related conferences described above, Canada's largest federally funded Network of Centres of Excellence (NCE) – the Canadian Water Network – has developed knowledge translation planning tools to help researchers get research findings and products into the hands of key audiences.[26] Further, federal regulatory processes (*e.g.*, the *Canadian Environmental Protection Act*), standards and guidelines at the provincial and federal levels, and EC's partnered large ecosystem initiatives (*e.g.*, Great Lakes Program, Northern Ecosystem Initiative, Atlantic Coastal Action Plan) are typically well informed by recent science. Nonetheless, in the environmental field, and in the water sector more specifically, documented experience on better brokering this linkage is scarce. In the next section, we highlight a 2002 Environment Canada initiative that grew into a broader departmental effort to strengthen the science–policy interface. The initiative focused on freshwater systems, but provides lessons for other science-based environmental issues.

3.3.2 Water Science–Policy Workshops

3.3.2.1 Rationale

In 2000, seven people died and hundreds more fell ill in Walkerton, Ontario, due to contaminated drinking water. Some five to seven thousand people – comprising about one-third of the total population – were affected by a similar outbreak the following year in North Battleford, Saskatchewan. As these two events unfolded, it became apparent that strengthened drinking water-related legislation and source-water protection rules at the provincial level would not be far off.

At the time of these events, the Water Coordinating Committee of the Canadian Council of Ministers of the Environment (CCME) – composed of senior water managers from federal, provincial and territorial environment departments – approached Environment Canada's National Water Research Institute (NWRI) to broker a series of national science–policy workshops.

As this was happening, the Science Liaison function at NWRI was also undergoing change. A January 2001 Speech from the Throne to open the first session of the 37th Parliament of Canada signaled a strengthened role for NWRI, and senior staffers were asked if the Institute had adequate support in liaison and science–policy interfacing to complement this growing role. Consequently, in the same year Science Liaison at NWRI was expanded to recognize the increased importance of knowledge translation and dissemination. A new Science Liaison Branch (SLB) led by a director-level position of equal status to NWRI research directors was formed and the science liaison function was augmented with additional staff with dedicated expertise in science writing and science–policy links. With this unique mandate, the SLB was well positioned to broker the meetings proposed by CCME.

The CCME workshops were intended to bring leading researchers together with policy and program managers to provide the recent science to practitioners (the policy and program community in all sectors), identify research needs and develop mechanisms for sustaining dialogue. The logic was that any new policy, regulatory or program initiatives would be stronger if informed by the latest aquatic science knowledge. Equally important was the objective to have policy and program managers articulate their research needs, and thereby influence the setting of research priorities.

In all, five workshop topics were identified by CCME as high-priority areas where science knowledge was critical to inform forthcoming policy and program initiatives:

1. effects of agricultural activity on water quality, Quebec City, January 2002;
2. groundwater quality, Toronto, March 2002;
3. water reuse and recycling, Calgary, May 2002;
4. water quality monitoring, Vancouver, October 2002;
5. wastewater treatment for small communities, Burlington, February 2003.

3.3.2.2 Workshop Approach

The workshops were sponsored by CCME and co-chaired by science leads from NWRI and a given host province with a particular interest in a specific subject matter. Nationally and, in many cases, internationally recognized science experts were asked to provide the state-of-the-science on a particular subject. Speakers came from various sectors (academy, government, industry, municipalities) and were asked to review current knowledge and identify research gaps and policy implications in their area of expertise. They were selected based not only on their research credentials but on their demonstrated ability to communicate effectively to an informed, but not necessarily expert, audience.

In addition to a range of scientific presentations, the workshops were also balanced by speakers reviewing key national, provincial or territorial policy and program initiatives, or international policy experiences relevant to the workshop theme. For example, the Water Reuse and Recycling Workshop had program leads from various provinces, Australia and the USA speak on policy issues and regulatory guidance. This helped raise the prestige of the event and secure appropriate participation.

Invitations were targeted to policy and program managers from senior levels of government, as well as key contacts from professional and industry organizations and local and regional government. The emphasis was on bringing research experts together with those individuals involved in developing and influencing water-related policies, regulations, programs, guidelines, strategies or related initiatives in a specific subject area.

Workshop moderators, typically selected from the policy and science program community in all sectors, guided session discussions to focus on broader science–policy issues, rather than scientific details. Throughout, science users or practitioners were encouraged to identify unique problems or research issues and needs. Combined with updates on related program and policy initiatives, these aspects of workshop design quickly became a main draw for research managers *and* policy personnel.

Workshop attendance (by invitation only) typically ranged from 40 to 70 individuals – an intentionally small number to ensure sufficient discussion on the recent science and to get feedback from policy and program managers on their respective research needs and ideas for sustaining dialogue.

Pre-reading and on-site products included resource materials such as presentations, key reference lists and expert contact information. Workshop reports were prepared and made available online, capturing not only the various presentations but a synthesis of both research and policy/program needs and options for sustained dialogue (see Environment Canada[27] for online workshop products). SLB staff co-authored four of the five reports, ensuring a further level of consistency and continuity among sessions. Finally, at the request of policy and program participants, attendees were also sent CDs containing all presentations after the workshop.

Continuity in leadership provided by SLB staff was key to translating the experiences from each workshop into effective modification of subsequent workshops. In some cases, agendas and discussion periods were shortened or lengthened, materials and reporting style modified, or attendee numbers or representation changed to better suit the specific subject matter – all in the interest of developing an optimal formula that would help make the workshops better and more flexible as time went on. As an example, the workshop on small wastewater treatment systems had particular relevance for municipal and regional government officials, and in response the agenda and discussions included more attention to management, regulatory and financing needs. On the whole, the CCME-sponsored workshops provided a unique opportunity in Canada to better bridge the science and policy communities working in the water sector.

3.3.2.3 Assessing Workshop Effectiveness

Anecdotal response after each workshop was positive, but to develop a better metric of effectiveness the authors subsequently surveyed workshop participants. An electronic survey targeting the roughly 300 participants engaged in the initiative was sent out to solicit feedback on the extent to which the workshops and their products had been useful in strengthening the link between policy/program personnel and the research community.

In all, 25% (78) of workshop participants responded – a large enough response rate to allow us to report qualitatively on general observations and perceptions of workshop effectiveness. Roughly two-thirds of respondents were policy/programs managers, and the remaining one-third were scientists and research managers (see Schaefer and Bielak[28] for more survey detail).

Participants were first asked whether the workshops were having any impact. Put another way, were the workshops effective at informing a given decision-making process? On this question, 90% of responding policy/program managers found the workshops and their products useful. In many instances, respondents indicated that the workshops directly informed decision-making – that is, the development of a specific policy, program, regulation, guideline strategy or some other related management decision. Some specific examples included:

- 'assisted in developing our province-wide monitoring program';
- 'information gained will be used in our jurisdiction's first-ever state of aquatic ecosystem report';
- 'will be useful in the development/finalization of our municipal wastewater discharge regulations';
- 'has assisted the province in developing draft regulations for groundwater management';
- 'helped develop consensus on need to pursue the development of a Canada-wide framework for water quality monitoring';
- 'helped to develop our provincial water strategy';
- 'helped make modifications to our provincial agriculture regulation.'

Similarly, 90% of responding scientists and research managers reported that the research issues/needs identified through a given workshop had been useful in refining their own organization's research priorities.

To obtain an indication of the extent of dissemination of workshop knowledge and, thus, the potential capacity to more broadly inform decision making, participants were asked if they had distributed workshop products (*e.g.*, binder, report, web address). Twenty-nine percent of respondents distributed the report/web address/workshop binder to their immediate work group, 24% to other sections in their departments/organizations, and 12% disseminated the materials to other departments/organizations.

Eighty-two percent of all respondents indicated the workshops were successful at linking research and policy. However, many noted that while the workshops were an excellent beginning they felt that sustained communication and networking would be key to ensuring science more routinely and significantly informs decision making and research priorities more effectively reflect policy and program needs. In fact, 96% of respondents indicated that sustained dialogue/networking was important. When asked about their preferences for the specific mechanisms for staying connected, 69% suggested a combination of electronic networking and face-to-face meetings as the main vehicle. The main sentiment from both research and policy communities was that they preferred to stay networked through some form of regular electronic contact (*i.e.*, web link, email lists), with occasional face-to-face meetings as the science evolves.

In sum, the sustained interfacing between these two groups was not only desirable but crucial to better bridging the science–policy divide.

3.3.3 Recent Developments

The science–policy workshop 'formula' has garnered considerable interest and continues under the Environment Canada banner. More recent sessions have been held on the themes of microbial source tracking and pharmaceuticals and personal care products in the environment. The policy-ready nature of the products from these sessions has fuelled a more aggressive campaign to get this knowledge into the hands of those that have policy or program oversight, in all sectors.

Reports have been posted online and articles written for various industry association newsletters to increase dissemination to relevant audiences. More broadly yet, knowledge gained from the pharmaceuticals workshop and captured in the report has been targeted to dozens of organizations in all sectors with related interests or mandates across Canada. The uptake and utility of this knowledge by these organizations will be the focus of subsequent work. Workshop results are also routinely presented at meetings and various events. Word has spread and Science Liaison staff are increasingly approached by research and program leads on potential topics for future workshops.

This kind of knowledge-brokering activity received further attention within Environment Canada. In 2006, the Department's first Assistant Deputy Minister of Science and Technology tasked the newly named Science and Technology Liaison (S&TL) Division to broaden its mandate beyond water S&T to represent the full breadth of Environment Canada's S&T and enhance knowledge transfer within and beyond the S&T Branch. S&T Liaison activities currently focus on targeted science writing, strengthened information sharing on research results, a renewed web presence with an emphasis on S&T user needs, and a more strategic use of science summaries, all of which has relevance for internal and external practitioners. As an example, a regular news-digest capturing Environment Canada's water science activities and progress is distributed laterally and upwards to senior program managers, including the Deputy Minister, within the Department, and externally through a subscribe mail list (with more than 2000 subscriptions to date).[29]

News-digests for other areas of the Department's research are in various states of development, but all are based on a 'one-pass' information gathering system that provides information for an internal-to-EC report distributed across the S&T Branch. This report has become one of the most downloaded documents on the Departmental Intranet.

Finally, under a new department wide publishing policy announced in mid-2008, Environment Canada's S&T publishing output (averaging 736 peer-reviewed environmentally-related publications annually from 2006 to 2008) will soon be better harvested through an automated system called EC SCIENCE ALERT. Designed and made workable, in large part, by S&TL staff, this policy will allow for more strategic use internally and improved dissemination externally.

Of particular note is the fact that all these initiatives have been conceived and executed while maintaining the trust of the research community, which, in our experience, takes considerable time to develop. All products are written collaboratively with science leads, and science–policy interfacing tools are developed and delivered in collaboration with scientists. This mode of operation enhances credibility and usefulness. This is particularly true of two series of research impact studies where the work of scientists (either individually or in teams) was, and continues to be, described and linked to the tangible benefits it had produced for Canadians.[30]

Despite its small size and modest budget, the S&T Liaison Division is continually looking to build the case for knowledge translation and brokering, and for innovative and practical approaches to better link scientists with policy analysts and program managers, though this is not without cost. Better assessing the uptake and impact of knowledge production is not a trivial exercise. Land and Water Australia, for instance, roughly quadrupled its previous communication budget to deliver on its new Knowledge and Adoption Strategy, arguing that 'funding the arrows' is as important as funding research itself.[18]

3.3.4 Conclusions

Our experience indicates that national science–policy workshops are an effective vehicle for bringing the science and policy/program communities together. Of equal value are the knowledge brokering or intermediary function and the skill sets this function requires if the exercise is to be successful. There is little substitute for face-to-face discussion and networking opportunities. Nevertheless, our observation is that sustained two-way communication is a precondition to more routine and effective science–policy interfacing. Our view is that a major institutional shift towards bridging the science–policy gap seems unlikely given a continuing climate of fiscal restraint; the Australian experience is a notable exception. Instead, a toolbox of regular and innovative translation and brokering activities holds the most promise to better connect science generators with science users.

Further research is required to better understand the preferences of science users. The science needs of a senior policy advisor may well vary from that of a junior policy analyst. Program managers may have yet different needs. A federal standards and guidelines official may require different information, in a unique format and at a different time than, say, a municipal wastewater manager or industry representative. In fact, there has been insufficient attention paid to science needs of practitioners. It is to address this problem that the authors have initiated a survey of the science needs of Ontario's watershed-based management agencies/conservation authorities, with further studies planned.

Finally, metrics are required that can demonstrate unequivocally when research and policy are better interfaced. Science users no doubt value relevant and understandable material in a medium that is accessible, user friendly and synthesized, has content that is credible and sound, and comes from a dissemination source that is competent and trustworthy. From the authors' experience with science–policy workshops, we can assert that researchers and research managers are keen on regular dialogue with policy leads and are anxious to shape research activities to respond to emerging program priorities.

With this desire to connect clearly evident on both sides of the science–policy divide, obviously we must now seek to develop practical indicators that will reveal when, in fact, the divide has been bridged.

References

1. M. Saner, 'A Map of the Interface between Science and Policy', Council of Canadian Academies, Staff Papers, Ottawa, 2007.
2. B. Steel, P. List, D. Lach and B. Shindler, *Environ. Sci. Policy*, 2003, **7**, 1.
3. G. B. Doern and T. Reed, (Eds.), *Risky Business: Canada's Changing Science-Based Policy and Regulatory Regime*, University of Toronto Press, Toronto, 2000, p. 385.
4. L. A. Plouffe, *Isuma*, 2000, **1**(1), 135.
5. J. Lomas, *Isuma*, 2000, **1**(1), 140.
6. Canadian Centre for Management Development, Creating Common Purpose: The Integration of Science and Policy in Canada's Public Service, CCMD, Ottawa, 2002, p. 29.
7. B. Jarvis, *The Role and Responsibilities of the Scientist in Public Policy*, A discussion paper on science and government, Public Policy Forum, September 1998.
8. Council of Science and Technology Advisors, *SAGE: Science Advice for Government Effectiveness* (May 1999), Retrieved 14 October 2008 from www.csta-cest.ca/index.php?ID=90&Lang=En.
9. Science and Technology Advisory Board, *Science Communications Framework for Environment Canada*, Environment Canada, Ottawa, 2000, Report #2.
10. Science and Technology Advisory Board, Finding Better Solutions: A Guide to Integrating Social Science and Community-based Knowledge into Planning and Policy Development, Environment Canada, Ottawa, 1999, Report #4.
11. Science and Technology Management Committee, *Management Framework*, Environment Canada, Ottawa, 1998, Report #1.
12. European Commission, *Scientific Evidence for Policy-making*, Directorate-General for Research, Socio-economic Sciences and Humanities, EUR 22982 EN, 2008.
13. C. Pohl, *Environ. Sci. Policy*, 2008, **11**, 46.
14. J. Holmes and J. Savgård, 'Dissemination and Implementation of Environmental Research', Prepared for the Swedish Environmental Protection Agency, 2008, Report 5681.
15. E. McNie, *Environ. Sci. Policy*, 2007, **10**, 17.
16. D. Cheng, M. Claessens, T. Gascoigne, J. Metcalfe, B. Schiele and S. Shi, (Eds.), *Communicating Science in Social Contexts New Models, New Practices*, Springer, Dordrecht, 2008.
17. MORI, *Science in Society: Findings from Qualitative and Quantitative Research*, Conducted for the Office of Science and Technology,

Department for Trade and Industry, Department for Innovation, Universities and Skills, London, 2004.

18. A. T. Bielak, A. Campbell, S. Pope, K. Schaefer and L. Shaxson, in *Communicating Science in Social Contexts New Models, New Practices*, ed. D. Cheng, M. Claessens, T. Gascoigne, J. Metcalfe, B. Schiele and S. Shi, Springer, Dordrecht, 2008, p. 201.
19. New Zealand Parliamentary Commissioner for the Environment, *Missing Links: Connecting Science with Environmental Policy*, Wellington, New Zealand, September, 2004.
20. House of Commons, *The Use of Science in UK International Development Policy*, Science and Technology Committee, London, 2004, 13th Report of Session 2003-04, Vol. 1.
21. J. A. Hutchings, *Can. J. Fish. Aquat. Sci.*, 1996, **53**, 943.
22. H. Krever, *Commission of Inquiry on the Blood System in Canada (Krever Commission) Final Report*, Library and Archives Canada Electronic Collection, 1997. Retrieved 14 October 2008 from http://epe.lac-bac.gc.ca/100/200/301/hcan-scan/commission_blood_final_rep-e/index.html.
23. Government of Canada, A Framework for Science and Technology Advice: Principles and Guidelines for the Effective Use of Science and Technology Advice in Government Decision Making, 2000. Retrieved 13 October 2007 from http://strategis.ic.gc.ca/pics/te/stadvice_e.pdf.
24. IOG, 'Public Science in Canada: Strengthening Science to Protect Canadians', Institute on Governance for the Professional Institute of the Public Service of Canada, Conference Summary, 2007.
25. Environment Canada, *Environment Canada's Science Plan: A Strategy for Environment Canada's S*cience, Science and Technology Branch, Ottawa, 2007, 21 pp + x.
26. Canadian Water Network, *Knowledge Translation Tool Kit*. Retrieved 24 September 2008 from http://www.cwn-rce.ca/index.php?fa=Research.KT.
27. Environment Canada, *Science-Policy Workshop Reports on Water Quality*, 2009. Retrieved 31 March 2009 from http://www.ec.gc.ca/scitech/default.asp?lang=En&n=5B7438C1-1#workshop (2009).
28. K. A. Schaefer and A. T. Bielak, *Environ. Monit. Assess.*, 2006, **113**, 431.
29. Environment Canada, *Water Science News*. Retrieved 31 March 2009 from http://www.ec.gc.ca/scitech/default.asp?lang=En&n=1890C965-1 (2009).
30. Environment Canada, *Science and Technology into Action to Benefit Canadians*. Retrieved 31 March 2009 from http://www.ec.gc.ca/scitech/default.asp?lang=En&n=4B40916E-1 (2009).

CHAPTER 3.4

ECOBAG – A Regional Science/ Water Policies Interface

PHILIPPE VERVIER,[a] JÉRÔME DEPASSE,[a] MICHEL COMBARNOUS,[b] HUGUES AYPHASSORHO,[c] MARTINE GAECKLER[d] AND MARC JARRY[e]

[a] ECOBAG, 5 rue Michel Labrousse, BP 42353, 3 023 Toulouse Cedex 1, France; [b] TREFLE, University of Bordeaux, Esplanade des Arts et Métiers, 33405 Talence Cedex, France; [c] CEMAGREF, 50 avenue de Verdun, 33612 Cestas, France; [d] Agence de l'Eau Adour-Garonne, 90 rue du Férétra, 31078 Toulouse Cedex 4, France; [e] Université de Pau et des Pays de l'Adour, IBEAS, Avenue de l'Université, BP 1155, 64013 Pau, France

3.4.1 Background of Scientific Support Needs for Integrated River Basin Management Implementation

The European Water Framework Directive (WFD) aims to contribute to the provision of sufficient supply of good quality water needed for sustainable, balanced and equitable uses. The WFD also aims to the significant reduction of groundwater and surface water pollution through 'daughter directives' (Directives 2006/118/EC and 2008/105/EC). Objectives of 'good status' are planned to be reached by the end of 2015 but the WFD has a much longer time span with two further river basin management plans scheduled for the periods 2015–2021 and 2021–2027. In this context, as specified in Article 20 of the directive, scientific and technical progresses have to be integrated within the

Water System Science and Policy Interfacing
Edited by Philippe Quevauviller

Published by the Royal Society of Chemistry, www.rsc.org

implementation of WFD, in particular at the reviewing periods of the river basin management plans (2015, 2021, 2027).

The WFD hence gives some indications of where and when, in the implementation process, the integration of new scientific and technical knowledge will be required (economic analysis, standards for monitoring of quality elements, review and updating of river basin management plans, identifying priority hazardous substances, and so on); however, there are no indications/recommendations on how this integration should be carried out.

This question of integration of scientific knowledge into professional sectors and into public policies has been considered differently within the 40 last years. During the 1950s and 1960s it was generally believed that useful results for industry and society would start flowing from basic research according to a so-called linear model of innovation.[1] The 1980s and 1990s saw a period of moving from the 'science push doctrine', corresponding to self-governing science, to 'society pull science'. Therefore, research started to be governed by the need to respond to problems confronting the societies and economies. However, it is one thing to require a science pulled by society and another to organize an efficient production of knowledge that can support environmental policies.

Gibbons *et al.*[2] have described the development from the dominant 'traditional' mode of production of knowledge (so-called Mode 1) to a much more complex and heterogeneous mode of knowledge production (so-called Mode 2). In Mode 1, knowledge is mainly created and communicated within academic institutions and within domains of specialised scientific disciplines. It is a response to problems of increasing complexity faced by modern society and, notably, to the water question linked to most of human activities and organizations. Mode 2 is said to be 'trans-disciplinary', since it mobilizes several scientific disciplines (not only from universities). Quality in Mode 2 is not only linked to intellectual dimensions but is also linked to social, political and economic support, still, however, taking place in a scientific arena.

In the late-1990s, it became important to develop cooperation between industry, public policy and research. It was what Etzkowtiz and Leydesdorf[3,4] called a 'triple helix' of science policy. This metaphor illustrates the fact that the development of each partner during a collaborative work follows a spiral related to the iterative interactions with the other partners. The global evolution of the working group between industry, public policy and research can be described as a helix, which is comparable to DNA.

Considering water policies, if a large-scale approach is indispensable (at EU or national level), local specificities (*i.e.* River Basin District scale) have to be taken into account. Therefore, scientific support to water policy implementation has to be considered at the two levels. This chapter describes the development of scientific support at a River Basin District (RBD) scale. The example given focuses on the Adour-Garonne River Basin (south-western France) and describes the science–water policy interface 'ECOBAG' (environment, ecology and economy of the Adour-Garonne Basin).

3.4.2 A Regional 'Science–Water Policy' Interface: ECOBAG

To support the implementation of integrative water policies, such as the WFD, it is necessary to create links between knowledge producers, water managers and water users to deliver information, knowledge and tools helpful for the policy maker.

ECOBAG[i] is a science–policy interface (SPI) established at regional level between the Universities and the Research Institutes of Bordeaux, Toulouse and Pau, the water managers, the regional representatives of the ministry of environment and the regional governments. These entities made a commitment to support the development and the transfer of knowledge and the tools, which are necessary for decision-making for:

- sustainable management of water resources and related aquatic ecosystems;
- sustainable development of territories for a supply of surface waters and groundwater in sufficient quantity and quality for uses of water that are sustainable, balanced and equitable.

To reach these objectives, ECOBAG:

1. is acting as a translator of scientific knowledge into clear, accurate, unbiased information to policy makers;
2. is supporting the production of specific methods, tools or knowledge that are required by socio-economic actors, by supporting research and development (R&D) approaches.

The methodology is based on three steps. The first step is to identify the major issues within the river basin for which water 'actors' have difficulties in implementing the corresponding policies. Starting from these difficulties, the second step is to identify the appropriate tools, methods or knowledge that are required to support the decision makers in the implementation of water policies. This second step is based on an iterative and collective process by involving representatives of decisions makers, water managers, water users, related water enterprises, innovative enterprises, technological platforms and researchers from all the fields that are concerned with the water system (from physics to human sciences). The third step is to build the appropriate consortium for the project that will have to deliver the identified supports.

[i]Members of ECOBAG: Région Aquitaine, Région Midi-Pyrénées, Direction Régionale de l'Environnement en Aquitaine et en Midi-Pyrénées, Agence de l'Eau Adour-Garonne, PRES 'Universités de Bordeaux', PRES 'Université de Toulouse', Université de Pau et des Pays de l'Adour, CEMAGREF, BRGM, INRA, ADERA.

This method, which crosses up-to-date research knowledge and issues related to the implementation of the water policies, has major advantages:

- Advantage 1 – a decrease of the misunderstanding between what the water managers and decisions makers are expecting from research and what the research is supposed to deliver; this is the consequence of:
  - a shared diagnosis of what are the difficulties met by the water managers, which can be tackled by already existing tools, methods and knowledge;
  - a clear definition of the expected deliverables (methods, tools and knowledge) from research activity that have to be usable by water managers and decision makers;
  - a shared anticipation of the difficulties that the actors and the researchers of water will meet in their own field;
- Advantage 2 – facilitating multidisciplinary approaches, interactions and synergies between fields of research and technology, participation of the deliverable users:
  - because the expected methods, tools and knowledge have been identified through a global approach by gathering different expertises, the fields of research required in the consortium to deliver appropriate outputs are identified throughout the process;
  - the facilitators in charge of the collaborative and iterative process of identification of the deliverables are the appropriate actors to create positive interactions between the partners of the consortium, between the various fields of research, between the various fields of technology, and between research and technology.

3.4.3 Examples of Successful ECOBAG Projects

3.4.3.1 Improving Exchanges of Information and of Knowledge between Researchers, Socio-Economic Actors, Managers and Decision Makers of Water Systems

ECOBAG is playing the role of translator as described in Chapter 5.2. Days of dialogue aimed at cross presentations of information and knowledge, and the connected difficulties, by researchers, decision makers, water managers and socio-economic actors have been conducted during a year and a half. The different issues that have been tackled are:

- water, matter and species fluxes within the Adour-Garonne River Basin District – role of discontinuities from upstream to downstream;
- methods and tools for survey of waters and the associated ecosystems;
- environmental and economic consequences of pressures within the Adour-Garonne River Basin District;
- agriculture and water quality.

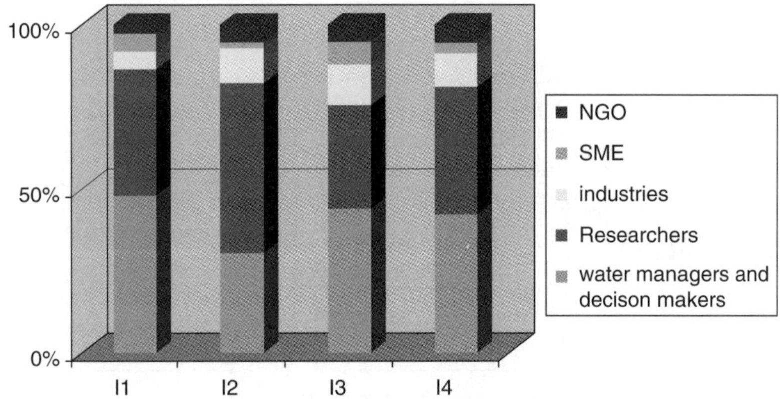

Figure 3.4.1 The 300 actors of the Adour-Garonne River Basin that participated in 'days of dialogues' represented non-governmental organizations (NGOs), small and medium enterprises (SMEs), industries, researchers, water managers and decision makers.

For their communication, the researchers agreed to work with socio-economic actors, with the objective of preparing a comprehensible message addressed to a diverse audience (Figure 3.4.1). This facilitated the dialogue between the participants.

Participants in the 'days of dialogue' were interviewed to evaluate what they acquired during the dialogue. The questionnaire identified the fact that, whatever the teaching competence of the presenters, the rate of comprehension of their presentation was increased considerably by the preliminary work proposed by ECOBAG.

3.4.3.2 A Collective and Iterative Process to Identify the Support Required by Water Managers and Decision Makers

The 'triple helix' method has been tested for ten years at the Adour-Garonne RBD scale by ECOBAG. One of the most important items of feedback for a significant improvement of research results integration within water policies is the importance of collective reflection to identify the needs for support of the integrated water management, by innovation and R&D.

Within the last four years, ECOBAG has developed a collective and iterative method, which is based on three steps (Figure 3.4.2). The first step was to identify the major issues within the Adour-Garonne River Basin District for which 'water actors' face problems in implementing the corresponding water policies. The second step was to organise an iterative and collective process of co-identification of problems that would require scientific support. This process involved, from the start, decisions makers, water managers, water users, enterprises of water, innovative enterprises, technological platforms, and

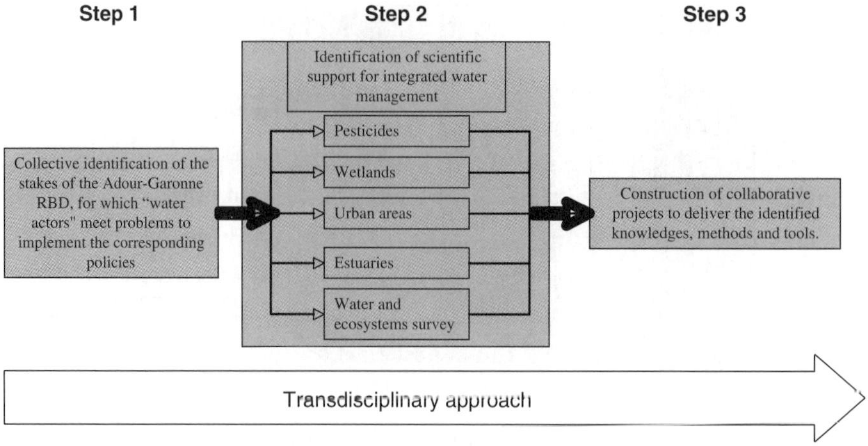

Figure 3.4.2 The collective and iterative process proposed by ECOBAG aims to identify the supports (knowledge transfer, multidisciplinary expertise, demonstration, research development) appropriate for integrated water management. This process involves, from the start, non-governmental organizations, small and medium enterprises, industries, water managers and decision makers, and researchers from all disciplines that are relevant for sustainable development.

researchers from all disciplines that are relevant for sustainable development. This process allowed the rewording of problems associated with water management or with the implementation of policies in the form of questions appropriate for scientific and technological purposes. It also allowed identification of the kind of support (knowledge transfer, multidisciplinary expertise, demonstration, research development) that would be required for a better water policy implementation. The third step was to build the collaborative projects, aiming to deliver the previously identified knowledge, methods and tools. These products must be usable by socio-economic actors.

This method was applied for the stakeholders identified at the level of the Adour-Garonne River Basin, in particular those associated with the problems involved in 'Pesticides', 'Wetlands', 'Urban areas', 'Estuaries and Water Survey'.

This method has permitted identification of 55 knowledge needs, methods and tools, thereby supporting problems associated with integrated water management.

Identified scientific support requirements were arranged in five categories without hierarchical order or exhaustiveness, as follows:

1. needs for state of the art of trans-disciplinary knowledge for given problems;
2. description of pressures-state-impacts relationships, especially for social and economic impacts;
3. operational tools and methods;

4. evaluation tools and methods to assess *ex ante* or *ex post* the programmes of measures;
5. innovative approaches to integrate social and economic dimensions in the formulation of responses to the measured or envisaged impacts.

The next step for these stakeholders within the ECOBAG process is to establish a hierarchy and to build up appropriate collaborative projects, an example of which is the project Concert'Eau described below.

3.4.3.3 Example of Concert'Eau (http://concerteau.ecobag.org/)

By running the collective and iterative process of identification of research needs in support of the WFD implementation, compliance with agriculture emerged as one of the priorities of the Adour-Garonne RBD. Within the Basin, 39% of surface water bodies risk not reaching the Good Status objective of the WFD by 2015. Nitrate and pesticides are one of the most important issues.

At European level, pollution of drinking water by pesticides and nitrates is acute for most EU Member States. The stake is linked to the compliance of the Water Framework Directive and of the Common Agricultural Policy by reducing the impact of agricultural pressures on environment without decreasing the sustainability of farming.

In these contexts, a Pilot River Basin 'Gascogne Rivers' has been proposed to the WFD Common Implementation Strategy (CIS).[5] Within the 'Gascogne Rivers' area (6800 km^2 representing 6 % of Adour-Garonne RBD and 263 000 inhabitants), 60–80% of the area is agricultural zones (maize, sunflower, soya and breeding). Nitrate pollution essentially originates from agriculture (87%) with a nitrogen surplus lying between 30 and 80 kg of nitrogen per year. Other origins of nitrogen are wastewaters from cities (6%), villages (4%) and industries (3%).

According to the regional policy on pesticides in place in the 'Région Midi-Pyrénées' (driven by the GRAMIP or Regional Group for Actions on Pesticides in Midi-Pyrénées Region), most of the territories of the Gascogne Rivers PRB have been classified as priority zones for reducing surface water pollutions. Within the Gascogne Rivers PRB, actors of the upstream part of the River Gers basin (entitled zone 'GersAmont') (http://www.gers-gascogne.com) have been mobilized for several years to face the degradation of the quality of surface waters. Quality of drinking water is one of the main concerns for 30 000 inhabitants. Within 'Gers Amont', 65% of water quality controls display at least one compound with a concentration higher than 0.1 $\mu g\ L^{-1}$. Total concentrations of all compounds are higher than 0.5 $\mu g\ L^{-1}$, which is the maximum value authorized for drinking water.

Conclusions from the work conducted in the PRB were that agro-environmental measures tested in four experimental small agricultural watersheds (from 100 to 1000 ha) have not been applied on a large scale. Within ECOBAG, analyses conducted by sociologists have demonstrated the importance of socio-economic dimensions in the process of co-operation and in the willingness of farmers to participate in the implementation of water protection projects.

To improve the situation, a project funded by the LIFE ENVIRONMENT programme and the Adour-Garonne Water Agency, and coordinated by ECOBAG, began in 2006, proposing to develop a collaborative technological platform (CTP) to support the water managers in their objective to efficiently apply measures on a large scale to reduce agricultural impacts on the quality of surface waters. This CTP is being developed and in the process of demonstration in the Gascogne Rivers PRB; it gathers scientists from a broad range of disciplines, decision makers, water managers, cooperatives and agriculture organizations. The CTP enables facilitation of a collaborative construction of scenarios with the stakeholders (farmers, local water managers, users and scientists) striving for changes to agricultural practices. In parallel, the Adour-Garonne water agency and the local political authority have been conducting, for 3 years, a broad programme of measures to reduce the impacts of pesticides on the Gers River upstream of the city of Auch. The CTP facilitates the *ex ante* evaluation of the environmental, social and economic consequences of these scenarios. It makes it possible to compare the scenarios proposed by the farmers and local agriculture and water managers, to define or to improve the measures needed for implementing European, National and Regional policies (Figure 3.4.3), including the WFD and the Common Agriculture Policy (CAP).

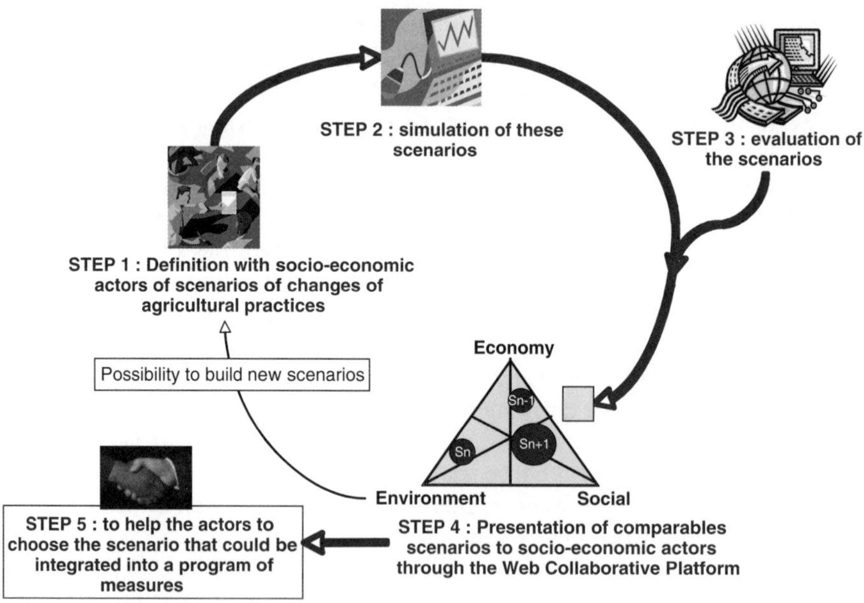

Figure 3.4.3 The CTP Concert'Eau, which includes a high-technology toolbox, supports a process of collaborative work between representatives of actors of agriculture and water. The process is organized in five steps.

The CTP includes five steps aiming to:

1. prepare and coordinate working groups of actors (cooperatives and agriculture organisations, administrative officials, scientists, NGO), producing different scenarios of changes of agricultural practices that would match the WFD and the Common Agriculture Policy objectives, and that present economic viability and social acceptability;
2. simulate the scenarios of changes of agricultural practices;
3. evaluate the social, economic and environmental consequences of these scenarios;
4. help the actors of agriculture and water to compare the different scenarios, through the decision support system (adapted from Mulino), in their social, economic and environmental components.
5. help the actors to choose from among the different proposed scenarios the one that could be integrated into their program of measures.

The Concert'Eau project is developed to be transferable to other territories or to other environmental issues. The potential of transfer is tested with the Region of Navarre in Spain and in the Alsace-Lorraine Région in France. Its efficiency has been demonstrated through the decision of the Adour-Garonne Water Agency and the Conseil Général du Gers to use its results to reshape their Programme of measures for 2010.

3.4.4 Conclusions

For 14 years, ECOBAG has tested several means to develop an efficient science–policy interface and demonstrated its feasibility at a large RBD scale (Adour-Garonne RBD has 7 million inhabitants), on the basis of collective identification of scientific support needs and of the development of effective supports by R&D activities.

The success of this process is based on various SPI components (as described in several chapters of this book). The activity of knowledge transfer is producing more fluxes of information between actors of a RBD, and, in addition, relationships are being developed between scientist and socio-economics actors that stimulate the collective and iterative process required to identify the needs of scientific supports.

Concert'Eau has demonstrated that the development of decision support systems has to rely on prior collective identification of needs, thereby ensuring the taking-up of developed tools in the process of integrated water management. The project's success is demonstrated by the fact that water managers will use it to improve their programme of measures. To reach this goal, the process of construction of the Concert'Eau Project has taken into account the different dialogue levels at the European, Adour-Garonne RBD, Regional and local scales.

We have also learnt that it is important to reduce, as much as possible, the period of collective reflection to identify the needs of scientific supports. This is one of the most important means of improvement, which ECOBAG has to reinforce. ECOBAG has also learnt that the acceptability and usability of developed tools, methods and knowledge needed by the actors of integrated water management require that activities of research and development be carried out in partnerships involving research, enterprises and socio-economic actors.

References

1. E. Erno-Kjolhede, *Managing Collaborative Research: Unveiling the Micro-dynamics of the European Triple Helix*, Copenhagen Business School Press, 2001, 345 pp.
2. M. Gibbons, C. Limoges, H. Nowotny, S. Schwartzman, P. Scott and M. Trow, *The New Production of Knowledge, The Dynamics of Science and Research in Contemporary Societies*, Sage Publications, London, 1994.
3. H. Etzkowitz and L. Leydesdorff, (Eds.), *Universities and the Global Knowledge Economy. A Triple Helix of University-Industry-Government Relations*, Pinter, London and Washington, 1997.
4. H. Etzkowitz, A. Webster and P. Healey (eds), *Capitalizing Knowledge. New Intersections of Industry and Academia*, 1998, SUNY.
5. M. Gaeckler and P. Vervier, *Report of the Pilot River Basin Group on Agriculture – Phase II – period September 2005–December 2006. – 2007–Experience in Analysis of Pressures and Impacts from Agriculture on Water Resources and Developing a Related Programme of Measures*, ed M. Cherlet, Institute for Environment and Sustainability, 2007, 217 pp.

Section 4: Communication and Education Needs

CHAPTER 4.1

General Introduction on Communication and Education Needs*

PHILIPPE QUEVAUVILLER[a, b]

[a] European Commission, DG Research, rue de la Loi 200, 1049 Brussels, Belgium; [b] Vrije Universiteit Brussel (VUB), IUWPARE, Building T, Pleinlaan 2, 1180 Brussels, Belgium

4.1.1 Introduction

Discussions at 'science-meets-policy' events[1] often highlighted difficulties of communication between scientists and policy-makers (*see* Chapter 1.2). The two communities have different codes and logics and hence the dialogue is far from straightforward. Policy-makers often think that science should be tailored to support policy, while scientists are often upset if their research is labelled as 'applied'. Furthermore, scientists are naturally striving to 'find the best approximation of the truth' and to focus on solving remaining uncertainties rather than working on 'compromises' usable by policy-makers. From the policy-maker's viewpoint, communication toward constituencies is aimed at convincing them that an issue is relevant at policy level, and they need 'strong science' to present evidence about the problem relevance, the costs of not addressing it, likely solutions and related expected implementation costs.[1] Data and facts supporting required arguments addressed to various stakeholders

*The views expressed in this chapter are purely those of the author and may not in any circumstances be regarded as stating a formal position of the European Commission.

Water System Science and Policy Interfacing
Edited by Philippe Quevauviller

Published by the Royal Society of Chemistry, www.rsc.org

cannot all be provided by researchers in an isolated way. The need for strong science goes hand in hand with communication that will enable to provide the evidence that policy makers are seeking for. As highlighted in Chapter 1.2, policy-relevant research generally focuses on 'real-world' situations, and communication of findings has to be conceived in such a way as to avoid the risk of feeling that research is abstract and irrelevant to society's pressing challenges. This is exacerbated by the increasing need for a greater inter-disciplinarity, particularly acute for integrated water resources management, which also requires proper communication within the scientific community and between different disciplines, which is widely developed in the present book.

Experiences linked to the implementation of the EU Water Framework Directive (2000/60/EC) have shown that contacts between researchers and policy-makers made at an early state of projects resulted in a better communication of scientific findings.[2] This is not solely concerning communication but also the identification of the right question to be addressed – at the right moment – by research, and this requires a proper dialogue between the scientific and policy-maker's communities. The active interactions of researchers with the full range of interested groups enhance their ability to 'see the whole picture' and identify salient questions, and provide independent assessments of policy options.[1]

4.1.2 Communication in the Light of Matching Agendas

Coordinators of projects that have been identified as relevant to water policies (see examples in Chapter 1.4) need to clarify specific issues to policy-makers, by describing the aims, milestones and technical challenges, and to get feedback from them so that they can understand what the policy expectations are over the duration of their project. Unfortunately, in practice such exchanges of information/knowledge rarely occur, which may lead to divergent directions being taken by the projects in comparison to policy orientations.[3]

Matching research and policy agendas is not easily achieved. Indeed, besides research agendas set specifically for policy development, other research activities are carried out in organisations such as universities, academies and research councils that are not necessarily driven by policy concerns. Research agendas may often have a poor match with the shape of the 'real world' social problems that policy is grappling with.[1] Concerns were expressed, for example, that many FP6 environment-related RTD projects are driven by the interests of the science community rather than the policy community (with the exception of projects funded under the 'Scientific Support to Policy'; *see* Chapter 1.4). This led to the question (largely addressed in Chapter 1.2) of how it is possible to ensure that the framing of problems for research takes into account the perspectives of both the policy and research communities.[1]

4.1.3 Enhancing Dialogue

All parties involved in science–policy interfacing need to create opportunities for dialogue, which should lead to the development of trust, helping to improve not only communication but also mutual understanding between the policy and science communities.[1]

Increased efforts are required to make research outcomes more accessible to policy makers. Different audiences need different outputs, and traditional forms of communication within the science community (*e.g.* in the form of scientific papers) do not meet policy-makers' needs. This is well illustrated by the situation encountered in the 'water world', and has a close link with the issue of translation/synthesis of findings for policy uses that is further discussed in Chapters 1.2 and 5.2.

Researchers now increasingly recognise the need to include dissemination plans and set a percentage of their budget for communication in project proposals. This does not mean, however, that scientific knowledge will be immediately transferable into policy. In other words, while emerging problems may be identified by science, policy-makers may not immediately see their relevance.[1]

4.1.4 Synthesis Needs

At the end of research projects, the most critical issue is the way the scientific information is 'digested' or 'translated' so that it may efficiently be disseminated to policy end-users and possibly applied (*see* Chapter 5.2 of this book discussing the role of 'translator' in this context). This is certainly the weakest link of the science–policy chain. Indeed, it is often considered that only a small percentage of research projects are known by policy implementers at national or regional level, which illustrates the need to improve awareness about scientific outputs but also to encourage policy actors to reflect on research needs linked to the policies they are responsible for.[1,3] The need for analytical synthesis and communication skills to assimilate large amounts of information, identify the essence and communicate this for non-specialists becomes increasingly acute.

One might consider that a scientists' job is about performing research and publishing that work in peer-reviewed scientific journals. Few efforts are made to 'market' specific users as the criteria for communication are rather linked to the international reputation of selected journals. Scientists might be dynamic communicators in conferences when presenting their work; they will often be less at ease when being in the situation of explaining their results to non-scientists. As suggested by Holmes *et al.*, others should take over the responsibility of translate scientific findings for wider audiences including policy makers.[1] This often requires a rewriting of published material to make it 'accessible' without changing the substance or over-simplifying. This translation process is discussed in Chapters 1.2 and 5.2. Let us repeat here that the issue is not to translate science into 'what is acceptable to decision-makers' but to consider different levels for different audiences: (i) extensive, detailed reporting for expert community, (ii) ten-page summary plus detailed annex for decision-makers and (iii) three-page folder plus video for general public.

4.1.5 Exchange Platforms, Networking

Participation of policy officers in research project meetings may not be practicable due to a lack of resources but efforts are needed to organise regular joint meetings focusing on specific themes. This is already happening in the WFD sector thanks to the Common Implementation Strategy platform[2] and should be systematised. In particular, science into policy can be promoted through working groups that include both scientists and end-users. The need for national 'receptors' was highlighted to ensure that countries are better prepared to act when new European legislation is coming up.[1] A factor that influences this is the existence of dedicated national structures responsible for liaising and communicating with the European level; examples of this are given in Section 3 of this book (French and Canadian experiences). We could go even further in recommending regional relays, as illustrated in recent examples.[3]

4.1.6 Education Needs

Finally, education through appropriate training (either at university level or through *ad hoc* training of professionals) is key to enhancing mutual understanding and communication of science–policy interactions. This aspect has long been absent from academic curricula, but awareness of the importance of proper interfacing is now increasingly recognised at national and international level (*e.g.* in the framework of Master degrees), an example of which is given in Chapter 4.4. Education needs concern policy-makers as well as scientists who need to acquire a 'transversal thinking'. We have to recognise, however, that this interfacing awareness, while important in the context of the research and policy works, will never be considered as a strong pillar by the respective communities. There is a strong need to develop *ad hoc* curricula responding to specific profiles of 'interfacing' people that should become a recognised (and rewarded!) function. This interfacing function is increasingly taken on board by consultancy companies or scientists 'thinking out of the box' but it is still not very high in the research and policy agendas. Awareness is rapidly changing, however, and this is reflected by specific services developed in environmental agencies/ministries (see, for example, Chapter 3.1). The next step will be to recognise the interfacing role as a specific function/job, which should go hand in hand with the development of appropriate education and training programmes.

References

1. A. Scott, J. Holmes, G. Steyn, S. Wickham and J. Murlis, Report of the 'Science-Meets-Policy, Conference, London, November 2005, www.theknowledgebridge.com.
2. Ph. Quevauviller, P. Balabanis, C. Fragakis, M. Weydert, M. Oliver, A. Kaschl, G. Arnold, A. Kroll, L. Galbiati, J. M. Zaldivar and G. Bidoglio, *Environ. Sci. Policy*, 2005, **8**, 203.
3. Ph. Quevauviller, B. Harris and P. Vervier, in: *The Water Framework Directive – Ecological and Chemical Status Monitoring*, ed. Ph. Quevauviller, U. Borchers, K. C. Thompson and T. Simonart, John Wiley & Sons Ltd., Chichester, 2008, pp. 411–444.

CHAPTER 4.2

WISE-RTD – A Portal for Science & Technology Transfer to Policy Making & Implementation in Integrated Water Resources Management

WILLEM J. DE LANGE,[a] JURGEN PLYSON,[b] PATRICK WILLEMS,[c] THOMAS VANSTEENKISTE,[c] FRANK PROVOST,[d] FRED HATTERMAN,[e] GUIDO VAES[d] AND PATRICK SWARTENBROECKX[d]

[a] Deltares, PO Box 85467, 3508 AL Utrecht, The Netherlands; [b] 2MPact, Kortrijksesteenweg 1007, 9000 Ghent, Belgium; [c] Katholieke Universiteit Leuven, Department of Civil Engineering-Hydraulics Section, Kasteelpark Arenberg 40, 3001 Leuven, Belgium; [d] ProvConsulting, Zakstraat 111A, 9112 Sinaai, Belgium; [e] Postdam Inst. For Climate Impact Research, Telegrafenberg A51, PO Box 60 12 03, 14412 Postdam, Germany

4.2.1 Introduction: Historical Perspective, Operation Environment

Bridging the science–policy gap has been of major concern to the European Commission since the beginning of this century and has led to several activities to support the issue. For instance, in the FP5 project Harmoni-CA

Water System Science and Policy Interfacing
Edited by Philippe Quevauviller

Published by the Royal Society of Chemistry, www.rsc.org

(Chapter 2.4) workshops have been held for water managers and policy makers to discuss needs and offers with scientists and technology providers. During the workshops it became clear that these different groups speak different languages, for instance: 'you need guidance' (scientists, meaning technical guidance) *versus* 'we already have guidance' (managers, meaning policy implementation guidance). Other Harmoni-CA activities made clear that managers are aware that 'there is much out there' and scientists show awareness 'that their developments are not ready to use in practice when a research project ends'. So, the need to transfer knowledge and experiences is apparent, but the time frame and the acting-worlds appear often to be too different to allow simple connection between offers and demands.

The WISE-RTD web portal aims to bridge the science–policy gap by transferring available information on guidance, state of the art technology and experiences in the field of integrated water resources management. The transfer consists of translating and linking this information to (in the vocabulary of) the planning process of the European Water Framework Directive and other related Directives. The WISE-RTD web portal connects the demands from policy making and implementation practice to technology providers and scientists in Europe and abroad working on water related problems.

The WISE-RTD web portal is part of the Water Information System for Europe – WISE (launched in 2007 under the auspices of the EC) – and as such it connects existing knowledge available in databases of the EC, national and river basin databases as well as databases on scientific knowledge and technology. The portal aims to support information finding at local implementation level as well as at river basin management level and provides information to strategic policy makers and policy implementation managers on solutions of the near future and on ongoing and planned research projects. The core of the system is an 'intelligent' procedure to map the scientific and technology information to water policy implementation tasks, and vice versa.

This chapter encompasses the aim and background of the WISE-RTD web portal. It describes how information can be made available from individual providers and how users are guided to find the relevant information. The support to information upload and how the system serves the science–policy interfacing is described in detail. Finally, the future sustainability of the system is examined.

4.2.2 Strategy and Scope: Towards a Sustainable Web Portal

The WISE-RTD web portal aims to provide information to actors at different levels. At the strategic level, science–policy interfacing is looking for the long-term prospects of research to support long-term policy issues like the sustainable ecological status of the WFD or the strategy on adaptation to climate change. Policy implementation by water managers requests information on process planning, concrete goals and guidance for operational management and stakeholder involvement. The technical and practical implementation needs support to solve concrete problems. The team that develops the portal aims to respond and adapt to user requests and to support both information providers and information finders at these different levels of operation.

The first developments of the WISE-RTD web portal primarily focused on support to the implementation of European Water-related Directives. Within the FP6 project SPI-Water, specific attention has been paid to enhance the usability of the portal in non-EC countries around the Mediterranean Sea (*i.e.* Morocco, Greece) and to improve the system based on experiences built up during workshops with actors in international river basins (*i.e.* Scheldt, Danube). During the workshops, river basin management actors have shown clear interest in the WISE-RTD portal and need for support in different languages. First steps have been made to provide WISE-RTD portal information in the major European languages.

The WISE-RTD web portal is aimed to be a system flexible for adaptations in which information is uploaded directly by the provider. The portal is based on open source software and is implemented such that the system is extremely flexible for adaptation and extension. Information can be uploaded on an individual basis instantaneously (Section 4.2.1) or by automatic linkage to larger sources (*i.e.* river basin databases) supported by the WISE-RTD Quality Assurance team (Section 4.2.2). All information undergoes quality checks with respect to content, availability and characterization.

The focus of WISE-RTD on the WFD context enables the definition of a complete set of keywords as the basis for mapping offers and demands. Though this means that the mapping (in support of the guided search, Section 4.2.4.3.1) is based on a closed set of keywords, WISE-RTD also provides an entry for searches based on freely chosen keywords (Section 4.2.4.3.2). As mentioned before, the context of WISE-RTD is planned to be extended, *e.g.* to other (European) water related Directives, and in that way it is an open system.

4.2.3 Background: Context, Science, Techniques

4.2.3.1 Context Scope

When the development of the WISE-RTD web portal started in 2003, numerous websites with relevant information could be found on the Internet. These include sites of research projects, sites on regional WFD implementation of specific river basin(s) (districts), on tools and technologies available in different areas of integrated river basin management, *etc.* A search based on related keywords on the web ('Google') gave thousands of results. A water manager looking for support to a specific WFD implementation problem could hardly find appropriate and useful results from such a long list. Words have different meanings in different contexts and determination of the appropriate combination of keywords is still not easy. In this context, the need for a tool that enables selection of information in support of the WFD implementation was defined within the science–policy bridging process started by the Harmoni-CA project (Chapter 2.4).

As pointed out above, the WISE-RTD portal's first aim was to transfer information in support of the implementation of the WFD. The WFD is a framework providing planning, goals and concepts and has set de facto standard terminology. Because of its clear context and the explanations in the CIS guidances,[1] their terminology is understood by policy makers, water managers and technical staff as well as by researchers and technology developers. Therefore, the WFD terminology is chosen as the basis for the linkage between offers and demands in WISE-RTD.

The WFD context and terminology enables the offer of a guided search in support of different types of users as well as helping information providers to characterize their items. However, the WFD focus limits the terminology used for the system, which gave rise to some negative impact. Therefore, the terminology in WISE-RTD has been expanded to areas related to integrated river basin management.

4.2.3.2 Scientific Background

WISE-RTD is in essence a knowledge base on where to find water science and technology and water policy implementation related information (currently in the context of the WFD). It supports the search for (WFD) specific information and provides links to related information. Key-issues derived from the WFD and its guidance documents (GDs) are chosen as the basis for the linkage between offers and demands in WISE-RTD. These key-issues are reflected by WFD implementation tasks, which are classified in different WFD-related disciplines or themes, *i.e.* economics, pressures and impacts, reference conditions. However, it is not the implementation tasks but keywords linked to these tasks that form the core of the mechanism used by the WISE-RTD web portal to identify related information sources. The keywords cover a list of water-related items and activities. They are classified in "Activity related" keywords (verbs) and "Context related" keywords (water related items).[2] The contextual keywords are structured in different levels, basically to make classifications. The level of the contextual keywords corresponds to different levels of detail: the lower the keyword level, the more detailed the information. Table 4.2.1 gives a few examples of activities and contextual keywords, including their classification.

In Table 4.2.1 'Water body status' is considered as a main keyword (level 1 keyword) and is further specified at higher level based on the keywords 'Water quantity status', 'Water quality status', 'Morphological status', *etc.* (level 2 keywords). Some of these level 2 keywords are further sub-classified at a next level (level 3 keywords) and sometimes at the deepest level (level 4 keywords). 'Water quality status', for instance, is detailed in 'nutrients', 'pesticides', *etc.* Activity related keywords are not structured in levels but instead were grouped

Table 4.2.1 Examples of activity related keywords and contextual keywords structured at their levels.

	Contextual keywords			
Activity related keywords	*Level 1*	*Level 2*	*Level 3*	*Level 4*
Identification	Elements in the water cycles			
Screening		Atmosphere		
			Precipitation	
Quantification				
		Water body type		
Selection			Surface waters	
Collection				River Lake
Scenario analysis			Groundwater	
Multi-criteria analysis			Transitional water Coastal water	
Cost–benefit analyses		Soil		
	Water body status			
Decision supporting				
Technical feasibility testing		Water quantity status		
Priority identification			Rainfall – runoff Evapotranspiration Water scarcity	

Critical path analysis			
		Water quality status	
Expert judgment			
			Nutrients
			Pesticides
Criteria definition			
			Toxic and hazardous substances
Design			
Development		Morphological status	
			River continuity
Harmonization			
			River profile
Relevance analysis			
	Driving forces or pressures		
Ability analysis			
		Pressure by human activity	
Physical modelling			
		Diffuse pressures	
Mathematical modelling			Nutrient loss
			Pesticide loss
Coding			
		Point pressures	
			Wastewater disposal

into common themes, like expert activities, water management activities, modelling disciplines, *etc*. (not in Table 4.2.1).

In the concept developed, the linking between information at the offer and demand side of WISE-RTD will be done through combinations of keywords. By combining an activity related keyword with one or more contextual keywords, a context is specified for the activity.[2] For instance, by combining the keywords 'Quantification' and 'Water Body Status', a more specific activity is considered, concerning the quantification of the water body status. When several water-related keywords are added, the activity becomes more specific. Moreover, depending on whether the linking is done only with higher level keywords, or also with lower level keywords, general or more specific activities can be covered. For instance, when 'Nutrients' and 'Surface water' are considered as keywords in the last example then one deals with the quantification of the nutrients (as specific components of the water body status) in surface waters. The activity is now specified at a more detailed level.

Each information source in WISE-RTD has to be linked with at least one activity related keyword and/or at least one contextual keyword and as such to one or more WFD implementation tasks. Information sources carrying the same keyword combinations are considered interrelated. This procedure for 'mapping' information forms an essential part of the WISE-RTD web portal system.

Clearly, a lot of links are possible by different combinations of activity and context related keywords. The WISE-RTD web portal handles the possible range of keyword combinations by organizing a multiple linking mechanism. This mechanism allows keywords combinations to be narrowed or broadened (starting from the default linking mechanism), resulting in less or more information that will be considered interrelated.

4.2.3.3 Technical Background

4.2.3.3.1 Database Structure, Available Categories, Link to Front and Back Office

WISE-RTD is open software in the sense that the source code is available for everyone to view. The system is developed around a well-organized information database, considered to be at the back-end of the WISE-RTD system. A database management system will establish links to the visible front-end of the system and provides the necessary operations to manage, map and visualize the stored data.

The WISE-RTD database structures information along two major categories; *i.e.* information and output from RTD projects *versus* policy implementation documents.

The output from RTD projects is considered as scientific information of use for supporting specific water policy implementation tasks. This scientific information can take the form of technical guidance or methodologies (as potential output of RTD projects); case studies and experience demonstrate the application or testing of specific methodologies and tools that can support policy implementation. On the policy implementation side, the information takes the form of guidance documents or implementation reports for specific cases. These policy implementation related documents are being faced with the output of RTD projects in the science–policy interface of WISE-RTD.

4.2.3.3.2 WPIS, Smart Link to Keyword Combinations

A tool has been developed to support the upload of information in the WISE-RTD system. This tool is called the WISE-RTD Web Portal Input System or WPIS. It provides an intelligent interface through which information sources can be uploaded to the WISE-RTD information database. The WPIS aims to retrieve all essential meta-information on the provided input with a minimum of effort by the provider. The upload procedure involves different steps, among which an important one is the selection of WFD items of most relevance to the uploaded information. WFD items are represented using a generalized terminology, and are associated to keywords. Through appropriate choice among the WFD items, correct keywords will be given to the input. Smart combinations of activity and context related keywords address directly WFD implementation topics and form the basis of the interrelations between different information items uploaded in the system. The WFD items are structured in levels; the lower the level, the more detailed the information. As such, the uploaded input automatically will provide information at a less or more detailed level. However, due to the limited focus of the portal (currently limited to the WFD), items other than WFD related are not foreseen in the input system. Input providers might need to suggest new items to extend the system.

The final step of the input upload involves providing a 15-line summary on the provided input. Based on this summary, the WISE-RTD Quality Assurance team decides to accept or correct the information input. More important is that this 15-line summary will appear at the front-end of the system. It might help the users of the system to quickly investigate whether the information is useful before it will be consulted in more detail. The information upload is completed with specifications on geography, URL, funding agency and project duration (if relevant).

Users do not need to register with the WPIS to be able to upload information. However, after registration the user is given the possibility to manage his/her input.

4.2.3.4 Guided and Non-guided Search Front End Link

To facilitate the search for information, the WISE-RTD web portal provides at its user interface guided and non-guided search engines. The guided search engines are user type specific.

4.2.3.4.1 Guided Search

The guided search is targeted to enhance the information search for several user groups (policy makers, operational managers, stakeholders, modellers, researchers, ICT-developers, *etc.*). The search items depend on the user group and aim to represent typical demands made by users. The search items are linked to keywords or WFD implementation tasks in the same way as is done for information items uploaded into the system. Through the mapping procedure, based on keywords, the user request will filter the available information to display the most relevant information.

4.2.3.4.2 Non-guided Search

The non-guided search (also called free search) allows users to search for information based on free text or through selection of specific keywords from the complete keyword list.

The free text search will look for the provided text in the information title and summary based on text-matching techniques. The text search will be performed for any text block present in the WISE-RTD system, but not in external web-sites. More details on the technical organization and implementation of the guided and non-guided searches can be found in Section 4.2.4.3.

4.2.4 Operation Process: Aim/Focus Management, Input/Output Support, QA, Adaptations, Extensions

4.2.4.1 Aim and Focus of Operational Management

The WISE-RTD web portal is supported by a team of specialists with different responsibilities. Sub-teams are responsible for the software-technical support (development and maintenance of the software and IT-infrastructure) (IT team), the conceptual development of the system (scientific team) and the quality assurance and user support (QA team). Expansion of the system is prepared and developed by the scientific team and implemented by the IT team. The QA team is responsible for the upload of information and the search for information by users and the quality assurance of this uploaded information. In the long term, user support is foreseen to be spread over different countries to provide support in different languages. Users can suggest improvements and expansions that will be forwarded to the scientific sub-team.

4.2.4.2 User Support 1: Upload of Information into WISE-RTD

The quality of the WISE-RTD system response to questions depends on the quality of the information provided in the system, on the performance of the procedure for mapping information and on the quality of the guided search entries. The WISE-RTD input is initially provided by the team members, but will be maintained by its users through the Web Portal Input System (WPIS; see Section 4.2.3.3.2). This input system has been developed in response to the experiences of users with the input of information (by project members or by external users as observed during the SPI-Water project, workshops and visits). It has been made as intelligent as possible, *e.g.* by a smart way of stepwise detailing the categorisation and associating keywords and by linking provided geographical information to all other geographical categorisation in the system (member states, river basins, *etc.*). The WPIS aims to retrieve information directly from the source, as opposed to the web portal team having to go treasure hunting.

The WPIS was designed to enable everybody in the field, who has any water related information that can be of use to someone else, to add their information to the web portal. To make the threshold as low as possible the system needs to be simple. Therefore, it is designed to be a simple-to-use system that leads the user automatically from one screen to another (Figure 4.2.1).

To keep all information up to date, users of the WPIS can create their own account to which all their input is linked. When they log on the system they can add or change their already submitted information. All information added into the system is categorised as 'to be checked', after this information has been quality assured by the WISE-RTD Quality Assurance (QA) team the information receives the 'checked' label. If the information is changed in the WPIS the classification will return to 'to be checked' (Figure 4.2.2).

Information added into the WPIS is immediately visible in the web portal. Nevertheless, the user is always warned that this information has not been checked yet by the QA team. This is important because the user of the WPIS is not always qualified to give the correct keywords to the uploaded information.

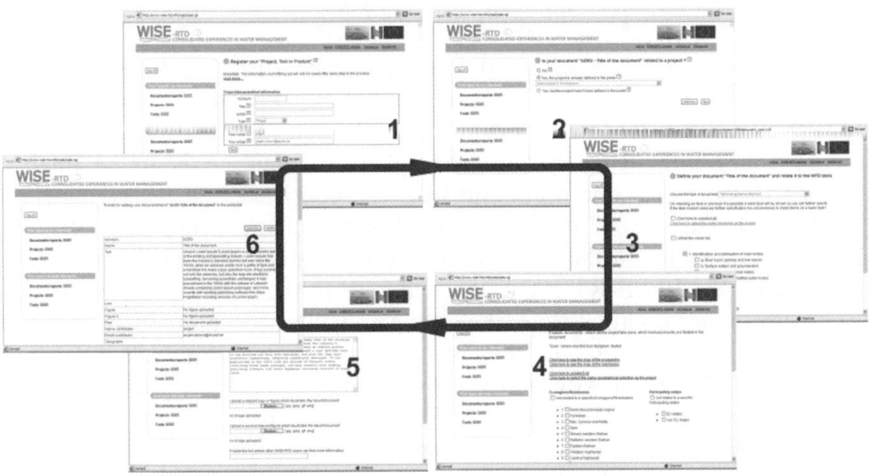

Figure 4.2.1 Adding information in six easy steps.

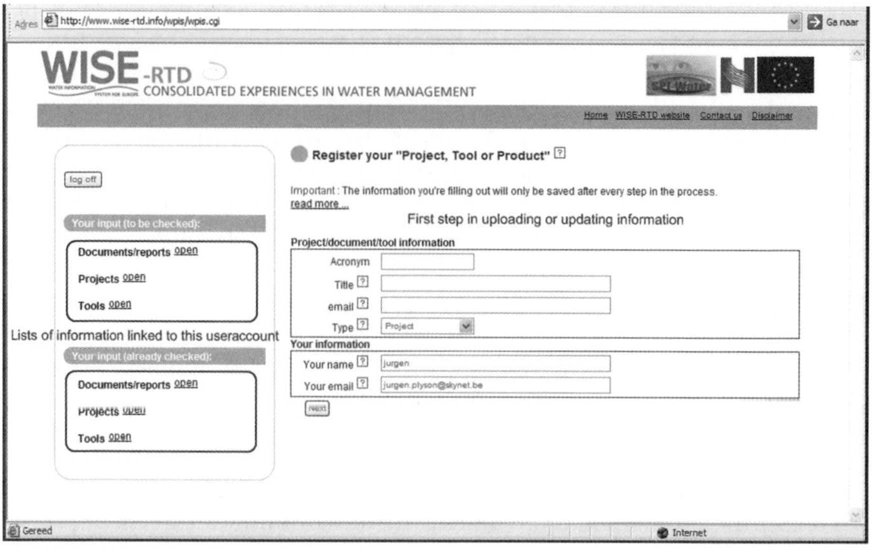

Figure 4.2.2 Account linked information is subdivided in 'to be checked' and 'checked' information lists.

In a preliminary version of the WPIS, users have been asked to select the applicable keywords from the vast keyword list. From experience we learned that users selected mostly too many keywords, such that their information would appear all over the web portal. To remediate this we replaced the keyword list with a set of predefined tasks and situations that can be selected. Every task stands for a group of logically combined keywords. With this way of presenting a selection we narrowed down the list and furthermore made sure that users select logical combinations of keywords instead of conflicting ones.

The WPIS tool ensures – through its guided steps – that the uploaded information becomes categorized and linked with keywords. However, practice has shown that interpretations of the WPIS steps still may differ between users. As such, the WISE-RTD portal cannot (yet) be considered a self-maintaining science–policy interfacing system. All uploaded input needs to be quality checked to maintain conceptual and technical consistency of the contents of WISE-RTD. This means that there is need of a permanent support, operating procedures to maintain the sustainability of the system. This task currently is executed by the WISE-RTD QA team. Based on the experiences of the QA team with the information flow, the QA team will also continuously improve the WPIS. The WPIS tool and the QA team play an important role in feeding the WISE-RTD portal with actual information. Clearly, for the short-term application and appreciation of the WISE-RTD web portal, it is important that the portal provides up-to-date information for actual issues in the present phases of the WFD implementation. The QA team has to support the retrieval of appropriate actual information for the different user groups of the portal.

This can be done by means of dissemination activities or user support in the upload of information.

4.2.4.3 User Support 2: Retrieving Results

Next to the linking of the scientific and technological outputs with policy implementation, the WISE-RTD science–policy interfacing instrument functions like a gateway, providing relevant information demanded by different user groups. Each user group has specific types of demands and wants to access and display the information in a way consistent with these demands. The WISE-RTD portal developed two major search entries at its user interface:

1. a guided search or user guidance, serving to guide specific user groups in finding relevant information;
2. a non-guided or free search application that searches for, and retrieves, related information based on free text or through selection from the predefined keyword list.

4.2.4.3.1 User-guided Search

User guidance by the WISE-RTD portal aims to meet the demands typically made by the different user groups involved in the implementation of water directives (in the current implementation of the portal limited to the WFD). The different user groups considered are:

- the water policy development and implementation community, covering the network of individuals and organizations involved in management;
- the scientific community, dealing with research, technology development and consultancy on water related issues;
- other stakeholders; representatives from industry, agriculture, NGOs, and so on.

The WISE-RTD user guidance will distinguish the above user groups and forward them to the customized demands for available information in a limited number of steps. The different entries are developed in close co-operation with representatives of these user groups. Therefore, the structure, presentation and terminology are adapted to the needs of these users.

Guided Search as Policy Maker or Implementer. The policy makers and implementers constitute the main focus group of the WISE-RTD portal. For this user group, information uploaded in the interfacing instrument needs to be filtered for specific policies or policy implementation step(s) of concern. As such, the entry requires the policy maker or implementer to screen/select specific policy implementation steps or tasks. In the current version of the web portal, only the tasks to be accomplished in the WFD are considered. These WFD implementation tasks are classified by sequence and milestones of the WFD as policy implementation related users like to have guidance and examples on tasks to be carried out by given deadlines.

The entry figure is presented by the flowchart of the (first) planning cycle required in the WFD and is taken from the WFD CIS Guidance Document 11 (Figure 4.2.3). The overall flowchart lists the major steps in the implementation of the Water Framework Directive, indicates linkages between the tasks, shows milestones and deadlines and includes recommendations on the overall timing for accomplishing the tasks. The figure identifies the main steps for the implementation of the WFD but within each step different implementation tasks exists.

The different policy implementation tasks are specified from the CIS Guidance Documents (GDs). According to these GDs, the tasks are classified in different water-related disciplines or themes: water economy (GD1), water bodies (GD2), pressures and impacts (GD3), *etc*. The different implementation tasks are, however, not sequentially described in different consecutive sections of the GDs.[2]

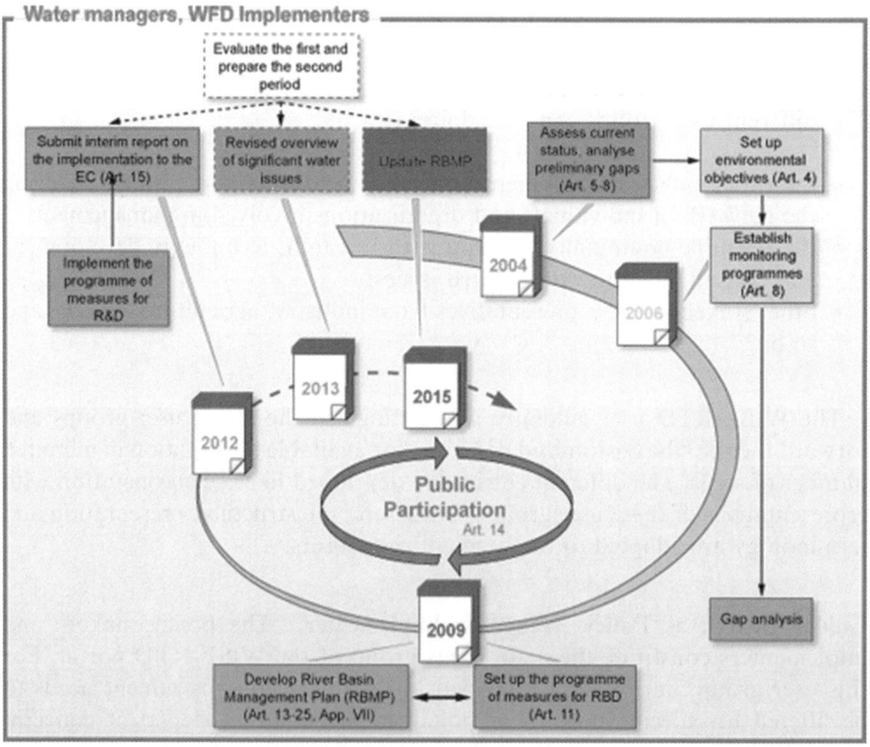

Figure 4.2.3 Milestones in WFD from CIS Guidance 11.

In the planning cycle, the individual tasks are grouped according to stages with deadlines by the key dates: 2004 for the characterisation and analysis tasks, 2006 for monitoring and compliance testing tasks, 2009 for the programme of measures and the river basin management plans to be developed, 2012 for the implementation of these plans, and 2015, 2021 and 2027 for evaluation and revision.[2] The planning guide in GD11 is specifically related to the implementation of the WFD. It contains and structures all steps needed for WFD implementation, sequentially according to time schedule and key dates. The implementation tasks are linked to the keywords, enabling the policy demands to be interfaced with the RTD results and the other information sources uploaded in the portal.

As the focus of the WISE-RTD web portal grows to a wider scope (beyond the WFD), this policy makers and implementers' entry will need further development. Policy implementation tasks covering the newly implemented water policies need to be defined. Also, additional keywords might need to be specified. Moreover, for each concerned policy, an implementation scheme corresponding to the WFD planning cycle needs to be implemented in the portal.

Guided search as Consultants and Modellers. Viewed in the context of the WFD, holistic modelling of river basin hydrology, hydrodynamics and water quality, taking into account the origin of pollution and ecology of individual river stretches may play an important role. River basin models can provide information about, for example, the chemical status of the waters bodies, being tools for development of measures to realise the defined objectives as well as to provide solutions for problems and protection measures for the aquatic environment. However, consultants and modellers are often not aware of the technical requirements of the WFD that necessitate scientific support. They often do not know much about the policies, the implementation phases, and the role models can play in the implementation.[2]

The 'modellers' entry of the WISE-RTD portal focuses on how research and consultancy on water related issues can be addressed in the WFD. The entry is organized in four sub-entries, each advancing 'modelling' in a different context:

1. Type of tools
2. The modelling process
3. WFD tasks related to modelling
4. Model-supported water management.

The sub-entry starting from the tool categorization enables investigation of how classes of tools and/or individual tools may serve for water management in river basins. The tool categorization essentially comes from experienced modellers and is both water subsystem and process based (*i.e.* catchment rainfall-runoff models, hydrodynamic river models, groundwater models, and so on).

The second subentry on the modelling process aims to facilitate Quality Assurance (QA) throughout the entire modelling process. The approach is based on guidelines to support the modelling process, developed by the HarmoniQuA project (http://harmoniqua.wau.nl). The idea is that each single modelling step will be guided, with emphasis on transparency and reproducibility, accuracy criteria, model validation and uncertainty assessment (beyond the technical, modelling related uncertainties). The complex process of modelling is therefore decomposed in five modelling activities:[3] (i) model study plan, (ii) data and conceptualisation, (iii) model set-up, (iv) calibration and validation and (v) simulation and evaluation. These main steps are applied in the 'modelling process' subentry (Figure 4.2.4).

After selection of a specific step of the modelling process, the information uploaded in the WISE-RTD web portal will be filtered. This is technically done by selection of specific keywords related to each step of the modelling process.

If technical users want to get access to the WFD tasks in which major modelling activities can be expected, the third subentry 'WFD tasks related to modelling' can be followed.

The WFD requires development and implementation of a wide range of technical tasks that relate *inter alia* to characterization of catchments, monitoring procedures, establishing relationships between catchment pressures and impacts on aquatic systems, and remediation measures where water bodies are considered to be at risk of failing to achieve their environmental objectives.

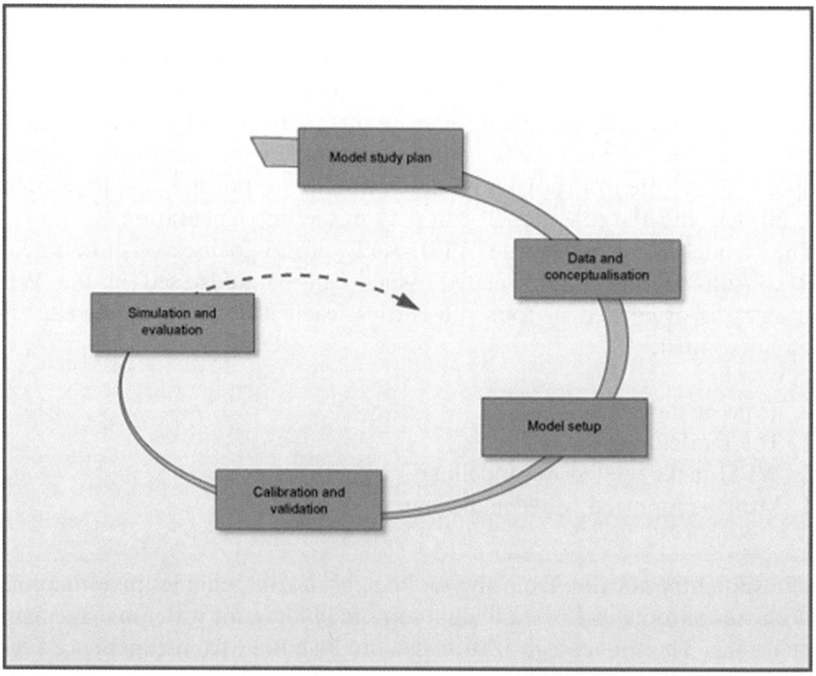

Figure 4.2.4 Main steps to structure the modelling process.

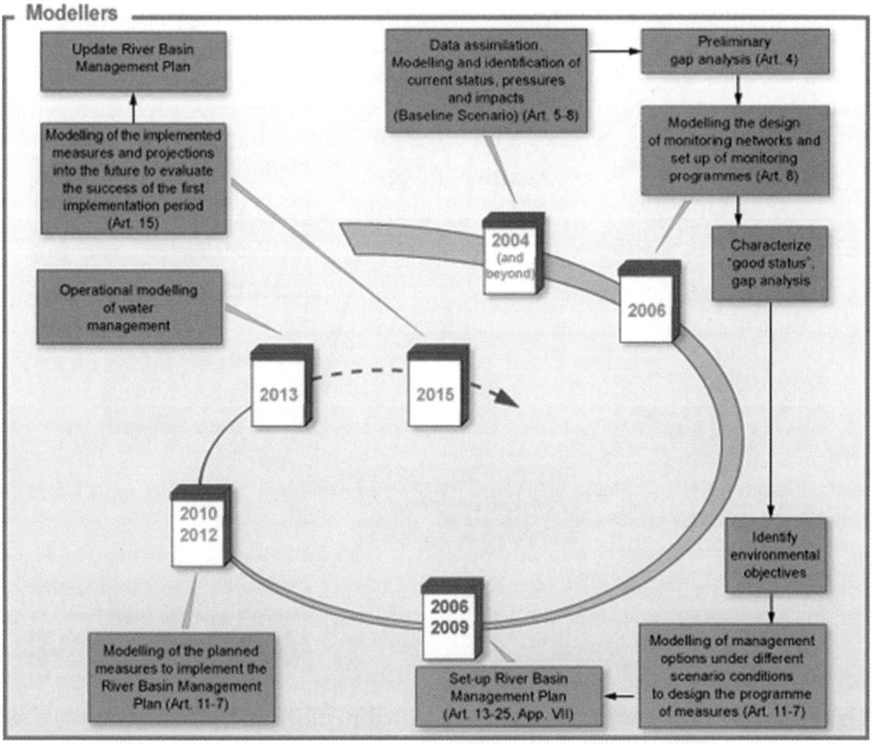

Figure 4.2.5 WFD's implementation timetable and the role of modelling.

For each step in the WFD implementation cycle (Figure 4.2.5) WFD tasks in which major modelling activities can be expected are specified. These implementation tasks will forward to uploaded information in WISE-RTD. Information about time frames and slots where model studies can play a major role in the WFD implementation are also related in this entry.

Though many guidances for water managers supporting the WFD implementation advocate the use of models, a guideline on the use of models in water management was lacking. The fourth and innovative 'model-supported water management' subentry intends to meet this demand and find elements of practical implementation of the WFD.

The planning framework applied integrates modelling and management in a way that allows adaptive management and modelling, driven by participation, facilitates social learning and supports decision making. The planning framework, shown in Figure 4.2.6 and developed within the Harmoni-CA project,[4] describes the main tasks that have to be considered to integrate participatory river basin planning and modelling along the main stages of the WFD implementation. The sub-entry is structured in six main steps (Figure 4.2.6): (1) problem description and goals definition, (2) conceptualization, (3) scenario definition and identification of management alternatives, (4) simulation and

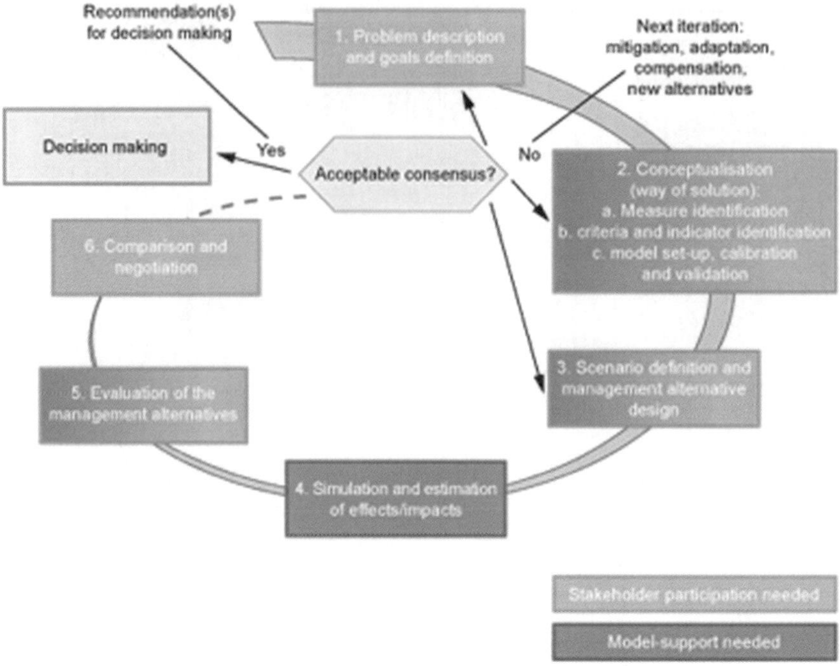

Figure 4.2.6 Framework for model-supported participatory planning of measures and integrated river basin management.[4]

estimation of effects/impacts, (5) evaluation of the management alternatives and (6) comparison and negotiation.

When selecting one of the boxes, additional information about the specific task is given as well as related guidance documents, tools, projects and experiences.

Guided Search as Stakeholder. Stakeholders are considered to be an important user group of the WISE-RTD portal as they are involved at all levels of water policy implementation. They may be representatives of industry, environmental, farming and community interest groups, and so on.

Available information of interest to stakeholder activities will only be provided by the WISE-RTD system after application of a filter, again based on keywords linked to each stakeholder user entry or filter item (Figure 4.2.7). Such information filters are foreseen for all sectors of the water industry (wastewater management, coastal management, *etc.*) and all forms of water use (drinking water supply, navigation, *etc.*).

4.2.4.3.2 User Non-guided Search: Open Questions 'Google'

Next to the user guidance to access information in the WISE-RTD portal, a simpler 'free search' alternative is offered. This entry requires no prior

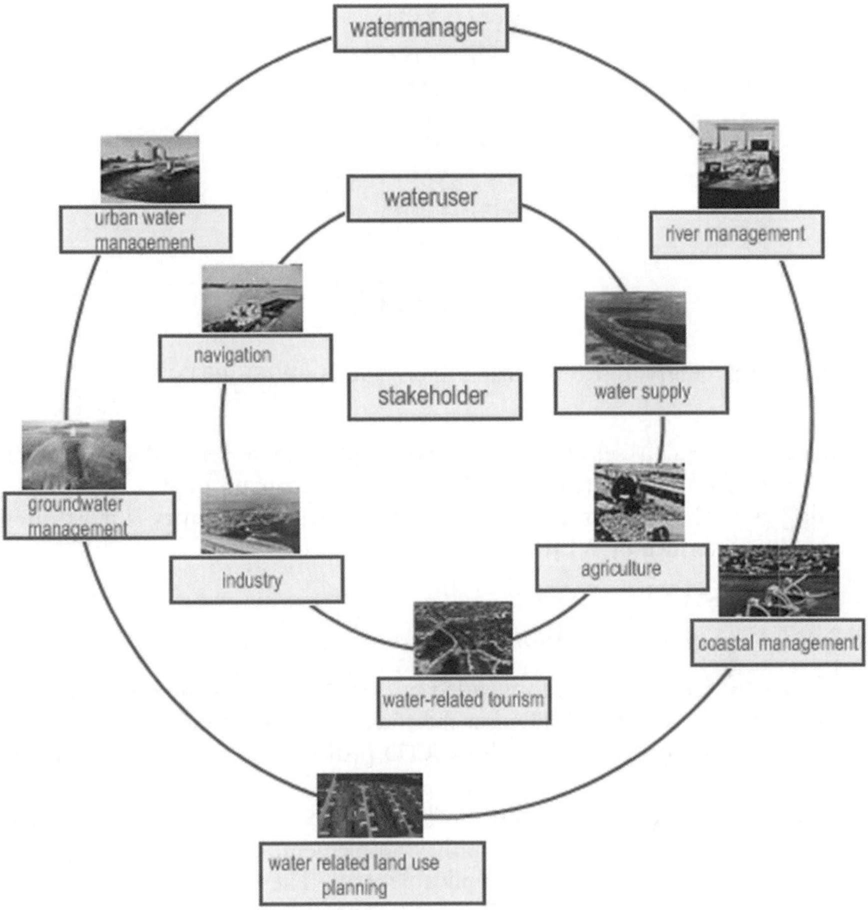

Figure 4.2.7 Stakeholder entry.

knowledge of the WFD and is developed to be understood and used by anyone. This 'free search' entry allows WISE-RTD information to be accessed and to transfer knowledge acquired in EU countries for helping the implementation of IWRM principles in other countries.

The 'free search' users will be able to retrieve information directly from the WISE-RTD database, without following a sequence of user guidance steps. The information may be searched in two ways: by a free text term or by a search in a predefined list of keywords. As explained above, the free text search will look for information containing one or more words that are specified in the free text.

The more advanced search on predefined keywords is based on the occurrence of the specified keywords associated to the information sources. Every single information source in WISE-RTD is associated to one or more (contextual) keywords, giving indications about themes or contents of information

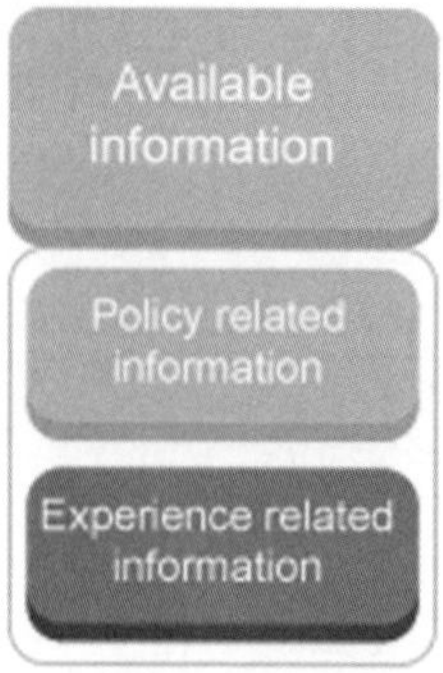

Figure 4.2.8 Entry to select available information in WISE-RTD.

sources. As such, a selection of keywords may be a very effective search manner, resulting in limited but highly relevant information links.

Analysis of the portal statistics showed this 'free search' entry is most often used at the interface of WISE-RTD.

4.2.4.3.3 Showing Available Information to the User

An organized collection of all uploaded information is presented in a final and central entry at the WISE-RTD user interface. It allows information on policy implementation documents as well as RTD projects and results to be directly consulted in the WISE-RTD system.

Through this 'information' entry information boxes on relevant outputs as developed through research and policy implementation projects are made accessible to users in a readily applicable way. These different information boxes at the opening page of WISE-RTD (Figure 4.2.8) show a clear difference between the policy implementation information (upper box) and the RTD project and results (lower box). Every box compiles corresponding information lists where information can be consulted from.

4.2.5 User Interaction: Evaluation and Adaptation

Bridging the science–policy gap is not a one-time effort. The development of WISE-RTD started in 2003 and has been evaluated in numerous workshops by all kinds of users (scientists and technicians, water managers, policy makers, *etc.*) ever since. The system has been adapted to the user comments in virtually all cases, in several cases even a major revision has been carried out. In this section, we illustrate the process of bridging the science–policy gap by addressing how the development team adapted the system to user comments and coped with some crucial issues.

4.2.5.1 First Phase, Before Launch 2003–2007 HCA Period

During the first workshops, it became clear that the position of the WISE-RTD portal is new in the way that it aims to support communication between different 'worlds of action'. The selection of science results and technologies and the translation of the applicability into language of water managers and policy makers is unprecedented.

Given that the WISE-RTD portal obtained a position on the Internet, the principal choice of 'linking without document storage' became apparent:

- The basis of linking information to requests (mapping) followed from the context of the science–policy focus being the WFD. However, numerous attempts have been made to accomplish a reliable mapping; several attempts to link available information to lists of keywords failed. The first successes came only after building a tool supporting the upload of information.
- A general philosophy behind the system has been that knowledge on information should be stored rather than the information itself. The provider of information has the best knowledge to specify the purpose and for what problems it can be used. By giving a short summary on the uploaded information and specifying the appropriate characterization in terms of keywords, users may profit from this knowledge that is made explicit in the portal. The information itself is maintained at the provider's website.

On the side of information requests, it appeared that different kinds of users have different understanding and needs. Water managers have a different request for an issue like 'tools' or 'guidances' than their technical staff (modellers), or than technical developers; same words have a different meaning. Therefore, different entries for guided search have been initiated. During different workshops it appeared that most European river basin managers work along the time frame and deliverables of the WFD. A graphical representation (for instance Figure 4.2.3) of this appeared to be very attractive for this group as guidance to find relevant information. This is one example of the observation that users have been extremely important in the development of entries on guided search to find appropriate selection schemes, lists, *etc.*

4.2.5.2 Second Phase, After Launch 2006–2008 SPI-Water Workshops

In the second phase, upload of large amounts of data and extensive evaluation in several river basin organisations have been accomplished within the FP6 SPI Water project.[5] More than 500 European funded research projects and their results have been uploaded. The QA team provided significant testing and generated improvement of the input support tool.

During this activity, a significant amount of (European) research and implementation projects and results were found that have vanished or do not have any place (website, document) to which users can be forwarded by WISE-RTD. For those results and projects that could provide some digital information, a (temporary) document store has been constructed, so that the information can remain available. What to do with vanishing information and how to ensure that existing information will stay available is still a subject of discussion in the European Commission. This is a major issue for the sustainability of the Web Portal.

The WISE-RTD web portal has been evaluated[5,6] in workshops in 2007–2008 at several river basins in Europe (Scheldt, Danube, Pinios) and in non-EU countries along the Mediterranean Sea (Morocco, Lebanon). In the context of the workshops, especially the one along the Mediterranean Sea, a French version of the portal has been developed. Later, Spanish, Polish, German and Dutch versions were also started to support the dissemination.

During the workshops, many users responded positively to the aim and principles of the WISE-RTD portal. The initiative taken at the European level to initiate this system is very well appreciated. However, users ascertained a lack of widespread, up-to-date information. Application of the WISE-RTD portal will only occur when the system provides information for actual issues in the present phases of the WFD implementation. Present information in the portal often does not yet cover tasks that are presently worked on. Most information on experiences and projects appears to connect to the policy implementation tasks of the past. Consequently, the actual support may be of limited magnitude.

The WISE-RTD user-interface was in general considered as 'technical' and 'complex'. Moreover, non-EU countries are not familiar with the specific terminology describing the WFD steps. Improvements to enhance the usability and user friendliness of the WISE-RTD web portal were strongly recommended. Based on the gathered suggestions and requests, a new look and feel of the user entries at the WISE-RTD front-end was discussed and a transparent and extended information structure for the database was established during the second phase of the SPI-Water project. As such, the WISE-RTD web portal evolved step by step to a sustainable and solid system, attempting to counteract the science–policy interfacing problem.

4.2.5.3 Evaluation in The Netherlands

A series of workshops in The Netherlands were held in 2007 at several water boards and river basin management meetings. At the applied level in the water boards, much interest was shown in information/reports on problem solving and practical decision making by similar organisations abroad. Technical issues in water boards often appeared to be very specific and not available yet on WISE-RTD. A clear need is encountered to upload this kind of information. The attendants of the workshops were positive with respect to uploading information on experiences and implementation practices, though that information is often in Dutch. This upload is stimulated by the translation of the WISE-RTD into different languages, one of which is Dutch. To open up this 'Dutch' information for persons abroad, it was suggested that a 15-line summary in English is provided.

The series of workshops in The Netherlands was held to derive support for the establishment of an automatic link between WISE-RTD and the renewed version of the national WFD database system in The Netherlands on which all parties are obliged to store there WFD reports. The water managers were convinced of the WISE-RTD strengths and gave support. Meanwhile, the technical implementation of the automatic link has been accomplished. It is also used as a prototype for automatic links with other databases (*e.g.* Danube river basin), which currently (2008) are in the process of being accomplished.

4.2.5.4 Concluding Remarks

The dissemination workshops held on these WISE-RTD prototypes have led to an overall good impression of user demands and the capability of WISE-RTD to respond to these. Many users responded positively to the aim and principles of WISE-RTD and were motivated to learn how the system works. During practical use, however, the system appears to be not intuitive, somewhat complex and the strength of it does not become clear within the time people are willing to spend. In addition, at the time the workshops were held, information loading activities were going on. This might cause a low search performance for some tasks due to a lack of information content. Aside from these feelings and remarks, users have found that the WISE-RTD database contained interesting documents, but it needs some time to find out the best way to approach them. Most often the 'free search' approach was consulted to enter the information database. This approach is preferred by users as it requires no knowledge of the WFD (steps) and because it appears very attractive and is easy to use (similar to a 'Google'-like search).

In response to the experiences and recommendations of the many evaluations, the system has been gradually improved and minor and major problems concerning lay-out, structure of the user entry and information content have been solved step by step. A future challenge will be found in the multiple demands to cooperate, participate and support projects and organisations with respect to linking their information systems on water-related issues.

4.2.6 Sustainability

WISE-RTD has been developed and supported by projects funded by the European Commission. Owing to its flexibility, extensions can be developed easily. This enables it to provide support to projects aiming at new areas of interest, such as flooding, twinning, adaptive management, *etc.* Continuation of support therefore is foreseen partly to come from research and implementation projects in support of European Directives.

The team supporting WISE-RTD will establish a sustainable organisation and will be guided by an advisory board consisting of the major stakeholders. Though the kernel of the team will be relatively permanent, it will be adapted to the needs of the projects demanding support, *e.g.* by specialists on specific subjects.

As mentioned before, user support is foreseen to be operated from different institutes in different countries and also may provide support to the upload of relevant information. Contact will be strengthened with (inter-)national river basin organisations to open up reports with technical practices and decision making experiences.

References

1. CIS (Common Implementation Strategy) Guidances 2004–2008, circa.europa.eu/Public/irc/env/wfd/library?l=/framework_directive/guidance_documents/.
2. P. Willems and W. J. de Lange, Concept of technical support to science–policy interfacing with respect to the implementation of the European water framework directive, *Environ. Sci. Policy*, 2007, **10**(5), 464–473.
3. H. Scholten, J. C. Refsgaard and A. Kassahun, *Structuring Multidisciplinary Knowledge for Model Based Water Management: the HarmoniQuA Approach*, ed. C. Pahl, S. Schmidt and T. Jakeman, iEMSs 2004 International Congress: 'Complexity and Integrated Resources Management', 2004, International Environmental Modelling and Software Society, Osnabrück, Germany.
4. F. F. Hattermann, Z. W. Kundzewicz et al., Model-supported implementation of the Water Framework Directive—A Water Manager's Guide, Guide in development under Work Package 3 of Harmoni-CA, E.C. funded concerted action, contract EVKI-2001-00192, 2007, http:/www.Harmoni-CA.info.
5. SPI-Water D30, Guidelines and recommendations on the transfer of knowledge on IWRM and WFD to the non EU countries, report D30 for SPI-Water Science-Policy interfacing in support of the water framework directive implementation, contract no. 044357, 2008.
6. Arcadis, *Verbeteren draagvlak WISE-RTD*, ARCADIS Den Bosch, The Netherlands, 2007, report 110402.001325.

CHAPTER 4.3

EUGRIS – More Than a Database

JÖRG FRAUENSTEIN,[a] PAUL BARDOS[b] AND ANTONY CHAPMAN[c]

[a] Federal Environment Agency, II 2.6, P.O. Box 1406, D-06813 Dessau-Rosslau, Germany; [b] r3 Environmental Technology Ltd, Room 120, Department of Soil Science, The University of Reading, Whiteknights, PO Box 233, Reading RG6 6DW, United Kingdom; [c] wca environment ltd., Brunel House, Volunteer Way, Faringdon, Oxfordshire SN7 7YR, United Kingdom

4.3.1 Introduction

The past decade has seen the development of numerous sources of web-based information across Europe, ranging from project websites, company websites, industry and regulatory bodies, online publications, best practice and other technical guidance. While the increased availability of this enormous range of information is, in general, a positive thing, allowing for better informed decision making to be made far quicker than previously, it has also brought with it a range of problems as well as solutions.

The development of web-based information resources has, until recently, been focused on the needs of a relatively small group of interested parties in each case. For example, websites for specific projects have generally been used as a platform for the exchange of information between project members, the storage of project deliverables and the announcement of events such as conferences and workshops. The outward looking section of such websites is often

Water System Science and Policy Interfacing
Edited by Philippe Quevauviller

Published by the Royal Society of Chemistry, www.rsc.org

limited to introductory information about the project and some general notes on the work. While this is useful to non-project members, the detailed project information is still relatively inaccessible. Other websites often focus their information towards a particular expert group of users who are most likely to use the information and for whom the website was originally conceived. Such a resource pays little heed to the potential user that non-core users could make of the information available. While these statements are clearly a generalisation, the result of much endeavour over recent years has been the production of a huge quantity of information that is not necessarily widely available, despite the fact that it is accessible online.

The development of multi-faceted European legislation co-ordinated over a large scale, beginning with the Water Framework Directive and continuing with other Directives for groundwater, waste and soil, among other things, has highlighted the need for information to be understandable for a wider audience as well as accessible to them. This has led to initiatives to help to improve the accessibility of information by creating online databases aimed at coordinating available information rather than generating information. This is the background to the development of EUGRIS, the soil and water information portal, which has developed over the last five years into one of the pre-eminent sources of technical, policy and research information and resources across Europe.

This chapter describes the history of EUGRIS, its value to the information market in Europe and its potential future as part of a wider consortium of information services. EUGRIS is accessible under www.EUGRIS.info and www.soilandwater.info.

4.3.2 Initiation of EUGRIS as an EU Funded Project

EUGRIS was originally developed between March 2003 and August 2005 in a 30 month project funded within the 5th Framework Programme of the European Commission (EVK1-CT-2002-80021). The German Federal Environment Agency co-ordinated the consortium of 11 partners from six countries, namely, Denmark, France, Germany, Hungary, Italy and the UK. An associated External Advisory Group accompanied the project and checked the user-friendly handling of the tool as well as the relevance of the information provided. The foundation version of EUGRIS was completed in August 2005 and has seen further substantial development since then (as described in Section 4.3.4).

At the time of the development of EUGRIS, web-based information on soil, water and other environmental issues was scattered over many websites and did not exist in a way that could be easily harvested by the full breadth of the scientific and policy community. Land and groundwater management requires an interdisciplinary approach and a considerable amount of supporting technical information and knowledge. The fundamental purpose of EUGRIS, therefore, was to provide access to information in a structured and contextually meaningful way. Therefore, EUGRIS helps information seekers to find the information they need and helps information providers to disseminate information.

4.3.3 What Does EUGRIS Offer?

The EUGRIS portal combines a technical and policy compendium, a search engine and a dissemination tool in one central place. It offers summary information and links to sources of more detailed information; it also enables registered users to disseminate their own information *via* an open and interactive system. EUGRIS provides technical information such as subject information, country information and research and funding information. It also provides service information such as a library and links to other resources, a 'who does what' directory, news and information on training courses. The modular structure is designed to enable both the continuous development of existing resources and expansion into other sectors in environmental policy and management, such as issues related to the soil thematic strategy. EUGRIS provides a nucleus that could be extended into a comprehensive European land and water management system; progress in this direction in collaboration with other information portals is currently in development and will be discussed in more detail later in this chapter. Last but not least, the new Member and Accession States benefit from this available information.

EUGRIS is free and provides a generally available, comprehensive and overarching information and innovation resource to support research and practical contaminated land, soil, water and groundwater management. It is aimed at the specific user requirements of the research community, legislators, regulators and planners, consultants and contractors, technology providers business and the public. Its philosophy is user directed. The user-friendly navigation allows target-oriented access to information according to different user's needs.

EUGRIS is designed as an open system, the majority of whose resources can be viewed by anyone, but it is not possible to post information anonymously. To post information (and see certain resources, such as directories of people) users must register themselves with EUGRIS, a process that also creates a personal entry in the EUGRIS directory. Anything posted by a registered user is immediately available to other users, but can be edited by the provider at any time. Registered users can also 'rate' and comment on postings as a form of external quality control; this option is only available to registered users on the same principle as the dissemination of information, namely that information cannot be added to the site anonymously.

EUGRIS is an information exchange, not a conventional search engine like Google, as it offers opportunities for users to post as well as view information. It also allows information providers to link their information to existing information on EUGRIS even if it has been posted by another source. Users can therefore search for all interlinked data on a certain project or organisation for example.

Much of the information stored on EUGRIS is meta-data (*i.e.* information about the information, rather than the information itself, which is stored on the source website). However, EUGRIS also contains a substantial series of technical summaries and a large glossary of technical terms.

4.3.4 Subsequent Development and Expansion of EUGRIS

The original EUGRIS project successfully developed and implemented the Information Management System (IMS) structure of the portal, prepared scientific digests on several topics including country information, and set up some European databases on research projects, training and people/organisations. Since the end of the initial FP5 project, EUGRIS has carried on and still exists today with the support of the partner organisations that created it and subsequent partners who use the portal as a basis for their own project work. As of November 2008 EUGRIS carried an extensive catalogue of 3257 resources (documents, software, hardware and website listings), 58 funding programmes listings, 284 R&D project listings, 378 organisations listings, 326 current news items as well as conference/training listings and a glossary of 506 terms, (often linked to national definitions and translations related to soil and water protection research). The expansion of EUGRIS since it became self-maintaining can be seen in Figures 4.3.1–4.3.3, which show the number of visits, users and resources, respectively, over the period from November 2005 to November 2008.

As well as continuing to provide its core service of a platform to view and disseminate information, EUGRIS has expanded to provide a platform for other national and European projects. It has undergone a major redesign during 2007 in response to feedback from the EUGRIS community for a more user-friendly, clearer, self-explanatory structure. The fact that the redesign was instigated reveals several positive aspects of EUGRIS. Firstly, it shows that it is

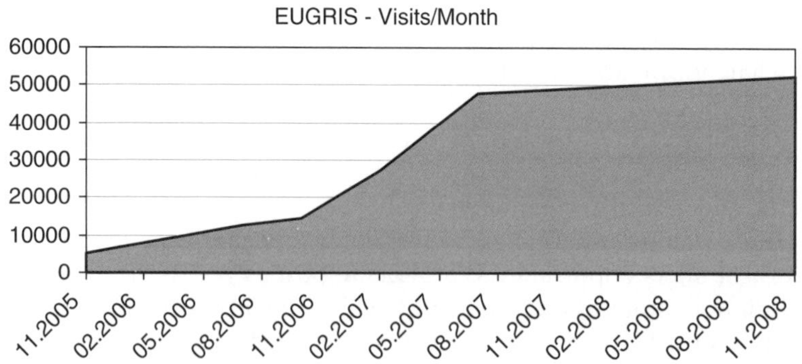

Figure 4.3.1 Visits to EUGRIS by month.

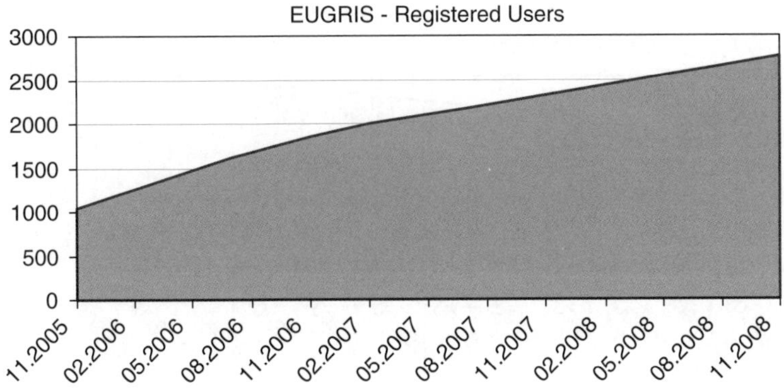

Figure 4.3.2 Registered EUGRIS users.

successful and has developed a core of users across Europe (and the world) who make use of the facilities; secondly, it shows that the users are sufficiently familiar with EUGRIS that they can suggest improvements that may not have occurred to the original project team; and, thirdly, that they wish to continue using the platform rather than develop a completely new one or divert their efforts to another one. The redesigned portal was launched in 2007 and has since continued to expand its user base and resource catalogue. Illustrations of the redesigned portal, and an example page from the Hungarian language mirror site, are shown in Figures 4.3.4–4.3.7.

With the upgrade and the new release of EUGRIS several notable features were included:

- powerful interlinked directories of projects, people, organisations and funding schemes;
- an advanced 'text' search;
- a library search for information by subject/topic, type and origin;
- an encyclopaedia of introductory summaries and country specific information;
- a directory of users and a directory of organisations;
- a developing multilingual glossary;
- new value-added features such as enhanced registration services are under development, *e.g.* easy tools to download and exchange links and listings within project teams.

Since its foundation as an output of an EU project supported by various organisations across Europe, EUGRIS has expanded into a community of people and organisations who co-operate to supply information for the benefit of everyone and also to promote themselves and disseminate their work. Projects that have joined the EUGRIS group since its inception include the FP6 projects AquaTerra, RISKBASE, SNOWMAN and EURODEMO, national

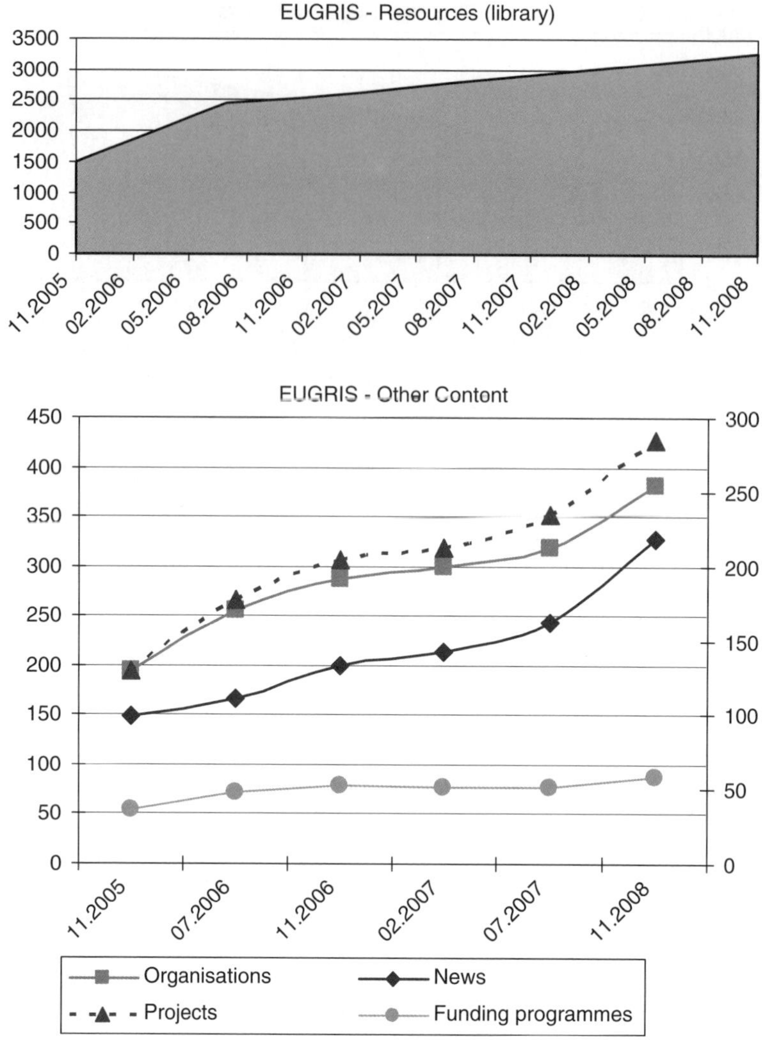

Figure 4.3.3 EUGRIS information.

projects such as Mokka (Hungary) and pan-European networks such as the Common Forum for Contaminated Land (www.commonforum.eugris.info).

EUGRIS now has a high level of use (with 10 000 to 20 000 sessions per week – see Figure 4.3.1), comparable with many commercial portals, which is a positive incentive for other organisations and projects to cooperate with the portal on further initiatives that expand the scope of the original project.

EUGRIS has demonstrated the need for and provided a means by which new and existing national and international gateways for information transfer can be developed and expanded. It has the potential to be a major influence on science and policy development in relation to groundwater, sediments, diffuse

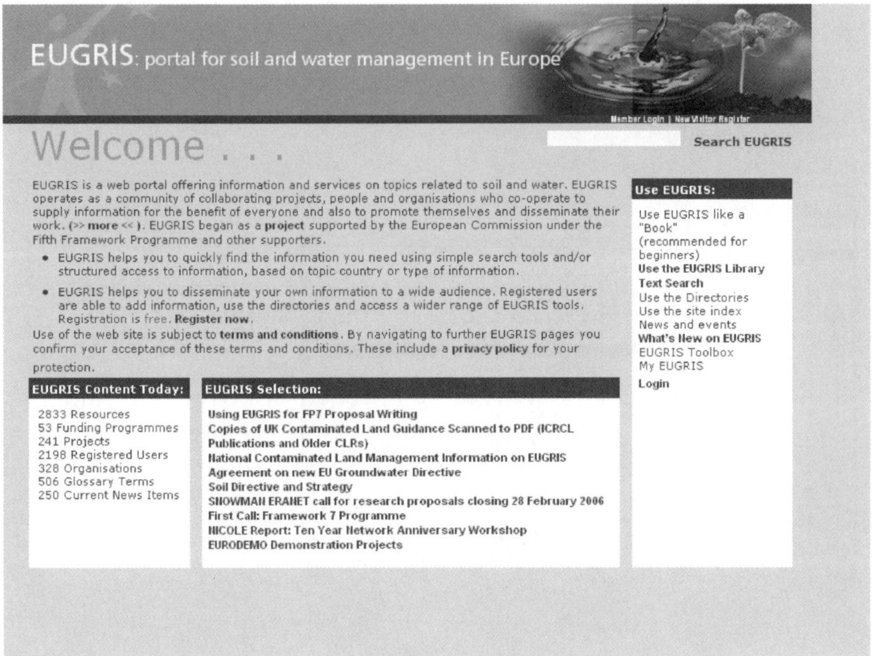

Figure 4.3.4 The redesigned EUGRIS homepage.

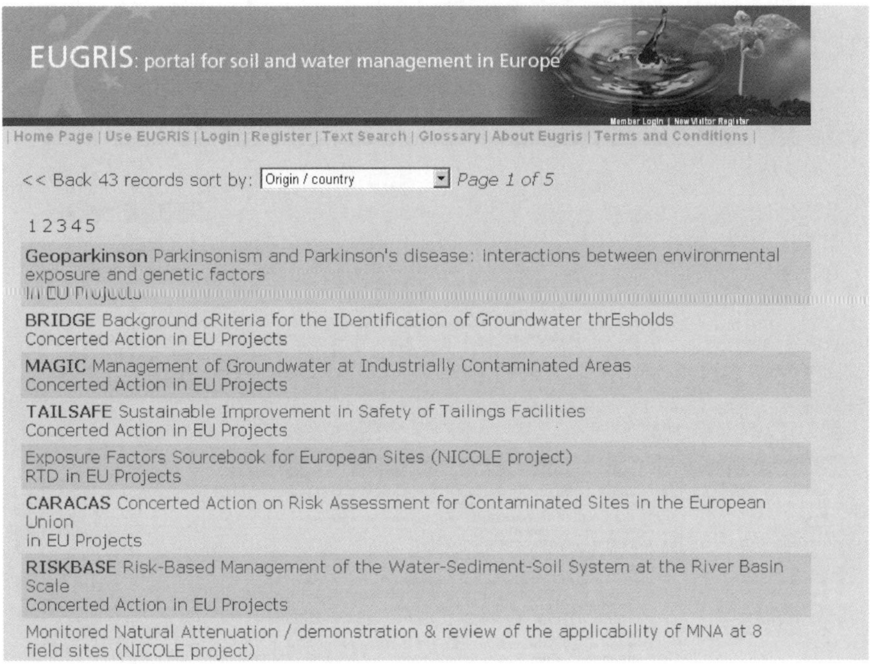

Figure 4.3.5 EUGRIS search page.

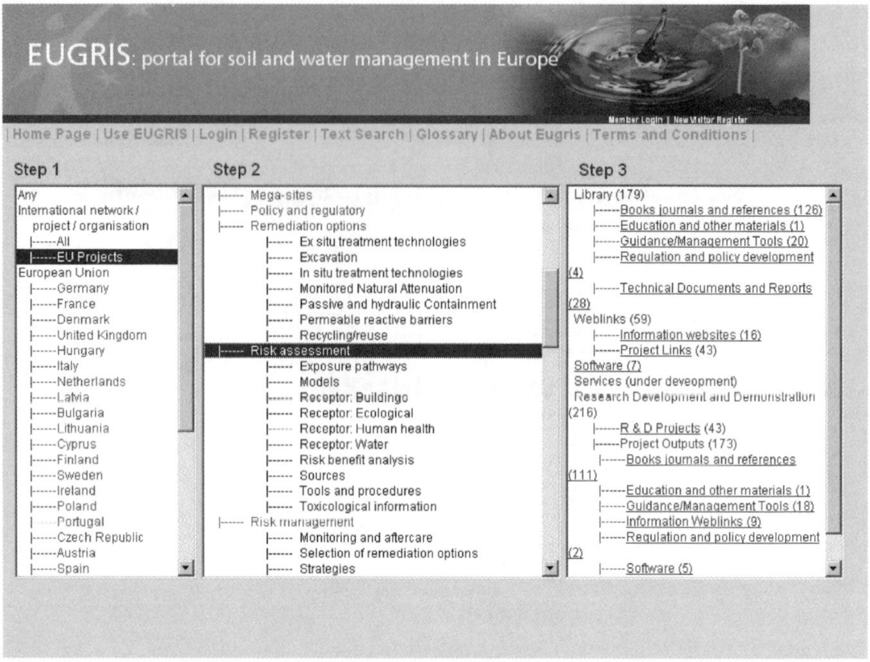

Figure 4.3.6 EUGRIS search results.

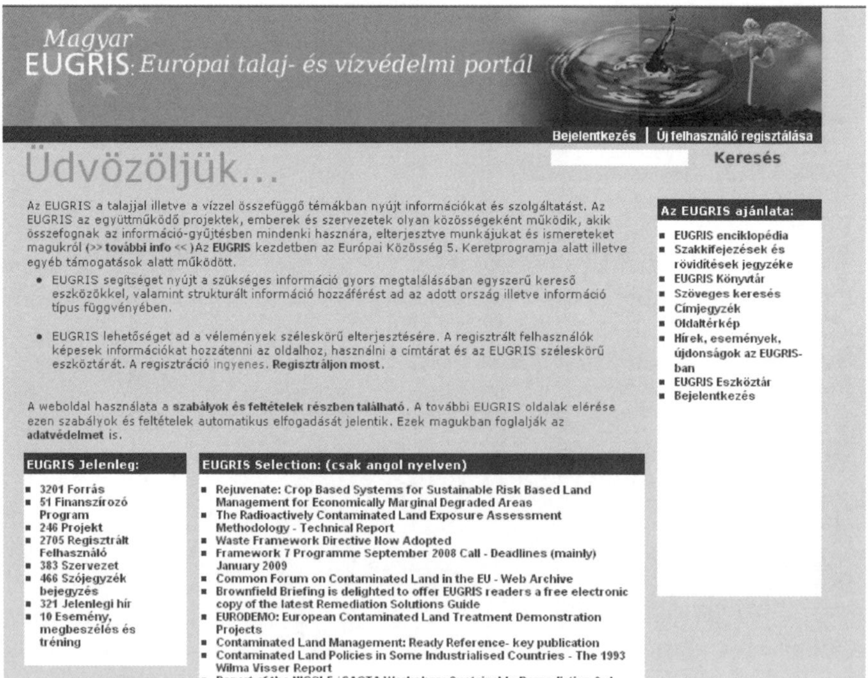

Figure 4.3.7 EUGRIS Hungarian mirror site home page.

pollution, river basin management, soil issues, brownfield regeneration and contaminated land management in the European Union. However, EUGRIS is not the only portal that supports the integration and dissemination of information in the field of soil and water. The Water Framework Directive encourages coordinated, integrated management of water bodies. A range of projects and initiatives have followed in its wake that aim to develop and improve the mechanisms by which scientific information is communicated to interested groups beyond the immediate scientific community in which it was generated. EUGRIS is playing a significant role in these developments.

4.3.5 EUGRIS, AquaTerra and Linking to Other Portals

AquaTerra was an Integrated Project funded under the sixth Framework Programme (*see* Chapter 2.2). Its core aim was to produce good quality science to improve understanding of the deposition, transport and degradation of contaminants within river basins. This was described in terms of the interactions between air, soil, water, groundwater and sediment, as well as the response of such environmental systems to wider external influences such as climate and economic pressures. A further aim of the project has been to ensure that this knowledge is disseminated to as wide a range as possible of the end users to whom it may be of value, such as those involved in the management of river basins, stakeholders within river basins and other members of the wider scientific community. The INTEGRATOR sub-project of AquaTerra aimed to distil the scientific output of the project and bring it to this wider audience. With this goal in mind, one task of the INTEGRATOR sub-project has been the development of an Online Information System (ATOIS), based on EUGRIS, to improve the accessibility of AquaTerra information through closer links with other portals.

Much time, effort and money has been invested in the generation of environmental information in Europe, which has resulted in a large and varied output, encompassing reports, book chapters, journal publications, hardware and software, websites, portals and online catalogues of information. A potential user needs to know what type of information is stored where in order to be able to make effective use of it quickly: finding information is time consuming and useful information could easily be overlooked, with the result that much of the information generated is underexposed and under-utilised. Recent initiatives focused largely on water and sediments, such as WISE-RTD (Chapter 4.2), SEIS, SPI-Water and RISKBASE, aim to overcome these issues in various ways, making a very positive step towards gaining more added value from scientific research across Europe. AquaTerra, through the development of ATOIS and in partnership with EUGRIS, is in discussion with a range of information holders in relation to the next logical progression, namely to improve connectivity and accessibility of information between portals as well as within them. Connecting ATOIS to other websites and making it possible to access information across different portals would mark a significant improvement in sharing information for all participating portals.

Initial methods to improve connectivity between websites will be simple measures such as embedded hyperlinks and keyword matching to allow further, targeted connectivity between sites. At present this is at a trial and investigation level only. If the initial linkages prove successful a further stage of more advanced connectivity using shareware on each of the sites may be undertaken. For such a cooperative exercise to be successful, several issues need to be recognised and appropriate protocols developed and agreed as necessary. Examples of these are listed below:

- the cooperation should be on the basis of sharing information between sites rather than donating it to other sites to avoid duplication of information and to maintain the integrity of each of the individual portals;
- ownership of data and meta data;
- data protection issues;
- intellectual property issues;
- updating of information;
- security issues;
- funding.

Progress in this area is ongoing and a definitive process of coordinating and linking the portals has yet to be developed. Some form of co-operation is expected shortly.

4.3.5.1 Soil, a Pressing Issue for Information Exchange

Soil protection policy at an EU level really began with the October 2001 Soil Protection Communication from DG Environment, which resulted in the 2006 Soil Thematic Strategy and the ongoing process to establish a Soil Framework Directive. Soil protection is therefore much 'younger' as an environmental policy issue, at least at an EU level, than water protection, where policies and Directives have been in place for many years. Contaminated land management, of course, has a relatively long policy history since its emergence as an issue in many countries in the 1970s. Many of the EU15 countries developed legal frameworks for contaminated land management, and work is in progress or finished for many new Member States. Several publicly funded European projects, networks and other initiatives like CARACAS, CLARINET and NICOLE have led to a more harmonised understanding of basic principles for contaminated land management and a general acceptance of risk-based decision making in contaminated land policy.

No similar emerging consensus exists for other soil threats. The proposed Soil Framework Directive is likely to drive consensus and stimulate the development of more detailed guidance for all Member States. Whatever the status of this Directive there remains a pressing need for information exchange of technical know-how, applied and policy solutions, decision making and guidance, not least to avoid duplication of effort between Member States, and ensure that research investment is really targeted at unmet information needs. There are two broad tasks in this process of information exchange: dissemination and communication. The first task focuses on data on soil status and monitoring across Europe, the second task focuses on knowledge and expertise.

The European Commission has been supporting the data task through the Shared Environmental Information System (SEIS), developed and managed by the European Environmental Agency (EEA). SEIS aims to 'tie in better all existing data gathering and information flows related to EU environmental policies and legislation. It will be based on technologies such as the internet and satellite systems and thus make environmental information more readily available and easier to understand to policy makers and the public' (http://ec.europa.eu/environment/seis/what.htm). From the original FP5 EUGRIS project, and through its ongoing operation and development, EUGRIS has established an online communication infrastructure to collect and provide information that can assist the exchange of knowledge and expertise. It can also support wider services, networking and joint research. While its scope related to soil threats has been largely focussed on point source contamination, it provides a solid foundation for an ongoing forum for the exchange of knowledge and expertise related to soil threats in general across Europe. Such a forum could:

- help researchers and science end-users identify relevant partners across Europe for future activities;
- ensure access to the state-of-the-art in research and technology to avoid duplication in mainstream non-competitive research;
- help a broad range of stakeholders to identify the real issues in a European context;
- assist end-users, researchers and other stakeholders, including the EC (DGs Research and Environment for instance), to identify gaps and needs in research within the areas covered and thereby provide input to upcoming research programmes in national, trans-national and EC contexts;
- support EC staff in developing policies;
- help authorities and administrators to liaise on specific topics that need to be dealt with on a European scale and to approach relevant experts and researchers for carrying out specific tasks connected therewith;
- remove barriers to information access for small organisations such as SMEs and local authorities.

4.3.6 Bottlenecks and Options for Future Development

Given the past compartmentalisation of soil and water research, it is not surprising that efforts to work together on common approaches are only just emerging. Most countries operating risk assessment policies use their own models and tools, and indeed several models may be in use at the same time. Protocols for exchanging data and information do not always exist even on a national scale. A broad variety of decision support tools exist, but their evidential basis and decision making approach is not always clear to users. Differences in approach are not only evident from regional or national perspectives, but also between scientific and technical disciplines and sectoral or 'stakeholder' needs. However, the interlinked nature of many environmental problems underlines the need for collaboration and working together across these barriers.

Making use of EUGRIS provides an infrastructure to facilitate the speed and efficiency with which shared approaches are developed, disseminated and exploited for the initiation of appropriate research and development. However, as with any information management system, EUGRIS faces several challenges at a European level: encouraging information content into the system, dealing with language issues, quality management and finding sustainable funding for example.

Within the EU-funded project period it was decided by the EUGRIS consortium to develop a lean system that minimised ongoing operational costs, and recognised that a central information collection activity would be hard to justify and fund in the long term. EUGRIS was therefore set up on the basis of information providers acting as a community who would deposit information into EUGRIS to bring Internet traffic to their own sites. Furthermore, the information input into the system was set up on a voluntary basis, not least because the EUGRIS project had no means of obliging information providers to use it. Consequently, the major challenge for EUGRIS has always been to encourage more visitors to disseminate their own information in the system. The main threat has always been a lack of interest in using EUGRIS for this purpose. Over the years that EUGRIS has been operating it has become clear that only a minority of its users deposit information, most use EUGRIS to collect information. EUGRIS' strategy in response to this has been to develop relationships with other projects, providing information management services in exchange for content (and resources). EUGRIS is now considering whether to offer these services more widely to research projects who can benefit from both having a tried and tested content management system and a low web service development cost, along with 'one step' dissemination to a proven large international audience. The more information providers who use EUGRIS, the better it gets and the more users will come to collect information – a kind of a positive feedback that EUGRIS is already beginning to see. However, EUGRIS has a difficulty in that it is now independently resourced and not an official EU initiative. This situation appears to make it difficult for EU research funders to recommend EUGRIS' use; on the other hand, as EUGRIS already exists, these same funders feel unable to fund its future development given the nature of the funding tools at their disposal.

Currently, EUGRIS is an English language system; although content in other languages can be posted into EUGRIS, they require an English summary.

This has been seen as a barrier to increasing the level of use across Europe. There are two broad problems: the first is that the system and its tools are presented in English, and the second that the content is posted in one particular language – usually English. One way forward that EUGRIS is exploring to address this problem is to work with national projects working in a particular country or group of countries who have an interest in linking their information exchange to the wider EUGRIS audience. EUGRIS has set up its first 'mirror' site in Hungary (www.magyar.eugris.info), shown in Figure 4.3.7, through its work in supporting a decision support tool for a nationally funded project (http://mokka.hu/). In this mirror site the system tools are available in Hungarian. Content summaries can be posted in English or Hungarian, but English is recommended for a wider dissemination impact. One possibility being considered is to allow users the option of 'local' or 'international' dissemination. EUGRIS is currently seeking partners for 'mirror' sites in other countries.

One concern about an open system like EUGRIS is moderation, *i.e.* how to deal with poor content being posted. A dilemma is that a 'live' system might attract poor quality information, but a system where information input requests are processed off-line is not only expensive but delays information exchange. It also runs the risk of being overwhelmed by input. EUGRIS' way around this dilemma is that *no* anonymous information can be posted, and all information posted is linked to the name of the person posted it, who must be registered with EUGRIS. Clicking on this name opens the personal entry of the author. This entry allows a user to determine the provenance of the information posted. It also allows EUGRIS' operators to remove privileges from users who consistently add poor information. In practice, over the 4 years EUGRIS has operated as a live system few unacceptable inputs have been made. EUGRIS also includes a comments system, which allows users to remark on the usefulness of posted information. Hence the site is intended to be self-moderating.

Naturally, most of EUGRIS' potential users expect cost-free access to its information. Unfortunately, as EUGRIS has found, public funding cannot permanently maintain all aspects of the portal; as a result, EUGRIS is using and considering other funding routes to maintain its service. Three broad funding routes can be identified, at least in principle:

1. Some infrastructure activities may be directly or indirectly fundable through future projects, using EUGRIS as dissemination and project management platform.
2. Some infrastructure activities may be fundable by revenue generated by sale of services and products.
3. A proportion of EUGRIS' operating costs may be recovered from advanced services offered to users such as page sponsorship, directory services, hosting services or registration services.

The future shape of the infrastructure will of course be determined by what is economically sustainable.

Initial discussions for a self-sustaining EUGRIS organisation have already taken place, but are not yet elaborated. Any organisation will need to take account of preceding investments in operating EUGRIS to date and EUGRIS partners' intellectual property interests. An ideal scenario would combine the EUGRIS infrastructure with a networking and content management approach.

4.3.7 Conclusions

The EUGRIS soil and water information portal provides access to technical, policy and research information in a structured way that facilitates its location and use. It largely provides metadata on information stored elsewhere but is also a source of a significant quantity of information in its own right. Since its initial development within the context of the EU FP5 funding programme, it has been self-maintained and has developed and expanded beyond its original boundaries of providing a source for searching for information and a place from which to disseminate information. It has reached the stage in 2008 where it is a widely used, significant source of soil, water and contaminated land information relevant to countries across Europe and is now used as a platform for a range of other national initiatives and EU funded projects. EUGRIS continues to expand and evolve to meet the demands and needs of the sector, with new content and powerful new tools under development. Its true usefulness depends on the participation and support from its users, who add their own information to the portal and make it a live resource rather than an archive.

The further development and the future of EUGRIS for water related information now appears to lie in ongoing developments to enhance both its own profile and those of other, similar portals such as WISE-RTD and RISKBASE, as well as the Online Information System developed within AquaTerra. The aim of these developments is to improve the accessibility of a wider range of information to a greater number of users by interlinking portals in various ways. This fits into the current European philosophy of greater communication between science, policy and stakeholder communities in the wider interests of the environment at large, and at this point in time no similar constellation of soil information interests exists. At present the discussion between the operators of different portals is ongoing and the exact nature of future cooperation is unclear. Nonetheless, the future of EUGRIS is clearly fixed in the direction of improved communication of science and policy information.

CHAPTER 4.4

EuroAquae and its Links to RTD (Research & Technological Development) & Management

PHILIPPE GOURBESVILLE[a] AND JEAN A. CUNGE[b]

[a] Université de Nice-Sophia, Polytech'Nice-Sophia, Antipolis, 930, route des Colles, 06903 Sophia Antipolis, France; [b] 31 Rue Doyen Gosse, 38700 La Tronche, France

4.4.1 Introduction

4.4.1.1 Water Sector Industry & Engineering: Demands & Needs for Hydroinformatics

The evolution of human activities, in the context of climate changes and growing population, induces situations that are evermore complex to manage. The sustainable development and, above all, management of water resources within the context of aquatic environment represents today, and will represent within the foreseeable future, a major challenge for post-industrial economy and social organisation. The essential aim of such management is to avoid, if possible, or at least minimize the risks of crises in water supply and wastewater treatment for the population, in water scarcity for irrigation, in management of consequences of floods, *etc.*

Initiated in Europe more than 20 years ago, hydroinformatics emerges as the central element for the progress of modelling activities and management of capacities on the theoretical side as well as in the operational field. The

Water System Science and Policy Interfacing
Edited by Philippe Quevauviller

Published by the Royal Society of Chemistry, www.rsc.org

Information and Communication Technologies (ICT) greatly extend the potential of the tools and, at the same time, modify the engineering activities in the water field. The development of ICT allows for synergetic use of simulation tools and communication technologies within a single methodological approach dealing with physical, social and economical aspects. The only possible way to manage successfully the problems evoked above implies consensual views and actions of decision makers and users (population, governments, administration, elected bodies, NGOs), on one hand, and an executive body: engineers, on the other hand. The very idea of hydroinformatics is precisely this: to make available and intelligible the results of engineering thinking to decisions makers and, in the other direction, to make clear for engineers the social requirements that they are to satisfy. Theoretically, this can be done thanks to the ICT but in practice it means that there is the need, within administrations, within consulting institutions, within contracting enterprises for professionals who understand this and have the knowledge and ability necessary to implement such a, still new, vision. In short, there is a need for 'ferment', for a catalyser, for engineering professionals who not only know how to build a dam or water supply system but also are able to talk to the outside world and modify by their ideas their own surrounding professional environment; and this is to be within the context of European Union reality, *i.e.* international context regulated by a common framework and interests.

According to all analysis and visions, intensive demand of hydroinformatics educated engineers and managers in public services and the private sector will continue for more than five decades (Implementation of the European Water Framework Directive[2]). In this context, the specialization and training of executive engineers and managers is the key issue. European knowledge, know-how and good practices in sustainable water management and hydroinformatics, although not always perfect, are nevertheless nowadays references and constitute the central elements of a technical common culture that is already widely requested and will be increasingly in demand in the future.

4.4.1.2 EuroAquae Project Description & Joint Degree

To meet the intensive demand for hydroinformatics educated engineers and managers in public services and private sector, five European Universities have decided to establish and promote a joint master's degree: 'EuroAquae – Euro Hydro-Informatics and Water Management'. The EuroAquae consortium is composed of:

- University of Nice–Sophia Antipolis (France)
- Brandenburg University of Technology Cottbus (Germany)
- Budapest University of Technology and Economics (Hungary)
- Polytechnic University of Catalonia (Spain)
- University of Newcastle upon Tyne (United Kingdom).

The EuroAquae consortium awards a joint degree defined as a Master of Science (Msc) that is recognized by the five participating countries and carrying the Erasmus Mundus (Section 4.4.1.3.) label. The partners of the consortium share the same vision and the same ambition for the future. Consequently, the main objective of the Master course is to prepare and train future scientists and executive engineers in charge of modelling and managing projects in hydrotechnologies and environment. These professionals have a vocation to assist local, regional, national and international bodies/entities, public services and to be involved in private companies.

Implementation of this new curriculum is based on co-operation and joint activities are organized around a specific mobility scheme. The coherence of the scientific contents and of the community of participants is promoted around the concept of a virtual university.

4.4.1.3 Erasmus Mundus Framework

Following the Lisbon declaration (2000), which focused on the development of 'the most competitive and dynamic knowledge-driven economy by 2010', the European Parliament and Council received in July 2001 a communication by the Commission on strengthening EU-third country co-operation in higher education. Following the positive reception of the Communication, the Commission adopted a programme proposal, Erasmus World, in July 2002. The programme was then renamed Erasmus Mundus. On 5 December 2003, the legislative process involving the European Parliament and Council was completed and the Erasmus Mundus programme Decision was adopted. It was published in the Official Journal on 31 December 2003[1] and entered into force on 20 January 2004.

The Erasmus Mundus programme has been very well received by the European higher education community. European academic institutions see the programme as a useful means to respond to the challenges that European higher education faces today, in particular the need to stimulate the process of convergence of degree structures and to enhance the attractiveness of European higher education world-wide. These are themes central to the Bologna process and to national reform in higher education in Member States.

The programme confirms the Commission's desire to encourage the opening up of European higher education to the rest of the world. It complements the European Union's existing regional programmes in higher education with third-world countries. Regional programmes, such as Tempus, ALFA and Asia-Link, will continue to foster international co-operation in higher education between the European Union and its partners.

Erasmus Mundus, however, is a new global scheme, providing a distinctly 'European' offer in higher education. It seeks, primarily, to enhance the quality and attractiveness of European higher education world-wide. Secondly, Erasmus Mundus Masters Courses and scholarships will provide a framework to promote valuable exchange and dialogue between cultures. By supporting the international mobility of scholars and students, Erasmus Mundus intends to prepare its European and non-European participants for life in a global, knowledge-based society.

The Erasmus Mundus programme is a co-operation and mobility programme in the field of higher education. It aims to enhance quality in European higher education and to promote intercultural understanding through co-operation with third countries. The programme is intended to strengthen European co-operation and international links in higher education by supporting high-quality European Masters Courses, by enabling students and visiting scholars from around the world to engage in postgraduate study at European universities, and by encouraging the outgoing mobility of European students and scholars towards third countries.

The Erasmus Mundus programme consists of four concrete actions:

- Action 1 – Masters Courses: they constitute the central component around which Erasmus Mundus is built. They are high-quality integrated courses at masters' level offered by a consortium of at least three universities in at least three different European countries. The courses must be 'integrated' to be selected under Erasmus Mundus, which means that they must foresee a study period in at least two of the three universities and that it must lead to the award of a recognised double, multiple or joint diploma.
- Action 2 – Scholarships: to give the Erasmus Mundus Masters Courses selected under Action 1 a strong external projection, a scholarship scheme for third-country graduate students and scholars from the whole world is linked to them. This scholarship scheme addresses highly qualified individuals who come to Europe to follow the Erasmus Mundus Masters Courses or to work for them.
- Action 3 – Partnerships: to encourage European universities to open up to the world and to reinforce their world-wide presence, Erasmus Mundus Masters Courses selected under Action 1 also have the possibility of establishing partnerships with third-country higher education institutions. These partnerships allow for outgoing mobility of graduate EU students and scholars involved in the Erasmus Mundus Masters Courses.
- Action 4 – Enhancing attractiveness: Erasmus Mundus also supports projects aimed at enhancing the attractiveness of and the interest in European higher education. It supports activities that improve the profile, the visibility and the accessibility of European higher education as well as issues crucial to the internationalisation of higher education, such as the mutual recognition of qualifications with third countries.

In concrete terms, during 2004–2009, Erasmus Mundus has supported about 100 Erasmus Mundus Masters Courses of outstanding academic quality and is defined as a joint activity between higher education institutions coming from the 27 European Union Member States and the EEA/EFTA countries (European Economic Area/European Free Trade Association). It has provided grants for some 5000 graduate students from third countries to follow these Masters Courses, and for more than 4000 EU graduate students involved in these courses to study in third countries. The programme has also offered teaching or research scholarships in Europe to over 1000 incoming third-country academics and for a similar number of outgoing EU scholars. Last, but not least, Erasmus Mundus will support about 100 partnerships between Erasmus Mundus Masters Courses and higher education institutions in third countries.

4.4.1.4 Water Sector Educational Environment in Europe

The water sector is a major economic domain in Europe with the public services it provides to communities. During the last decade, water utilities like Suez Environment and Veolia have strongly invested and recruited numerous professionals. In parallel, it is surprising to see the limited investment made by the educational environment to answer the demands of the companies and more generally of society. The educational sector in Europe is still offering 'traditional curricula', focused on a disciplinary approach like hydraulics and fluid mechanics and has difficulty promoting the multi-disciplinary and technical changes needed by the professional environment. This adaptation process is not achieved yet. On the contrary, the number of curricula is globally decreasing despite a real increasing demand. This situation has been pointed out by several national entities, underlying the urgency to redevelop curricula in such fields. Of course, EuroAquae cannot pretend to answer to all of the demands but it can at least provide a number of qualified graduates for the specific international hydroinformatics field.

4.4.2 EuroAquae Project History and Achievements

4.4.2.1 Joint Degree Agreement

According to the Bologna process, a EuroAquae master course is based on a two-year programme with four semesters of 30 ECTS. The course is a joint activity run by the five partners of the consortium and is organized in a pedagogic continuum to provide:

- an introduction and common knowledge/soft skills to the participants offered by each of the consortium members (semester 1);
- the acquisition and use of hydroinformatics concepts, methods and tools, carried out at a single location during semester 2. During this semester, each participant identifies and chooses a specialization for semester 3;
- during semester 3 a thematic specialization following four main options, according to the partners expertise and, moreover, the deep level of co-operation through the consortium members for the development of the master course and hydroinformatics: hydroinformatics systems (software engineering and modelling), urban waters management, inland waters management and decision support systems;
- during semester 4 a research project or a professional internship.

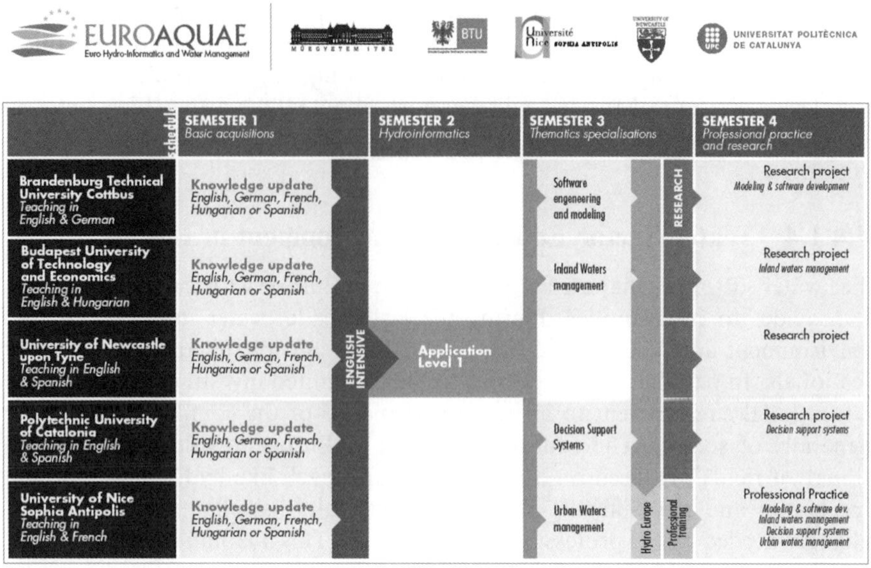

Figure 4.4.1 Organization of contents and mobility for participants.

The course provides one research and one practice oriented specialization. The choice of a professional practice internship for the last semester answers the demands of participants who then have the opportunity to finalize their professional skills and acquire good practices that are essential for a good integration in the international professional community.

Participants must follow, at least, 30% of the curricula in a different institution from their 1st Semester 'home institution'. The mobility scheme applied in semesters 2, 3 and 4 is presented in Figure 4.4.1. The master's course welcomes about 40 participants each year from both non-European countries and European Union members. The consortium considers it a priority to mix participants from third countries and from Europe to achieve a common technical culture and real promotion of good practices through a community of practitioners and professionals.

Since May 2006, the Msc has been officially accredited as a joint degree according to the national rules of each partner institution of the EuroAquae consortium.

4.4.2.2 Knowledge Skills of Graduates

Knowledge is the ability to understand; the capacity for critical, rational, scientific and strategic thinking. It involves reflection, selection, application, and consolidation of all kinds of information. Skills mean the mental ability to perform routine tasks. It is generally a local and time-dependent characteristic, strongly linked to the technologies available in a given environment. These two definitions should be used to analyse the content of the EuroAquae master course to answer the question 'How does the profile of this MSc graduate fit with water sector professions?'

The most general and official definition of the final professional profile and of the openings that the EuroAquae MSc diploma is supposed to give is the following:

The main objective of the Master is to prepare and train future scientists and engineers in charge of modelling and managing projects in hydro-technologies and environment. These professionals will assist decision-makers of local, regional, national and international collectivities, of public services, and of utilities, or they will be involved in consulting activities with private companies. Their technical expertise will be accompanied by the understanding of the social and economic context in which techniques are implemented

The target is their 'good integration within the international professional community'.

The above description is sometimes misinterpreted on three different points:

1. It may give an impression that the graduates have in-depth knowledge ('technical expertise') and skills of basic engineering in the area of hydraulics, hydrology, urban hydraulics and techniques of management of water resources. The duration of the course does not allow the acquisition of such in-depth basic knowledge from the course lectures.
2. The title of the diploma ('hydroinformatics and water management') sometimes implies to professionals that the graduates know *all* ICT technologies as well as numerical modelling that are useful and applied in the water sector. Consequently, it could be natural to some future employers to ask our alumni to be in charge of a modelling group within a company, or to develop a piece of modelling software. Again, for the same reason as above, the alumni who has just graduated does not know enough to carry out such tasks.
3. The course is supposed to 'train future scientists and engineers in charge of modelling and managing projects . . . '. The terminology is ambiguous: for some people and organisations anybody with an engineering degree is a scientist. Notably, in French the word 'scientifique' does not mean scientist. The EuroAquae curriculum and its mobility makes it difficult (with few exceptions to the rule) to train future researchers. In addition, the curriculum in itself does not pretend that the alumni can, after graduation, only in virtue of EuroAquae diploma be 'in charge' of anything.

After this warning it is important to underline that the EuroAquae course is currently unique in Europe and even in the world, although, as stated above, it may not fulfil all possible expectations of the profession. The course curriculum as defined by content of the modules leads to the following overall profile of knowledge and skills.

The EuroAquae Master Course objective is to prepare engineers:

- having reasonable knowledge of ICT enabling technologies and possibilities, as well as of modelling tools, both currently applied within water sector;
- capable of assuming an active role within companies and institutions involved in design, modelling and management of projects in hydro-technologies and related environment;
- able to act as *ferment* within their professional environment and also to radiate, through their attitude, basic characteristics of European ideas concerning technological culture and professional ethics;
- having a capacity of personal judgement, a sense of independence and autonomy.

This profile is forged during three semesters of University studies and one semester spent on a Master Thesis that typically stems from professional experience gained during last semester 6-month internship that consists in working in an industrial organisation as a qualified engineer; it may also result from last semester research activities at one of the Partners' Universities. The graduated alumni should acquire, during the first three semesters, the basic knowledge and skills listed below:

- a clear picture and basic knowledge of the mathematics and physics that are useful for enabling ICT and for numerical modelling in the water sector;
- basic knowledge of hydraulics of open channel flow, pressurized networks, groundwater flow – not at the level of specialisation but definitely enough to understand the physical laws and their mathematical formulation; basic knowledge of hydraulic engineering structures and their purposes (weirs, dams, dykes, *etc.*); understanding the differences between maritime, river and urban problems;
- basic knowledge of hydrology – not on the level of specialisation but definitely enough to understand the physical laws and their formulation as well as basic probabilistic techniques;
- basic knowledge of enabling technologies (GIS *etc.*) and also skills based on hands-on application experience of these technologies;
- fair understanding of modelling systems applied to the water sector and currently available on the market; this understanding covers the role of mathematics, algorithmic, numerical methods as well as a good understanding of limitations; skills gained through personal hands-on experience of application of at least two industrial tools;
- basic knowledge of methodologies and understanding of the tools applied for integrated management of water systems, including descriptive knowledge of some economic tools (understanding of linear & dynamic programming, understanding of the principles of DSS);
- capacity to organize and run engineering WEB collaboration projects; skills gained through hands-on and participation in such projects;

- ability to listen and to present the problems and difficulties in a clear and concise way in intelligible English;
- understanding that there are ethical problems in the engineering profession and that it is essential that stakeholders to any project understand this; having acquired an overall intellectual attitude of somebody who looks for integration of various technologies, interests and attitudes to achieve better result that satisfies most stakeholders;
- understanding that there are specific European attitudes in this context and possible differences between these attitudes and other (*e.g.* his own country's) attitudes.

4.4.2.3 Innovation in Education: Virtual University and HydroEurope

Coherence within the course program and the consortium is established through Web-based collaboration in the spirit of a 'Virtual University'. During the 1st semester teachers and students, independent of residence, participate in a Module 'Web-based collaborative engineering' thus becoming acquainted with each other through the Internet before meeting face to face in the 2nd semester. During the 3rd semester students from all locations will be trained extensively on collaborating over the World Wide Web on an engineering project within the Module 'HydroEurope Working as virtual company'.

Experience and skills of Web-based collaboration and communication enable the participants to better follow open distance learning within lifelong learning programs. These aspects are addressed within pre-professional training and research methodology.

The availability of teaching material on the home-pages of the participating teaching institutions and on a specific collaborative platform (www.euroaquae.org) supports the development of coherent teaching material as well as quality control and enables students to follow up the courses taught at other places.

Modern Information and Communication Technology (ICT) offers today the opportunity to explore and to promote a new dimension in engineering activity: collaborative work of team-partners residing at different locations and working at different times using the Internet as a common working and communication platform. Collaboration in this environment demands new skills and a new 'technological culture' to be generated simply by doing. This is a worldwide challenge in which, in future, experts and engineers from different countries with different languages, different mentalities as well as different specialization and professional experience have to collaborate in research, teaching and practice. To promote the methods and tools of the collaborative engineering in the hydroinformatics field, it is essential to implement this approach in the educational and training processes. Several experiments have been initiated in recent years and one of the most significant is Hydro-Web run by the Brandenburgische Technische Universität in Cottbus (http://www.hydro-web.org) since 1999. Since 2002, a new project called HydroEurope – an Intensive Programme within the Socrates/Erasmus framework (http://europa.eu.int/comm/education/programmes/socrates) of the European Commission – has been initiated by the EuroAquae consortium.

HydroEurope (www.hydroeurope.org) is following the way opened by Hydro-Web and is dedicated to students involved in six master courses specializing in hydroinformatics, water management and hydro-technologies. This activity is now a compulsory part of the EuroAquae master's curriculum. The objectives of HydroEurope are to promote, in a global European vision, the key concepts, methodologies, tools and good practices in hydroinformatics. Today, the problems related to water are increasingly complex and must be analyzed in a global way and with the right tools. The sophistication of the situations forces a reconsideration of the way of working: new practices, new tools, new technologies and new ways of collaboration.

The HydroEurope concept is developed through a case study – the 1994 flood of the Var River (France) – and welcomes each year about 100 participants – EuroAquae and other European master degree students – from more than 35 different countries.

The recognized success of HydroEurope was extended in 2006 with the creation and development of a similar approach in Asia with HydroAsia (www.hydroasia.org). HydroAsia runs a similar concept to HydroEurope on a case study based in Incheon (South Korea) and based on urban flooding analysis.

4.4.2.4 Internship of 4th Semester and Employment of Alumni

For the 4th semester, EuroAquae students have the possibility to join a company for an internship of 6 months. During this period, they participate, as professionals, in an ongoing project with joint supervision from academic and industrial tutors. During this period, the students have the possibility to understand the reality of industrial and professional life and, at the same time, to demonstrate their knowledge and skills.

After 4 years and similar experiences carried out in different master or engineering degrees, it is obvious that the internship represents the best option to obtain employment for the Alumni. In fact, most of them – over 90% – receive an offer to stay with the company after graduation. This approach confirms both the needs of the professional sector and the coherence of the curriculum developed by EuroAquae.

4.4.2.5 Statistics of Four Generations of Students

In September 2008, the fifth generation of students entered EuroAquae with the clear ambition of obtaining at least the same results as the previous groups. For the first three generations, 100% of the graduates are now involved in a professional life directly linked to the water field and hydroinformatics. Some of them are now developing an academic career, back in their home countries (about 12%), while others follow their career plan in public services (about 10%) in their home countries, but most of them (78%) join private companies. Noticeably, the vast majority of European students are now established outside Europe, in branches of European companies. Most of the non-EU students are employed by European industry in locations in and outside Europe. The quantitative elements demonstrate the validity of the initial vision and efficiency of the course in answering industrial demands. Interestingly, nearly 50% of EuroAquae students are young women.

4.4.2.6 EuroAquae Alumni Association

Since 2006, an Alumni Association has been implemented for the EuroAquae participants. The Association, defined under French law (Association Loi 1901), has the following objectives:

- to promote a beneficial relationship between EuroAquae, its alumni and other interested persons for academic, professional and social reasons;
- to represent the views of the alumni and to assist and advise the EuroAquae consortium on several key activities, including student recruitment and career development;
- to raise the status of the alumni in the environment of EuroAquae and the wider hydroinformatics community;
- to conduct social and cultural activities that are of interest to alumni;
- to provide a career network for graduates of EuroAquae;
- to promote, advance and protect the academic reputation of EuroAquae.

More than the traditional objectives for an alumni association, it is important to underline here the very successful development of the professional network after a few years. This network now provides real support to students and could really optimize career development by providing opportunities and even contracts to participants.

4.4.3 Why is the Profile a Success?

4.4.3.1 Hydroinformatics – Links with RTD

What is hydroinformatics? The term was introduced by Professor M.B. Abbott in 1992 and the inventor of the then neologism defined it as the integration of computational hydraulics and of artificial intelligence. Computational hydraulics is the fusion of numerical methods of applied mathematics, hydrodynamics and hydraulics, the latter being originally very much experimental science. At these beginnings it meant that numerical simulation and modelling is integrated with artificial intelligence in single tools. The meaning has evolved since. Probably the most consensual definition today would be that hydroinformatics is about making the best use of information technologies to manage water in the environment. The term 'manage' is used here in a very wide sense: certainly not limited to the ordinary understanding of 'water management practice'. Indeed, management here means dealing with whole gamut of information, from purely engineering subjects, on the one hand, such as data collection, measurement, interpretation, such as design of river basin management strategies including civil engineering structures, such as modelling (simulation) of river, groundwater and coastal flows and water quality to forecast natural events (floods) as well as the impacts of human activities. On the other hand, it is concerned with *information management*: the time when society left the engineer with the decisions concerning water problems is over. Decision makers today are numerous, are not from engineering professions and they cannot be asked to understand engineering language or argumentation. Conversely, the engineering profession is still far from an understanding of what is desirable to decision makers (citizens, elected bodies, NGOs, *etc.*). Hydroinformatics gathers a range of tools that, based on information technologies, may be used to explain exactly what these various groups wish and how engineers can realise these wishes and, especially, what various impacts of realisation could be. Moreover, these tools allow for participation of stakeholders in the process of execution of the projects; there are examples of such participation in recent years. Thus, hydroinformatics is also about movement, flow of information, not only about its content; it is at the heart of the process of communicating information. Here we come to the meaning of another term in the above definition, *i.e.* ' . . . the best use of information technologies to manage . . . '. What does the word 'best' mean? Here we enter the domain of social rightness, and also ethics, in a general sense as well as of engineering ethics. Thus, hydroinformatics became a technology in itself, a very specific technology that asks for a specific state of mind (or culture) and for specific knowledge.

How are hydroinformatics and, consequently, the EuroAquae Project related to the Research and Technology Development? Formally the EuroAquae Project, as part of the first Erasmus Mundus programme, is not concerned about research. It is a Master of Science course, not a PhD programme. Nevertheless, by its specific mobility character it creates *ipso facto* the inspiration towards research on hydroinformatics basic science and technologies. It does so through the links between the teaching-research teams of five universities, on the one hand because of personal exchanges among researchers-teachers and, on the other hand, through competitive thinking: teachers at each partner's university have to face mature master students coming from other universities carrying in their minds subjects and problems that are unsolved, that they met as research subjects on their way, that have not been solved so far. At the same time the teachers and researchers of each university have open access to full and detailed lecture notes of their colleagues from other partner institutions and this is a considerable incentive to compare research subjects or unsolved problems from other places with their own perspectives. Links to the Technology Development are very different. It has two specific facets. First is the process, described above, of exchange of information and incentive for each partner to raise its own teaching and education to the same level of technological information diffusion as others. Thus the latest technological innovations and innovative approaches of hydroinformatics are being propagated as taught subjects across Europe through five partner Universities. The second facet is that of the 4th semester internship of graduating students and also the professional activity of alumni employed by the European water sector industry after graduation. It is very much an interactive phase. On the one hand, students and alumni are learning a lot about technologies used in real life while, on the other hand, they are bringing into industry many revolutionary ideas and knowledge of a number of most recent or ongoing research technological developments learnt in several European universities. Although they are entering leading European companies to learn, there are two aspects of which the EuroAquae alumni are carriers and that are understood only by a few managers:

1. Most professionals employed by industry simply have no time to follow the technological developments going on and thus apply routine methods. An example is the use of market-available software codes for simulation of hydraulics and hydrological events. Once acquired, such codes may be used for decades by a consulting company without modification or improvement. EuroAquae alumni arrive with ideas that there are limitations to the application of various codes, that new developments have been made, that the introduction of technologically new tools is a good investment and brings a competition advantage. Thanks to their professional profile they are able to promote such ideas.
2. EuroAquae alumni gain insight into the fact that where the technology (hydroinformatics) is concerned one must think in terms, evoked above, of going beyond frontiers of civil engineering or environmental engineering. During their EuroAquae years they were taught, and in the majority they retained, that the development of technology follows user requirements. The users are understood here in a wide sense: coming not only from engineering sciences but also from all domains related to water, from chemistry, biology, ecology and up to social and political sciences concerned with our basic resource. Companies caring for their competitive advantage should push the development of pluridisciplinary technology in synergy with information management. EuroAquae alumni's profile is again a tool that makes such thinking more common and popular within companies and thus helps technological development.

4.4.3.2 Mobility, Intellectual Profile, Industry Needs (Research or Industry)

The experience gathered during 5 years of running the Project, and also the lessons gathered from three generations of EuroAquae graduates, have shown that the Course is producing a profile that is extremely useful to the water sector industry. Indeed, when the first generation of students went for 4th semester internship to industry, the latter expected trainees with a 'normal' engineering profile with some ICT knowledge applied to hydraulics and hydrology. After only a couple of weeks, however, the companies found that they had a specific profile that had been lacking within their personnel and that has proved immediately to be surprisingly useful.

The profile sought by the EuroAquae Project teachers and Management Board can be defined as follows:

- engineers educated for work in companies or governmental institutions and capable of proposing and developing frameworks integrating engineering projects of various water sector domains, environmental problems, and two-way communication with decision makers (politicians, citizens, elected bodies, investors);
- innovators and integrators: leaders (not in a hierarchical sense of the word) who help to develop and implement new technologies in both design and construction to create competitive advantages;
- professionals capable of integrating immediately engineering teams in foreign countries, in a foreign environment throughout Europe and beyond, and capable of proposing a European approach to engineering problems but also able to adapt themselves to local conditions;
- professionals that know the physical bases of water sector sciences as well as basic technologies of ICT (GIS *etc.*); knowing enough from hydraulic engineering to be able to judge what kind of physical and normative constraints a designer of a project is submitted to; knowing enough of ICT technologies to be able to propose communication bridges between designers and decision makers;
- engineers knowing enough about numerical simulation and correlative enabling numerical technologies to ask appropriate questions and be an efficient aid in the choice of software based on such methods by their employer; capable (in the context of a project or large development) of explaining to laymen why such or such software/method may be accepted or rejected, what are typical inadequacies of numerical models, why one cannot believe everything that comes out of the computer and why one cannot find everything on the Internet (at any rate not everything that is right);
- engineers able to decide about their own capacities and having the will to choose the way to develop their own personal qualifications beyond acquired knowledge.

Indeed, typical EuroAquae graduating students have their own personal judgement, sense of independence and autonomy, capacity to make their way within strange or foreign groups that they have to work with. They show the open-mindedness that allows for the capacity to be 'interfaces' between often narrow technologically or even completely different culturally (*e.g.* engineers and environmentalists, engineers and decision makers) professional groups. This comes partially from the selection process but essentially from student's overcoming the personal challenge posed by having to change location every 6 months. From the beginning they have an autonomous mind (otherwise they would never apply) but then mobility is an extremely powerful educational factor in forging their character, outlook and intellectual profile. A profile that proved to be in complete accord with the new needs of the management of European companies. These companies, until recently, based their activities on in-depth specialised engineering personnel of civil engineers, of hydrologists, of software programmers, *etc.* But now they have to work in different environments, to take into account the influence of new groups of stakeholders and decision makers, of multi-disciplinarity of projects, of globalisation making it necessary to co-operate with design and supervising teams as well as contractors coming from various countries, cultures, speaking different languages. In this context, new recruits like EuroAquae alumni – who know the basic activities and difficulties of each specific profession but at the same time have an autonomous and open-minded co-operative profile, with moreover a good knowledge of ICT – were immediately recognised as exceptional input by managers.

Industry has been offered the profile it needed, the profile that could not be shaped by the classical curriculum of existing faculties. This has been proven further by the fact that nearly all students are offered employment immediately after graduation by the companies where they spent their last 6 months of internship.

What about research? This question arises naturally because in traditional universities an MSc degree is also a prerequisite to enter a research career through the three or more years of PhD studies. Preparation for a research career is not a specific purpose of the Course, *i.e.* it is possible that a graduate continues for a PhD but such a situation can only be exceptional. Indeed, with the mobility approach, moving every 6 months, the chances of becoming a researcher, integrating into a research group, finding a professor who can guide a PhD thesis, financing it, *etc.*, are limited. The candidates are aware that EU objectives in creating the Course are not limited to radiating European excellence in education. There are also other aims: to familiarise the students *through practice* with the European water sector industry, to make them propagate the image of excellence of this industry in their countries and also to help European companies in exporting their activities as well as find associates in third countries for common projects and activities. Obviously, these objectives ask for professional profiles acquired through work in European companies rather than through research.

This being said, the EuroAquae project and course have built up solid bases on which to promote future co-operation between its partners and industrial partners towards the definition of research subjects and towards carrying out research PhD programmes. But this is another story.

4.4.3.3 Link with Management Capacities

Within the last decade the way that decisions are taken in the water sector has been greatly modified and, consequently, the managerial requirements that industrial activity, as well as planning of projects and administration, asks for have changed. The nature of the problems has changed with the integration of water quality problems, social demands and ecological issues. More specifically in Europe, most basic technical problems, such as hydropower structures, optimization of hydraulics works, conception of sewage networks, *etc*., have become routine and 'classic' in the absence of potentiality of great dams, river regulation and similar projects. The managers of companies or administrations now have the difficult task of understanding the technical aspects as well as social and environmental demands. Moreover, and possibly in the first place, they have to share the decision making process with the bodies and organisations representing interests of the public. Thus they face a double challenge:

- on the one hand, to communicate within their organisations or business consortia groups of specialists such as hydrologists, civil engineers, biologists, chemists, economists, *etc*.;
- on the other hand, to make operational an interactive information–communication process with the politico-socio-economics groups who are decision makers and, ultimately, investors and hence their clients.

To require that managers do this alone is obviously absurd. Moreover, no management course or MBA diploma can provide them with the necessary knowledge. The solution is to create within the companies collaborative processes, and, outwardly, to follow a global communication strategy. To implement both inward and outward strategies the managers need specific personnel who understand well-integrated water management problems, are educated in the spirit of collaborative processes and can weave an interactive network of communications between various professional groups.

The EuroAquae course does not aim to educate managers. However, its graduates satisfy most of the requirements concerning the above described managerial capacities. It supplies industry with people who can employ ICT within and outside the companies, who have the experience of collaborative engineering, and who know well enough the basic problems and ways of various professions to be able to set up internal (formal or informal) networks of information exchange. Along similar lines, they have the capacity to reach outside, towards non-engineering decision makers. Thus, some of them will undoubtedly become managers in the future! Interestingly, 60% of students of the 4th generation of EuroAquae declared, at the end of first year of the course, that they intended to create their own companies within several years after graduation!

4.4.4 Developments Thanks to the EuroAquae Project

4.4.4.1 EuroAquae Consortium – Links between the Partner Universities

4.4.4.1.1 Personal Links Allowing for Other Co-operations

The partners of the consortium share the same vision and the same ambition for EuroAquae. They built the vision, through synergy, from the major competencies of each academic partner and their professional environment, with the ambition of creating a programme able to cover successfully requested and innovative content, to maintain and promote the essential research and development activities, at the highest level.

With the co-operation between the different partners, the need for gradual integration and harmonization of contents has emerged. This procedure has been greatly helped by the specific Quality Assurance scheme established since 2004. The regular synthesis – produced every semester from questionnaires sent to students, teachers, industrial and academic supervisors – has clearly defined the improvements to implement and facilitate discussion among the partners.

Following this first step, the partners have extended the co-operation into the scientific field to set up the EuroAquae research group, by merging the research teams and with industrial collaboration. A major goal is the development of PhD activities under a joint supervision – '*cotutelle*' – of EuroAquae consortium members.

The personal links established among the teaching community involved in the EuroAquae have contributed to extend the co-operation and to develop new activities like joint proposals for research activities or expertise with industrial partners. This process contributes to strengthen the community and increases the quality of the communication among the actors of EuroAquae.

4.4.4.2 Links between EuroAquae Course and Industry

From the very beginning the EuroAquae course has been willing to establish a close partnership with industry. This objective solves the normal and basic need for a curriculum to be able to answer a real need of the profession. At the same time, this partnership established for the curriculum development and permanent improvement is also a key strategy for the professional insertion of the graduates. The proximity with the main companies involved in the water field at the international level brings to the curriculum an essential added value that allows for an evolution and adaptation of the contents. The professionals have not only been involved in the curriculum definition but take an active part in the teaching activities, with full responsibility for several technical courses like wastewater treatment, water supply system design and maintenance, SCADA systems, and so on. Professionals deliver also, in the different locations, several 'professional conferences' focused on practical aspects of engineers' life to train and initiate progressively the participants.

One specific aspect of the project is the creation and activity of the 'Club of Friends of EuroAquae'. This Club was set up at the very beginning of the Project. Its members are both institutions (consulting companies, utilities, administrations) and individuals working in these institutions. Currently, there are 32 registered members of the Club from a dozen institutions from six European countries. All Club members are well-known water sector professionals. Initially, most members were invited to the Club because of personal links with members of the Management Board or teaching-research personnel from the partner universities. Since then, however, several members have joined after having hosted (and often subsequently employed) EuroAquae graduates. Club members are not asked for financial support. They engage in helping EuroAquae through criticism and suggestions concerning curricula, through meetings with the students (there are two student meetings a year, during which members of the Club make presentation concerning the profession and their activities; these meetings are followed by informal discussions with students), through suggestions concerning placements for internship. The members of the Club have access to the EuroAquae Web site and can follow careers of the alumni as well as read students' CVs. One member of the EuroAquae Management Board is charged with the animation of the Club activities and, under the direction of the Project Co-ordinator, prepares a 'Club of Friends Bulletin', supplying Club members with important news and developments concerning the Course. The Bulletin appears every 1 or 2 months (ten issues during last 2 years).

4.4.5 European Added Value

4.4.5.1 Traditional Academia Concept and EuroAquae Mobility Paradigm

European Universities have become mass education institutions. The EC expects 60–70 out of every 100 pupils beginning school to receive a university education. This ambition, together with the objective of introducing a harmonized system of degrees, led to the Bachelor-Master System. It is believed that this system is flexible enough to respond to the needs of engineering education with an output of about 75% of enrolled students finishing with the Bachelor degree and 25% of the enrolled students going for Master level. Transition from Bachelor to Master level is possible either directly after Bachelor Award or by continuous education and life-long-learning programmes.

The Bachelor-Master programme is a consecutive programme; the Bachelor programme takes 3–4 years of education and the Master Programme 1–2 years. The differences of about 1 year are due to transition from former traditional university education/degree programmes in the different countries of the European Community.

The objective of the Bachelor Programme is the first profession qualifying degree; the Master degree is the second qualifying award for professional engineering. Education for research is provided in independent PhD research and courses after the Master graduation.

Implementation of the Bachelor-Master-Programme requires changes to the traditional programmes of 5 years of consecutive education. As most enrolling students aim for a Bachelor degree, all activities are concentrated to achieve the objectives of the Bachelor degree, which means the qualification for the engineering profession. Basically, the main objective is to acquire skills. This situation leads to a reduction of theory in the curricula during first 3 years in favour of training in engineering applications and mobility semesters spent as internship in companies or with universities abroad. The objective of a Bachelor Programme is to produce an engineer who needs the minimum, additional education is in practice provided by a company of first employment.

Implementation of the Master-Programme is intended as a consecutive programme on top of the Bachelor-Programme. This seems to be close to former national degree programmes of 5 years duration but, in fact, is very different. While within the former 5-year degree programmes a broad theoretical background was taught during the first years of curriculum as a basis for a 5-year programme, now the theoretical background is increasingly reduced to the extent that is needed for the Bachelor qualification, *i.e.* for most students of the first 3–4 years.

The consequences of this development have been considered for the design of the EuroAquae Master Programme. Knowledge about theory of methods and techniques to be applied during the Master course cannot be anticipated to be a prerequisite for students. This holds even more because the Master Programme 'Hydroinformatics and Water Resources' is not an integral part of any consecutive Course Programme with an underlying Bachelor Programme and recruits its students from different fields from engineering, natural sciences and informatics with various scattered basic knowledge on the Bachelor level. Moreover, taking in international students from all over the world makes it awkward to define any preconditions for curriculum access because of the variety of education programmes, especially in countries in Africa and Asia.

The EuroAquae Master Programme 'Hydroinformatics and Water Resources', aiming to be a Programme of Excellence, relies on the fact that enrolling students learnt during their individual Bachelor programmes 'how to learn' and 'how to work efficiently'. It gives all kinds of support to the students, from curriculum content presentation on the Web to entrance mock-up examinations to best prepare students for successful course participation. This provision of material is supposed to be an integral part of the curriculum and should be used by students for best individual preparedness for selection of the curriculum and to fill individual gaps before arriving for the first course.

4.4.5.2 New Education Paradigm

As stated above, EuroAquae partners share the same vision and the same ambition for the future: this vision was built, synergistically, from the major competencies of each academic partner and their professional environment, with the ambition of providing real added value to the participant and to society. The approach is supported by the concept of a 'Virtual University', which is based on the use of Web technology. The technology is not only used to share resources like lectures notes, documents and books but also to initiate collaboration and communication among participants, teachers and professionals involved in the EuroAquae activities. A communication platform is used for such purposes and combines a Content Management System and several environments like chat rooms and forums. A specific Web Video Conference environment based on the Connect platform developed by Adobe is also implemented and used by all. One of the added values of this technology is to support the development of coherent teaching material as well as quality control and enabling students to follow up the courses taught at other places. The availability of the resources contributes to establish an equal treatment for all participants and to strengthen the links of the teaching community by sharing implicitly good practices and contents.

The concept of a Virtual University, developed in a scarcity context of financial support for education institutions, answers the growing demand for global knowledge, able to address the diversity of situations that are the reality of engineers involved in the water field. In such a new and dynamic environment, the technology is simply used to connect the different experts and, at the same time, to develop a new knowledge able to handle the complexity of real situations. This dimension is today one of the most important aspects of the professional life of young engineers involved in the hydroinformatic field.

4.4.5.3 Water Sector Industry, Interrelations, Exportation

The part of the European water sector industry that the EuroAquae programme is concerned with is composed of national and regional administrations, consulting engineering companies, water utility companies and contractors. These four groups are targets of internship of EuroAquae students and, eventually, their employment. The first group, administrations, are of lesser importance numerically because they are concerned essentially with EU students, which are in the minority on the Course. Contractors, so far, have not proved to be interested in EuroAquae graduates. The situation is the complete opposite when Consulting companies working in Europe, and also beyond the EU, and water utilities are concerned. These two groups are involved in exportation of services, work, design and policies from EU towards Third countries and have been, from the beginning, extremely interested in the graduates with a EuroAquae diploma. The latter have been employed in these organisations either in Europe itself or in subsidiaries throughout the world. In that sense the added value objective of 'radiating European ways and helping European companies to export' is being reached with each new graduated generation of students. Another important point concerning European added value for the Water sector industry is the development of a new space for interrelationship between companies that otherwise may well be competitors or unknown to each other. The interrelationships help in development of common attitudes and is aimed at widening the market and exportation but also in the development of new links between clients (*e.g.* utilities) and suppliers (*e.g.* consultants). This occurs thanks to several components: EuroAquae Alumni association, members of which are working with various companies but maintain close personal contacts; EuroAquae Club of Friends activities; creation of a EuroAquae Research core group.

4.4.5.4 Radiating European Engineering Approaches and Ethics

Such radiation is obvious from a glance at the Third Countries where the EuroAquae Alumni of the first three generations are currently working: from South America, through Russia, Africa, India, Laos, China, Singapore to Australia and New Zealand. What they learnt from engineering approaches and ethics in Europe influences their work there.

4.4.6 Future Common Activities: EuroAquae Research Group

The partners in EuroAquae have successfully developed a unique experience of 'virtual university' that is recognized as a new paradigm by the community involved in hydroinformatics and water management. The concept has been extended in some parallel international projects as:

- HydroEurope (http://www.hydroeurope.org)
- HydroAsia (http://www.hydroasia.org)
- HydroAmericas (http://www.hydroamericas.org)
- HydroIndia (http://www.hydroindia.org).

These projects attract, every year, about 180 participants from more than 45 different nationalities. They have contributed to provide a worldwide dimension to the EuroAquae consortium, which is now engaged in many international activities.

The EuroAquae consortium builds this vision, synergistically, from the major competencies of each academic partner and their professional environment, with the ambition of creating and developing education activities, to maintain and promote essential research and development activities, at the highest level.

To support the education actions, the EuroAquae consortium in 2007 took two major decisions concerning its future:

- to merge the research activities of the five teams from five Partner Universities to create a joint European research group;
- to strengthen existing, and create new, links with the water sector industry.

The objective of the new EuroAquae research group, composed of more than 60 multidisciplinary scientists and researchers involved in the water sciences and technology, is to develop and promote innovative research in hydro-informatics and water management to answer some of the urgent needs and requirements of society. Impressive diversity within the research group allows the formulation of innovative and new approaches in various sectors involved in water management; for example, data acquisition, sensor definition and conception, hydrological and hydraulics modelling, software engineering, decision making, policy development and implementation, communication related to engineering problems and impacts of structures and systems implemented are some of the aspects where the research group can bring innovative ideas with a clear European dimension.

At the same time, the EuroAquae consortium, with its international industrial partnerships, will have the capacity to develop solutions that could be tailored to various regions of the world and fit the regional requirements: low energy consumption water supply systems, alternative urban drainage systems, impacts of climate change on design standards, simulation of new coastal protection, development of decision support system, implementation and development of new sensors for rivers and networks for real time management and so on, and also accompanying educational activities, including Life-Long Learning.

The research group has the vocation to increase the internal collaboration between the different institutions and also to promote new partnerships with industry and companies involved in water management. These new collaborations will strongly support the joint research activities by exchanging experience, knowledge and know-how.

Indeed, the EuroAquae research group has already started to develop some visible activities, such as in October 2008, when a joint working group was established with the Chinese authorities to prepare the definition of the new standards for urban storm drainage design and management in China, and with support of the initiative of the World Meteorological Organization (WMO) – Help Desk – for the development of Integrated Flood Management in developing countries.

References

1. European Parliament, 'DECISION No 2317/2003/EC OF THE EUROPEAN PARLIAMENT AND OF THE COUNCIL of 5 December 2003

establishing a programme for the enhancement of quality in higher education and the promotion of intercultural understanding through cooperation with third countries (Erasmus Mundus) (2004 to 2008)', *Official J. Eur. Communities* (31-12-2003).
2. European Parliament, 'DIRECTIVE 2000/60/EC OF THE EUROPEAN PARLIAMENT AND OF THE COUNCIL of 23 October 2000 establishing a framework for Community action in the field of water policy', *Official J. Eur. Communities* (22-12-2000).

CHAPTER 4.5

Inserting the Multi-lingual Urban Designer into the Water Discussion

SKYE DUNCAN

Graduate School of Architecture, Planning and Preservation, Columbia University, 1172, Amsterdam Avenue, New York NY 10027, USA

4.5.1 Introduction

4.5.1.1 Inserting Urban Design into the Science and Policy Discussion

As the awareness of water and its role in our current global status of increased surplus and scarcity is increasing, we find ourselves at a critical time in which our collective expertise and insights must be shared so we can re-inform each other and enforce the success of a common global water goal.

While scientists continue to conduct research and provide the world with new facts and figures regarding our current water status, predictions of our future water conditions, and ideas for new technologies to help manage various crises, policy makers also have the power to influence the way in which our water future is shaped at local, national and global scales of legislation. So surely the working relationship and interchange between two such important professions is vital in ensuring that one is able to inform the other to guarantee the most appropriate decisions are made, while areas of unknown are highlighted for future research.

Water System Science and Policy Interfacing
Edited by Philippe Quevauviller

Published by the Royal Society of Chemistry, www.rsc.org

As each individual's brain is of course wired uniquely, our interests and therefore chosen fields and professional directions vary. Our effective learning techniques are diverse and, along with our personal and trained communication skills, they often divide us, preventing us from uniting in trans-disciplinary education. We know there is a mutual interest regarding the subject of water across the professional spectrums and it is now time for us accept and embrace our differences, respect the knowledge and skills of our complementary counterparts and pull together to generate the most innovative and progressive approaches to managing our water status for every corner of the world.

After discovering that very few people have even heard of Urban Designers, let alone understand what they do and how it might relate to water and other environmental considerations, this chapter is an attempt to both inform the reader of the multiplicities and complexities that urban designers embrace, and to insert ourselves into this vital discussion of the state of water in our world. It began as a discussion with a group of people from URGe (Urban Research Group) in New York and continues to be an on-going research project. It is by no means a claim that we are expert hydrologists, or that we can manage this topic on our own, but quite the contrary. The aim is to offer the perspective of urban designers and to request our inclusion in this discussion. The intention is also to highlight the interdependent relationships of science, policy, economics and design in the shaping of our physical environment and to promote the advantages of trans-disciplinary efforts in solving our current water crisis.

4.5.2 Urban Design, our Pedagogy and our Focus on Water

4.5.2.1 What is Urban Design?

Urban design is a relatively new discipline and difficult to concretely define as it draws from architecture, urban planning and landscape architecture. It attempts to tackle and embrace, then visualize, synthesize and communicate, through design, the increasing complexities that exist in our various global systems, including economic, social, political and, of course, environmental languages. It does not require complete fluency in each of these languages, but fosters a base knowledge of each, and an ability to juggle them simultaneously in a 'multi-lingual' manner, so that suitable experts can be brought to the table at the appropriate phase of any project. This helps to encourage holistic approaches to urban form and infrastructure, and to place urban designers in a role where through their skills and actions they can facilitate in the loss in translation between policy making, scientific research and environmental impacts.

4.5.2.2 Urban Design Pedagogy

Because of the relatively new nature of this profession, steps to better handle the challenges ahead must begin, like any other field, with education. We must continue to foster and further push an ability to think, question, research and design at diverse and inclusive scales, from person to planet, and encourage a beneficial relationship between academia and the profession. We must repeatedly ask ourselves if our current skills are enough to deal with the complexities at hand in our ever changing environment, and continuously readdress and adapt our pedagogical efforts accordingly. One vital consideration that we know will not be going anywhere anytime soon, water, becomes an appropriate lens through which to discuss the field of urban design and our pedagogical goals as it weaves its way in varied states through every scale of our existence.

Urban design students almost always come with a previous professional degree in architecture or landscape architecture. The student work discussed further in this chapter is from the Master of Science in Architecture and Urban Design program at New York's Columbia University, part of the Graduate School of Architecture, Planning and Preservation (GSAPP). Here the students, who come from every corner of the world, are bombarded with crash courses in considering the impacts of environmental science, policy making, economics and social conditions in every design move they make. They must understand the past, present and potential future conditions of their sites, work simultaneously in multiple scales, and do so almost always in groups rather than individually. They are taught to thrive on the nutrients of scientific research for design inspiration and new ideas, and to in turn conduct their own research through questioning and generating original data. Science helps to provide something tangible to work with in this field otherwise known for its lack of concrete answers. Unlike science, however, each 'test' through a design proposal is a unique experiment, differing in client, site constraints, location and program, making it difficult to generate conclusive results. Similar to water samples taken for analysis from different points along its continuous journey, the conditions and influences vary, producing concentrations of political, cultural, scientific and environmental values that inevitably differ each time. (Figure 4.5.1 and 4.5.2)

How an urban design proposal manifests itself in reality is of course often heavily molded by policy legislation. When relating to water specifically, for example, policy ensures we must adhere to: how much water runoff certain land parcels must collect and treat on site, how much a ground floor commercial development on a retail corridor must sit above the 100 year flood line, whether local policies restrict a new innovative building design to recycle grey water for adjacent landscape irrigation, and so forth. The world of academia, however, allows the luxury of time and focused effort of using scientific and policy research to develop a design proposal and suggested results that can re-inform policy making, and eventually shape the future of our urban environment through practice. It is our responsibility within the academic world to push existing policies to their limits, find hidden potentials and interpret them in creative ways.

With greater exposure of our skills as a discipline comes the increased opportunity for the urban designer to facilitate the translation of languages

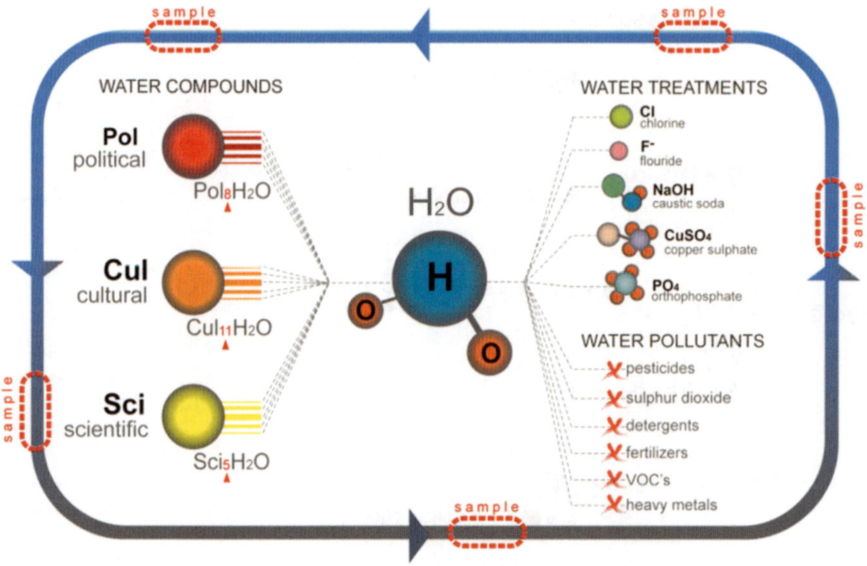

Figure 4.5.1 'Water Chemistry' (A) (S. Duncan, M. Williams and A. Casani). This project began as an attempt to understand and frame an argument for an urban design scheme that worked with water both at local and regional scales of the New York area. It asked the questions: How does one value water? Is this even possible? Surely this value continually changes, resulting in different outcomes depending on who's completing the appraisal, and on where the sample was taken from? After extensive analysis of the specific water cycle, and academic readings from the likes of Mathew Gandy and Erik Swyngedouw, the figure provides just one example of the many visual diagrams communicating the idea that water has more than just physical additives in it, some considered good, and others devastatingly bad (demonstrated by the list on the right-hand side). It also aimed to demonstrate that depending where a water sample was taken along its path (or where a localized design project was to be situated, represented by the exterior ring) there would be varying concentrations of non-physical political, cultural and scientific value adhered to the water compound (the list on the left), developing a new sort of 'water chemistry' within the built environment.

between professions. We must be careful, however, to provide enough time within each education program to teach the basics of all of these languages, while leaving enough space to further enhance the already fluent skills of design. On top of the multitude of influential disciplinary languages that must be covered (including policy making, environmental sciences, economic support and social science) we must also be sure to cover the multiple scales of design (the object, the building, the city, the region and the globe), and the almost infinite number of additional subject matters that impact urban design, from water to energy, and from land to infrastructure.

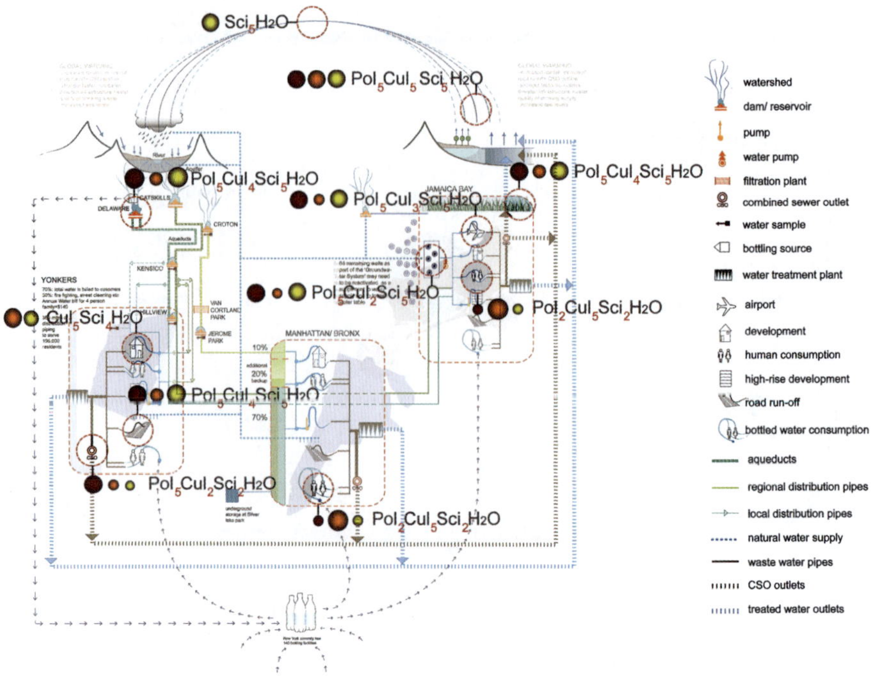

Figure 4.5.2 'Water Chemistry' (B) (S. Duncan, M. Williams and A. Casani). This image takes the Water Chemistry concept and demonstrates the concept on the actual water cycle of the New York area. The areas highlighted take a series of water samples along the water journey and show the concept that the non-physical concentrations vary at each given point of this continuous path.

4.5.2.3 Understanding Water and the Multiple Scales of Urban Design

When it comes to designing our physical environment, we see water play a role in every one of the extreme scales mentioned above. The examples are vast and for the purpose of this chapter have been kept brief, but whether designing a water efficient faucet, a fog collector in the desert or an ancient water harvesting vessel, we note the power of the smallest scale of industrial design to profoundly affect human lifestyles and urban environments. The more recent design of small scale objects such as the PlayPump (Playpump: http://www.playpumps.org/site), or the Q-drum (Q-drum: http://www.qdrum.co.za/), can ease dire conditions in areas such as Africa, reducing long arduous journeys for water collection through designing the shape of a container, or capturing the kinetic energy of children playing on a roundabout to pump a water-well. At the scale of the building we typically do everything we can to keep water out. We traditionally used it ritually, or for ornamental atmosphere, but will continue to rely on collaborative efforts of science and policy so that in the future we can capture, clean and reuse the millions of gallons we let flow through the carefully designed pipes in our homes and offices.

Human settlements have historically located themselves within close proximity to water bodies, for both survival and transportation purposes. Considering the scale of water and our cities, we have seen over a great deal of time the development of physical infrastructures drastically alter the relationship between the rural and the urban. As increased potable supplies are displaced further and further from their natural contexts, cities have flourished, the spread of disease has decreased, and our expectations of infinite, and often largely underpriced, water has grown exponentially. While some places have no choice but to carefully manage every drop they can access, most cities see perfectly clean water fall from our skies, run off our roofs and our roads where we often mix it with contaminated supplies and dump it in our local water bodies. With the powers of policy, science, design and economics combined we know we have the ability, and already the technology, to adjust these practices of managing water in and around our cities.

At the global scale we recently saw the year 2007 tip the balance so that more than 50% of the world's population resided in urban environments, developing an increased interest and importance to the role of those who design those urban environments, the Urban Designer. More people living in our cities means we need more space for them to live, and better conditions for them enjoy. It means more water must be cleaned and delivered to each city's inhabitants and more water will be consumed and taken from its natural cycle to run through our anatomy, our architecture and into our waste management systems. It means more of our local water supplies will be transported in embedded form through the products we produce and increasingly trade around the globe as our consumption patterns escalate. Efforts around the world are cleaning up waterfront properties to provide for new residential and supporting developments to house the new bodies, but simultaneously desert areas such as Las Vegas and Dubai are continually expanding at the enormous cost of nearby water ecosystems.

Science technologies have been embraced to achieve this expanding growth, policies set to enforce and allow it, and designers have shaped how this growth manifests itself in physical form. It is absolutely critical that together we begin to take these various professions and use their collective expertise to make some critical decisions and strategically enhance our urban growth in a sustainable direction. Should the science technologies for recycling wastewater to white water be developed and embraced in priority over large-scale infrastructures that remove potable supplies from its natural location? Should we address policy making and the challenges of water management at multiple scales simultaneously, remembering that while jurisdictional borders can generate differing local regulations, water knows no geographical boundaries, and its watersheds, rivers and aquifers abide to no such rules? Surely we should be restricting areas of growth to where future water supplies are adequate rather than allowing suburban sprawl in the desert areas, since together designers, scientists and policy makers have the power and ability to do so. This year in California in fact saw a delay in certain new developments as there was an inability to prove there was adequate water supply for the next 20 years.[i] An example that while undoubtedly troublesome for some parties directly involved, perhaps provides a precedent when understood as part of the bigger picture, and as responsible growth management, we could learn from to be able to ensure clean water remains a basic human right in our future.

In almost any project there are obviously numerous players involved, both affected by current water conditions and, in turn through their actions, affecting the state of water conditions in the future. While architects, engineers and designers design the appropriate buildings or spaces, construction workers translate the drawings on paper to physically generate the final form. Developers make sure the finances are in place and often provide pressure to make it as cost efficient in the short term as possible, and policy makers set the ultimate parameters (height, bulk, site coverage regulations) in which all parties must work. There is therefore an opportunity for policies to be proactive in requiring or incentivizing sustainable water practices in our built environment. With appropriate legislation in place, it would give designers the leverage to push water sensitive designs to those who are financing the project.

4.5.3 Managing Multiple Languages

4.5.3.1 Language of Urban Designers

While scientists succeed with numbers and graphs, and policy makers thrive and articulate themselves through the world of words, designers at large utilize both of these means and interpret them to express their work through varied languages of visual media. Urban designers are trained to deal with a multitude of professions, and must be able to present their concepts and ideas clearly both to someone off the street, as well as let's say an important political leader. Within their programs they must work with graphics, films, media and moving images, clear and creative data visualization, technical and accurate scale drawings, while also developing articulate verbal presentation skills. They are taught to understand the advantages of each form of media, and to use them for their strengths in effectively communicating varied and complementary types of information. These developed communication skills can be, and should be, used to help bridge the gaps between various disciplines, to translate and share information from one to another.

The 'Water Expo' held in Zaragoza, Spain, in 2008 demonstrated an example of taking complex scientific facts, along with a vast number of individual nations' water relating policies, and communicating this information through

[i] http://www.nytimes.com/2008/06/07/us/07drought.html?scp=6&sq=development%20limited%20due%20to%20water&st=cse#

built form, urban space, film media, graphic design and other visual performances to the general public. It was through these efforts that 5.6 million more people around the world are now more informed about the current global water conditions, and can take that information back to their own professions, or use it to adjust their daily lives. It was essentially an exercise in creative translation and communication of water relating information from various disciplines, through design, and one that may not have been so successful had it been only through the written word, or scientific analysis. (Figure 4.5.3).

Figure 4.5.3 Photographs from Zaragoza Water Expo, Spain, September 2008. (Photographs taken by S. Duncan.) In the images we see a large water drop generating the floor plan for the tallest building on the site, small television screens presenting short films dripping from the ceiling, minute droplets forming mist to cool the spectators in the public spaces on the hot Zaragoza days, bold graphic text lining the walls of rooms, and quantitative figures representing virtual water consumptions carved into wooden panels and assembled to generate an enclosure. The displays provide an opportunity for interaction that allows a visitor to absorb, to interpret and to, hopefully, retain the new information presented to them.

4.5.3.2 Language of Science

The sciences always seem to discover or discuss issues well in advance of the rest of the world – effects and causes of global warming, virtual water, etc. – and this is obviously partially due to the need to prove and confirm facts and theories before they're ready for general public consumption, but perhaps it is its general communication techniques or marketing skills that can be improved.

Let's take the subject of virtual water. A term thought to be coined in the 1990s, it defines the volume of water that is required to produce various industrial or agricultural products, virtual in the sense that it doesn't physically exist in the final product, but increases along the process. In some countries large volumes of potable water supplies are utilized for production for exportation goods, enhancing economic growth for an area, but at the cost of adequate human consumption supplies, or natural ecologies polluted with pesticides and fertilizers. Think about it, you are actually smelling water every time you drift past the flower arrangement on your table, consuming 16000 L of it every time you taste your tender steak cooked to perfection, sleeping on gallons of it each time you slide into your 400 count Egyptian cotton sheets, and staring at highly purified supplies of it each day as you sit at your computer containing complex electronic chips. As you purchase an imported garment in NYC for example, you are consuming the volumes of water from another country required to grow the cotton, to produce the fabric, to dye and wash it, to create the piece of clothing and to transport that item to NYC. You personally may have then had a hand in various global ecological destructions such as the almost inconceivable reduction of the Aral Sea in Asia due to cotton farming. Water is simply everywhere around us in an undeniable presence, but there is only a limited volume of fresh supplies of it. Although the concept is almost 20 years in age, it is only recently reaching the consciousnesses of many relating professions. Most of its exposure only in scientific magazines and academia means it only reaches so many. Scientific American, the longest continuously published science magazine in the United States,[ii] have drastically improved their graphic images and diagrams over the last 150 years, making their information more accessible to the general public, but is still only really available to those with enough interest to go out and purchase the publication.

While it is not too late today to consider the importance of the virtual water topic, if our communication lines between disciplines were more open and effective earlier, we might already have global policies in place that aid in managing water rich products between water poor countries. Our local policies might be more stringent on manufacturing and industrial processes, utilizing recycled grey water rather than drawing on white water supplies wherever possible. Recent student projects at Columbia University's Master of Science in Architecture and Urban Design demonstrate urban design as a tool for drawing on such scientific research and suggesting appropriate policy changes (Figures 4.5.4, 4.5.5 and 4.5.6).

[ii] http://www.actionbioscience.org/announcements/scientific_americanmagazine_na.html

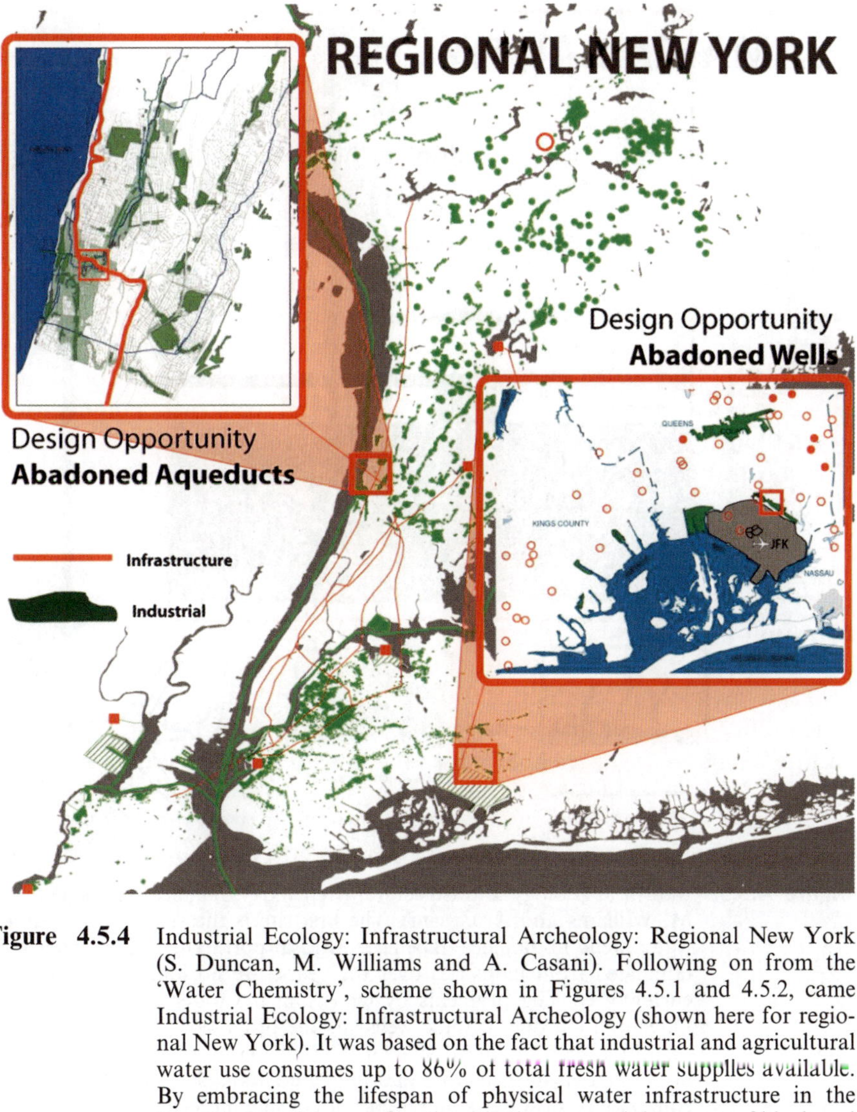

Figure 4.5.4 Industrial Ecology: Infrastructural Archeology: Regional New York (S. Duncan, M. Williams and A. Casani). Following on from the 'Water Chemistry', scheme shown in Figures 4.5.1 and 4.5.2, came Industrial Ecology: Infrastructural Archeology (shown here for regional New York). It was based on the fact that industrial and agricultural water use consumes up to 86% of total fresh water supplies available. By embracing the lifespan of physical water infrastructure in the New York region to define the sites, it proposed the re-use of black and grey water for industrial and agricultural production to reduce the overall white water consumption and manage the water footprint for an ever-expanding population. The regional scale shows the land zoned as industrial and some of the physical water infrastructure. Although there were numerous design opportunities to be found, two local sites were investigated in more detail. The first maps show the locations of the Old Croton Aqueduct in Yonkers and the abandoned water wells in Jamaica Bay.

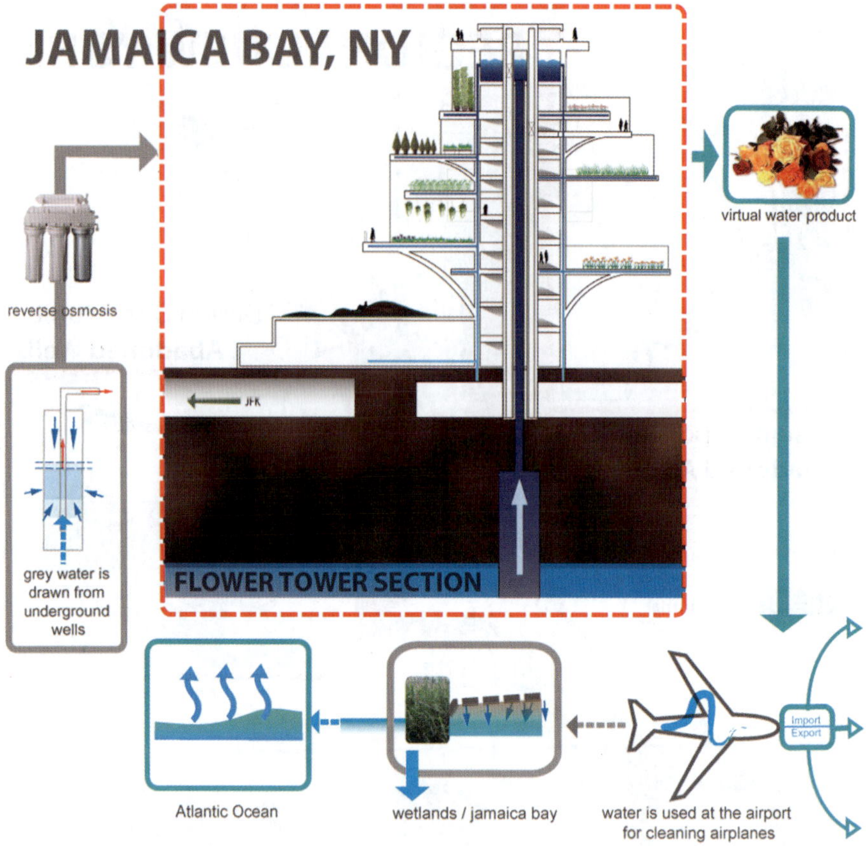

Figure 4.5.5 Industrial Ecology: Infrastructural Archeology: Jamaica Bay (S. Duncan, M. Williams and A. Casani). The first site is the series of abandoned water wells in Jamaica Bay near JFK airport. These have been contaminated from industries over the years, are no longer used and are currently causing problems by increasing the water table in the area. Directly upon one of these wells sits the proposed 'Flower Tower' – a vertical public botanical garden. This water intensive production of flowers avoids the need for importing internationally (consuming another country's fresh water supply) and of consuming local white water supplies. It instead draws up the contaminated water directly from the well below, and cleans it to then irrigate the urban horticulture platforms. The design proposal provides a wonderful new public amenity for the local community, where people can go to welcome the planes in and out of JFK airport, while increasing public awareness as they learn the details of the agricultural process.

4.5.3.3 Language of Policy

When it comes to the language of policy makers, we see the documents and codes often seem to take the local spoken and written language and use it in such

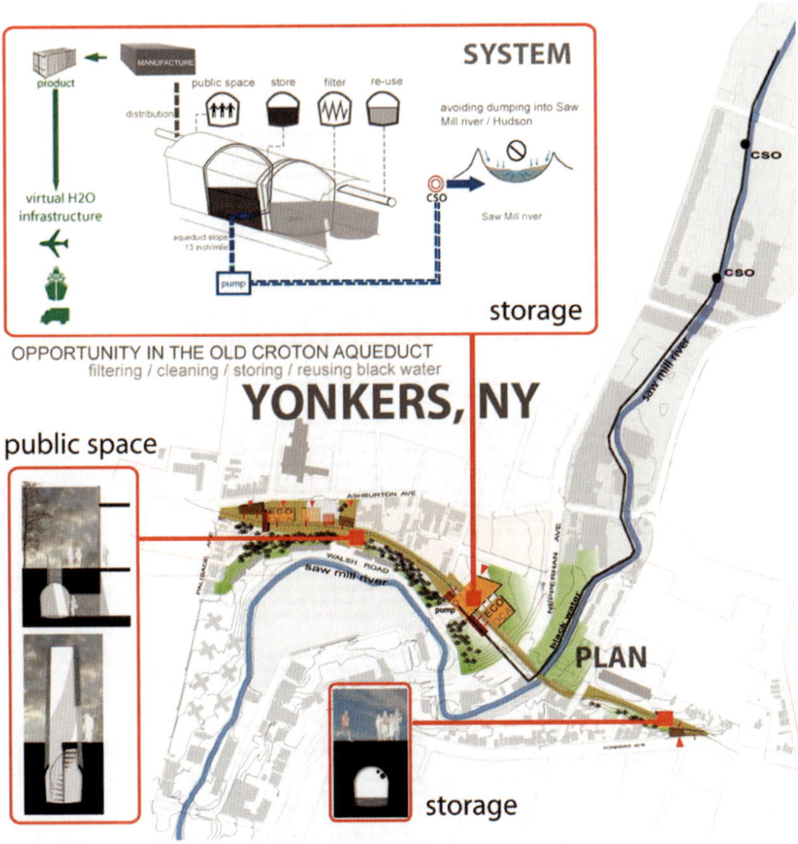

Figure 4.5.6 Industrial Ecology: Infrastructural Archeology: Yonkers (S. Duncan, M. Willams and A. Casani). The second site is the currently unused Old Croton Aqueduct (OCA) running through Yonkers that first delivered potable water to New York City from the Croton Aquifer in the 1840s. It was decommissioned in the 1950s and now has a long narrow public park that runs above it. The park, however, is in a tragic state in certain areas that it runs through the city of Yonkers with no supporting activity overlooking the area, generating an unsafe and polluted public space. It is here the proposal reinstates a life and purpose to the OCA, using it to foster the development of a new garment district that runs along the edge of the aqueduct line. The aqueduct is used in parts for storage of black water (diverted from its path to the CSO – combined sewer outlet) where it is cleaned to a state that it can be used for direct access for the industrial processes above. In other parts the aqueduct is opened up to provide public spaces for the district, where people can come and enjoy the history of their area, and relish in its prosperous and sustainable future while educating themselves of the openly displayed industrial processes. Once the local interventions consolidate through time, they would hope to ensure that resulting products will have global repercussion in virtual water form, since fresh water levels would maintain or increase in water scarce countries. Through managing sustainable production and trade internationally, New York can set a global precedent of industrial ecology and reduce stress on water poor countries.

CHAPTER 32
ENCROACHMENTS INTO THE PUBLIC RIGHT-OF-WAY

SECTION BC 3202
ENCROACHMENTS

3202.1 Encroachments below grade. Encroachments below grade shall comply with Sections 3202.1.1 through 3202.1.4.

3202.1.1 Footings. Exterior wall and column footings may be constructed to project beyond the street line not more than 12 inches (305 mm), provided that the top of the footing is not less than 8 feet (2438 mm) below the ground or sidewalk level. Foundation walls required to support permitted projections may be constructed to project not more than the permitted projection beyond the street line.

3202.1.2 Vaults. Vaults may be permitted in accordance with the *New York City Charter* and Chapter 19 of the *Administrative Code*. Such vaults shall comply with the provisions of this code and other applicable laws and rules.

3202.1.3 Areaways. Areaways, subject to approval by the Commissioner of the Department of Transportation, shall be protected by grates, guards or other approved means.

3202.1.4 Tunnels between buildings. Tunnels connecting buildings, and projecting beyond street lines, may be constructed subject to the approval of the Commissioner of the Department of Transportation. Such tunnels shall comply with the provisions of this code and other applicable laws and regulations.

3202.2 Encroachments above grade. Encroachments into the public right-of-way above grade shall be prohibited except as provided for in Sections 3202.2.1 through 3202.2.3.

3202.2.1 Encroachments subject to the area limitations. Encroachments that are subject to area limitations are those elements listed in sections 3202.2.1.1 through 3202.2.1.8, generally of an architectural character, that form an integral part of the building facade. The aggregate area of all such elements constructed to extend beyond the street line shall not exceed 10 square feet (0.93 m2) within any 100 square feet (9.1 m2) of wall area, except that a veneer may be applied to the entire facade of a building erected before December 6, 1968, if such veneer does not project more than 4 inches (102 mm) beyond the street line. The area of any such projection shall be measured at that vertical plane, parallel to the wall, in which the area of the projection is greatest. This plane of measurement may be at the street line, the line of maximum projection, or any point in between.

3202.2.1.1 Entrance details. Entrance details, including steps, and doors when fully open, may be constructed to project beyond the street line not more than 18 inches (457 mm). Entrance steps that project beyond the street line shall be guarded at each end by railings or check pieces at least 3 feet (914 mm) high or by other members of the entrance detail providing equivalent protection.

3202.2.1.2 Architectural details. Details such as cornices, eaves, bases, sills, headers, band course, opening frames, sun control devices, rustications, applied ornament or sculpture, grilles, windows when fully open, air conditioning units, and other similar elements may be constructed:

1. To project not more than 4 inches (102 mm) beyond the street line when less than 10 feet (3048 mm) above the ground or sidewalk level.
2. To project not more than 10 inches (254 mm) beyond the street line when more than 10 feet (3048 mm) above the ground or sidewalk level.

Exception: Architectural details that are more than 10 feet (3048 mm) above the sidewalk and that project more than 10 inches (254 mm) may be permitted subject to the approval of the Commissioner of the Department of Transportation.

3202.2.1.3 Balconies. Balconies, including railings and supporting brackets, no parts of which are less than 10 feet (3048 mm) above the ground or sidewalk level, may be constructed to project not more than 22 inches (559 mm) beyond the street line. When permitted by the provisions of this code, fire escapes that are part of a required exit may be constructed to project not more than 4 feet 6 inches (1372 mm) beyond the street line provided no part, including any movable ladder or stair,

3202.2.1.5 Light fixtures Light fixtures that are supported entirely from the building may be constructed to project not more than 2 feet (610 mm) beyond the street line, provided no part of the fixture is less than 8 feet (2438 mm) above the ground or sidewalk level.

3202.2.1.6 Flagpoles. Flagpoles that are supported entirely from the building may be constructed to project not more than 18 feet (5486 mm) beyond the street line, but not closer than 2 feet (610 mm) to the curb line, provided that no part of the flagpole is less than 15 feet (4572 mm) above the ground or sidewalk level.

3202.2.1.7 Wall signs Wall signs may be constructed to project not more than 12 inches (305 mm) beyond the street line when conforming to the requirements of this code and Section H111 of Appendix H.

3202.2.1.8 Projecting signs All permitted projecting signs may be constructed to project not more than 10 feet (3048 mm) beyond the street line, but not closer to the curb line than 2 feet (610 mm), when conforming to the requirements of this code and Section H112 of Appendix H, and provided that no part of the sign is less than 10 feet (3048 mm) above the ground or sidewalk level.

3202.2.3 Awnings. Awnings constructed in accordance with Section 3105 and supported entirely from the building may project beyond the street line as follows:

3202.2.3.1 Store front awnings. Store front awnings may project beyond the street line not more than 8 feet (2438 mm), provided no part of the awning is less than 8 feet (2438 mm) above the ground or sidewalk level, except for a flexible valance which may be not less than 7 feet (2134 mm) above the ground or sidewalk level, and provided that the awning box or cover does not project more than 12 inches (305 mm).

3202.2.3.2 Awnings over windows or doors. Awnings over windows or doors may project beyond the street line not more than 5 feet (1524 mm), provided that no part of

Figure 4.5.7 Paragraphs of written text from the New York City Department of Buildings Building Code.

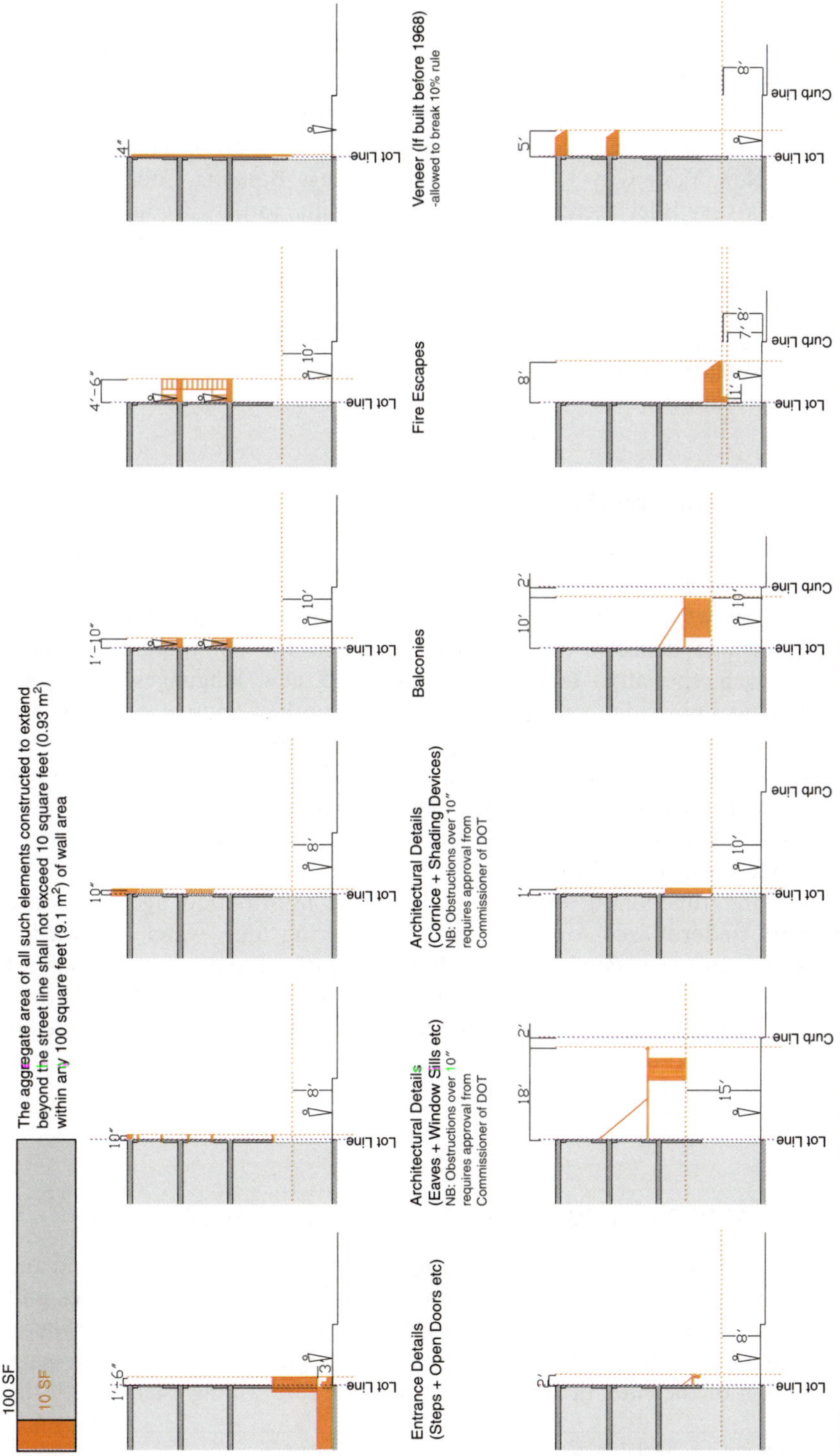

Figure 4.5.8 The same information as in Figure 4.5.7 presented visually.

a specific way that it almost develops into its own new dialect. The legislation, mostly all words with no images, becomes specialized in a way that even those within the profession can find it difficult to digest, leaving little hope for others to understand. In Figures 4.5.7 and 4.5.8 we see two examples of communicating the same piece of information. The paragraphs of written text are from the New York City Department of Buildings Building Code,[iii] and the diagrams, a very brief exercise to visualize this data so that it could be easily understood by those more visually inclined. It is by no means to say that one method is better than another (as mentioned previously we all successfully digest new information in different ways) but suggests that the additional effort required to present information in more than one manner may just be worth it, and be necessary to foster the successful transfer of information from one profession to another.

4.5.3.4 Combining the Languages

As we all know, languages present the constant challenge of varying dialects and are ever developing along with our societies and technologies. There is also the danger as we enter more specialized areas of expertise, and therefore differing language dialects, that we further widen the communication gaps between each specialty. Increased exposure to new languages of course improves our knowledge and understanding of another, and we must find the opportunities in allowing our varying professional languages to overlap, to fuse and to re-inform each other. We should embrace the challenge of looking at our subject matter such as water, not in isolation, but from the differing perspectives that allow complexities of different influences and constraints to transform into unique successful solutions.

Living and functioning in the middle of the information age, there are numerous underutilized opportunities for sharing our water knowledge through the Internet, conferences, online publications and so forth that we should be taking advantage of. We must, however, find a way to do this while developing the editing and clear communication skills that do not leave us dumbfounded and oversaturated with information to the point where we are paralyzed to inaction.

4.5.4 How Science, Policy and Design Interact

If we were to search long and hard enough, we would most likely find the scientific research and developed technologies for making the changes and improvements we desire for our water world already exist out there somewhere. The economics of utilizing these advanced technologies, however, often remain in the realm of the elite, or the designer with the rare willing client. With the

[iii] http://www.nyc.gov/html/dob/downloads/pdf/cc_chapter32_sbs.pdf

power to prohibit, discourage, allow, incentivize or require certain practices, policy makers are able to foster the shift over time to main stream use of water sensitive designs, eventually dropping the often astronomically high initial costs to within the affordable reach for designers to argue and justify their importance in their work. Below are a few examples of this interaction of the multiple disciplines impacting a situation, discussed only briefly, but in order to give a taste of the numerous scenarios out there.

While science is (and has been for a long time) telling us to use less water, policies must follow by enforcing it. Without the legislative power of regulations, architects and designers are often fought by developers on their good intentions of water saving designs, with the arguments of increased financial costs. Local or national policies should be enforcing more restrictive use, or at the very least fine tooth-combing their legislations to remove current penalties for good practice. Until people are forced to use less, or have to pay for excessive use, individual behaviors will not change. An appropriate value must be given to this precious resource.

To encourage green, blue, or water conscious design in our built environment, we must make the translation and transition from what science tells us is possible, to what policy tells us is required. Designers can aid in this translation and transition from one to another.

Take the issue of Combined Sewer Overflows (CSOs), an antiquated water management system (used in New York City and elsewhere) designed originally to save the cost of building two water systems. While in dry weather the sewage and wastewater from buildings go to water treatment plants, heavy rains means the roadway runoff combines with the wastewater and is discharged into local water bodies. Through the policies introduced by the US Environmental Protection Agency (EPA) in 1994,[iv] and amendments to the Clean Water Act[v] by Congress a few years later, efforts were required to focus on eliminating, or at least reducing, the pollution caused by such systems. While policy makers clearly worked from specific scientific evidence of polluted water ways to instigate their policy changes, they pave the way for other scientists and designers to further develop the concepts and technologies for managing water runoff by incorporating roadside swales, plantings and water detention and retention systems into their work.

Similarly, but less advanced in terms of public acceptance, science tells us we have the realistic technologies to clean and recycle wastewater supplies for human consumption, (our space shuttles can have wastewater regeneration systems to recycle urine into drinking water![vi]) but until policy permits, incentivizes or legislates it, designers will struggle to incorporate such practices into the buildings and systems they design.

Science tells us that global water levels are raising and occurrences of flooding increasing. Certain policies tell us that a ground floor development in

[iv] http://cfpub.epa.gov/npdes/cso/cpolicy.cfm
[v] http://www.epa.gov/oecaagct/lcwa.html
[vi] http://thelede.blogs.nytimes.com/2008/11/13/tasting-nasas-recycled-water/?ref=space

certain at-risk zones must be above a specified flood line. This therefore presents an opportunity to urban designers to generate innovative solutions in which a required raising of six feet above curb level for a shop to have its goods laid out does not destroy the desired vitality and activity of a main corridor shopping street. Should the sidewalk be raised and parking wedged beneath it? Should the entire roadbed be raised to meet the new level required for ground floor retail? Can the sidewalk be stepped and just half of it be raised? How would a 20 m ramp for accessible access impact the frontage of the store? It is most certainly time to begin thinking outside the box (Figure 4.5.9).

The Vertical Farm,[vii] a concept developed by Columbia Professor and microbiologist Dickson Despommier, takes scientific research and imagines a new world in which urban farming supplies a city with its fresh fruit and vegetables in inner-city skyscrapers. Able to utilize processed wastewater for irrigation, it could reduce white water consumption for an area, while reducing the risks of loss of crops to nature's increasing events of floods and drought. While the concept began in the sciences, it will be with the cooperation of policy makers, who will legally permit this to happen (both through construction and food standard regulations) and designers, who have the skills and creativity to develop it into a successful physical manifestation within the urban jungles, that this concept will be able to find its full potential.

An interdependent relationship each of our professions must keep in mind is that of water to power. Water is needed to generate power in the traditional sense, and power is in turn required to provide potable supplies of water. This relationship is not privy to water poor or water rich areas, whether fresh water must be imported, contaminated supplies cleaned or local sea water desalinated for consumption. As urban designers we find opportunities in this relationship to enhance our work. We try to design cities to be less car dependent by encouraging enticing pedestrian environments, adequate bicycle lanes and alternative transportation means. The long-term hope is to help reduce our carbon footprints, the effects of global warming, and in turn to help lessen the frequency and impacts of increased extreme water events that can destroy our urban and rural environments, droughts and floods.

4.5.5 Future of Hybrid Approaches

In understanding the relationships between science, policy and design, it is vital not to forget the critical role of economics. Just as this chapter is an attempt to insert urban designers into this discussion of water and trans-disciplinary approaches, economists must do the same. For no matter how innovative the technology, how progressive the policy, how successful the design, if the numbers don't add up, and the money behind the scenes is not taken care of, nothing will happen or change. While the world contains numerous well

---

[vii] http://www.verticalfarm.com/Default.aspx

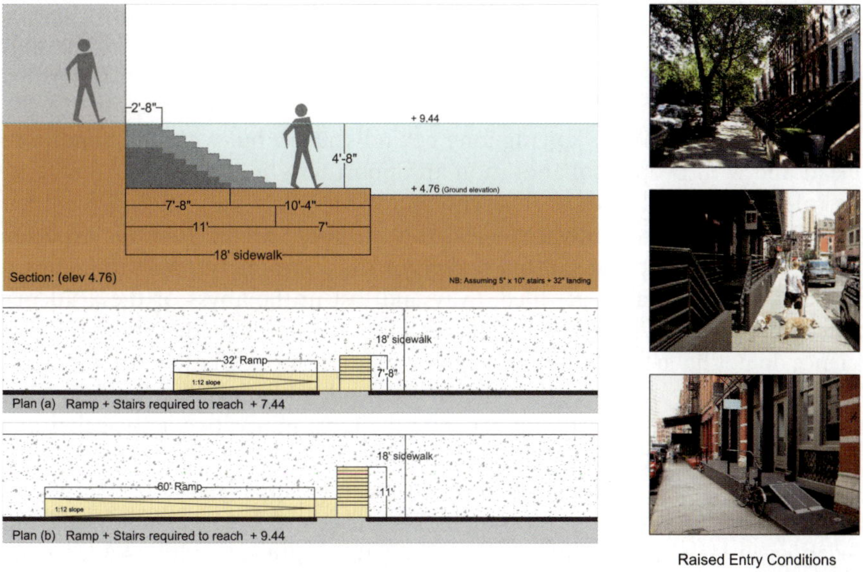

Figure 4.5.9 Flood levels on a retail corridor. This diagram demonstrates the impacts of the requirement to raise a ground floor retail establishment to a level safely above the base flood elevation for the area. Here, while the existing road level sits at +4.76, the ground floor must be raised 4′8″ to reach an elevation of +9.44. While this model is commonly seen in New York for residential developments, its impact on a main shopping street could drastically affect the porosity and accessibility of the store fronts, and the vibrancy of the street itself.

intentioned souls, if a less water sensitive project can go ahead and be constructed at a much lower cost, chances are it will.

It is often not until people feel the impact of their actions on their wallets, or see the blatant facts in front of their eyes, that awareness will increase and behaviors will change. Like the recent increased price of oil encourages people to bring out their bicycles or walk to their local shops to save money, we must find the equivalent triggers to alter our personal relationships with water.

Scientists and designers can only do so much to change behaviors and while they can aid in increasing awareness of the subject matter, only policy has ability to legislate to dictate certain responsible behaviors.

Some of us are privileged to have the power of knowledge with regards to this subject matter, but it's the processing and interpretation of this knowledge that becomes vital. We need to know what to do with it, and who to pass it onto to see the changes we need come to fruition. Increased knowledge, and therefore power, will only be enhanced through the creation of hybrid teams. We of course do not want economists to design our cities, or policy makers to conduct our scientific experiments, but we must foster a respect for what each profession does. To reiterate, this chapter in no way claims that urban designers have the power or know how to successfully complete any (water) project on their own, but merely requests and suggests that we be brought to the table to be a part of the discussions, like just one single food group contributing to an overall balanced diet fostering healthy development.

We see policy typically lagging long behind science, but with dire conditions ahead and with proper collaboration an opportunity is presented to shake up the roles and for policy to push science and design, and vice versa. Design has an ability to generate innovative ideas, but without specific scientific expertise to help develop these ideas into a reality, or the economic support to finance a new concept, they often get filed away and lost in the abyss of the academic conceptual realm. With a more trans-disciplinary approach, design, policy and science do not have to maintain a stagnant relationship with each other, but can flip, rotate, invert and strengthen the interdependent connections between them to push each other forward and generate the most innovative approaches and solutions to our water crisis.

We can aid this dynamic rhizome of relationships and cross fertilizing learning through shared professional conferences, through re-analyzing our pedagogical structures to offer cross disciplinary projects while students are in university, and by each ensuring appropriate experts are at the table when contemplating our future water world at various scales.

The images used in this chapter and their explanations might have been all that some people looked at, others may have skimmed only the subtitles for each section. If you are reading this final paragraph you might have read some or all of the written text, and it may or may not have been helpful coming from the perspective of someone outside the professions of policy and science. The hope is, however, that this chapter provides a little insight not only to the world of urban designers, how we think and what we do, but how the subject matter of water plays a critical role in encouraging us to share our expertise with each other, to talk, to read and to respect that we all have something complementary to offer when gathering around the table to discuss the vital conditions of water in our current environment, and that our more well educated decisions and practices can only have a positive effect in turn on our future water conditions.

Acknowledgements

This chapter is the further development of an on-going research project completed in collaboration with Mojdeh Baratloo and Marissa Gregory, as a part of URGe (Urban Research Group), New York, USA. It has been written from the perspective of an urban designer with an architecture background, and focuses on the subject of WATER, and its role of influence from both academic experience (while studying and then later teaching at Columbia University's Master of Science in Architecture and Urban Design program) and from professional experience while working in the New York City's Department of City Planning's Office of the Chief Urban Designer.

Related Reading

M. Barlow, *Blue Covenant, The Global Water Crisis and the Coming Battle for the Right to Water*, The New Press, New York, 2007.

J. Feyen, K. Shannon and M. Neville, *Water & Urban Development Paradigms*, CRC Press, Taylor & Francis Group, London, 2008.

M. Gandy, Concrete and Clay Reworking Nature in New York City, MIT Press, MA, 2002.

R. J. Goldstein, *Ecology and Environmental Ethics*, Ashgate Publishing Company, VT, 2004.

T. Hiss and C. Meier, *H_2O: Highlands to Ocean*, Geraldine R. Dodge Foundation, 2005, Gale Group, Farrington Hills, Michigan, USA.

K. Lanz, L. Muller, C. Rentsch and R. Schwartzenbach, Who Owns the Water? Lars Müller Publishers, Swiss Federal Institute of Aquatic Science and Technology, Switzerland, 2006.

B. Latour, Politics of Nature –How to bring Sciences into Democracy (English Translation by Catherine Porter), Harvard University Press, MA, 2004.

S. T. A. Pickett and P. S. White, *The Ecology of Natural Disturbance and Patch Dynamics*, Academic Press, New York, 1986.

E. Swyngedouw, Social Power and the Urbanization of Water. Flows of Power, Oxford University Press, New York, 2004.

E. Tufte, Envisioning Information, Graphics Press, Cheshire, Conn., 1990.

E. Tufte, The Visual Display of Quantitative Information, Graphics Press, Cheshire, Conn., 1983.

K. Watkins, Human Development Report, Beyond Scarcity: Power Poverty and the Global Water Crisis, The United Nations Development Plaza, 1 UN plaza, New York, NY 10017, 2006.

M. Webber, Catch-22: Water vs. Energy, Scientific American Earth 3.0 (Special issue) vol. 18, no. 4; www.SciAmEarth3.com (2008).

S. White, On Site Review 17: Water, Calgary, Canada, www.onsitereview.ca (2007).

Section 5: Summary, Way Forward and Conclusions

CHAPTER 5.1

Concept of Interfacing and Perspectives*

PHILIPPE QUEVAUVILLER,[a, b] PHILIPPE VERVIER[c] AND MARIE-PERRINE DUROT[d]

[a] European Commission, DG Research, rue de la Loi 200, 1049 Brussels, Belgium; [b] Vrije Universiteit Brussel (VUB), IUWPARE, Building T, Pleinlaan 2, 1180 Brussels, Belgium; [c] ECOBAG, 15 rue Michel Labrousse, 31023 Toulouse Cedex, France; [d] ONEMA, the French National Agency for Water and Aquatic Environments, Hall C, 5 square Felix Nadar, 94300 Vincennes, France

5.1.1 Introduction

Environment policies require a sound scientific evidence base to be developed and implemented at international level. Salient examples can be found within the Climate Change community, which managed to efficiently transfer inputs from scientists to policy makers through the Intergovernmental Panel on Climate Change (IPCC) process. Similarly, the European Platform for Biodiversity Research Strategy (EPBRS) is a forum within which natural and social scientists, policy-makers and other stakeholders identify structure and focus the strategically important research that is essential to address major challenges for halting biodiversity loss.

European water policies also require a dedicated interface between Member States, which are confronted with similar Water Framework Directive (WFD) issues, and with the European commission, from where a great deal of water legislation originates. Discussions are on-going between scientific and

*The views expressed in this chapter are purely those of the authors and may not in any circumstances be regarded as stating a formal position of the European Commission.

Water System Science and Policy Interfacing
Edited by Philippe Quevauviller

Published by the Royal Society of Chemistry, www.rsc.org

policy-making communities in the water sector to better organise science–policy interactions.[1] In this context, the need to develop a conceptual framework for a science–policy interface related to water was highlighted, which would gather various initiatives and knowledge.[2,3]

This chapter follows up considerations expressed in various other chapters of this book, providing some insight into the concept of science–policy interfacing as applied to water policy implementation and opening perspectives linked to the WFD Common Implementation Strategy (CIS).[4]

5.1.2 Science–Policy Interfacing in the Context of the WFD

The Water Framework Directive (WFD)[5] provides a stable platform enabling the build up of communication and best practice exchanges among different players (policy implementers, technology providers, scientific community, industrial stakeholders, NGOs, *etc.*). As described in other chapters of this book, this is reflected in clear improvements within the last 5 years, with plans for developing joint initiatives among research and policy-maker's communities. In particular, it is recognised that efforts are needed for presenting results of research and demonstration projects in an easily usable form by policy-makers (this is extensively discussed in Chapters 1.2 and 1.3). Adversely, the way research results are considered by the policy-making community is not straightforward, mainly for political reasons and difficulties in integrating the latest research developments in legislation. The difficulty is enhanced by the fact that the policy-making community is probably not defining its role as 'client' sufficiently well. In other words, the dialogue and communication are far from being what one would hope to ensure an efficient flow of information.[2] Improvements were thought to be possible through the development of an operational 'science–policy interface' based on a coordination of relevant programmes/projects with direct relevance to the WFD implementation,[2,6] which might include:

- a screening phase for evaluating needs linked to specific policy milestones (this may concern gathering background scientific information or tailor-made research, R&D and demonstration);
- a mechanism to identify those research projects that are most relevant with respect to policy steps, and which might be 'validated' through demonstration activities with the aim of being disseminated and applied at the most appropriate level (regional, national or EU);
- an exchange platform enabling both scientists and policy makers to discuss the corresponding research and policy agendas from the very beginning to ensure a more structured communication at all appropriate levels of policy formulation, development, implementation and review.

This interfacing concept should be more than fostering dissemination and application of research to respond to policy needs. It should actually strive to establish strong links among different funding mechanisms existing at the EU and national level and thematic policies. In particular, pilot projects should be promoted, combining the implementation of results of successfully completed research, R&D or demonstration projects with the implementation of related policy steps. Such a mechanism should look at the possibility of developing new partnerships by combining various types of projects [*e.g.* RTD and LIFE (Chapter 1.4) COST, Structural and cohesion funds (Interreg projects), agricultural funds, *etc.*, as well as projects funded at regional/national level] and promoting the establishment of innovative partnerships involving scientists, policy makers, managers and other stakeholders for the effective integration of science outputs into policy and management decisions. At present, such coordination is not yet even contemplated, as illustrated by different catchment-related projects funded by various mechanisms that were developed in parallel without building up synergies.[6]

In addition to the above described necessary means to set up efficient science–policy interfaces, it should be anticipated that:

- For the required projects to deliver appropriate scientific support, new skills should be developed by the coordinators. Since most of the research projects are gathering different scientific disciplines (physics, chemistry, natural sciences, economics, social sciences, *etc.*) and different socio-economic sectors (water managers, policy makers, enterprises, NGOs, *etc.*), project coordinators with 'translating' expertise (Chapter 5.2) are key resources to drive these projects.
- Since water is a global problem but a local resource, it would be relevant to set-up in each River Basin District of Europe, at least, one science–water policy platform playing the role of interface that could be involved in a larger European network strongly supporting implementation of European water policies.

In response to the above considerations, a concept is under development in collaboration among the European Commission' Directorate-Generals for Environment (including the LIFE programme) and Research (DG-RTD), aiming to establish an efficient and sustainable science–policy platform linked to WISE (Water Information System for Europe) that has a direct relevance and link to monitoring development and data.[7] This initiative is a continuation of exchanges developed over the last few years (Chapter 1.4), which led to the development of the so-called WISE-RTD web portal,[8] as briefly outlined below.

5.1.3 Operational Features

5.1.3.1 Harmoni-CA Initiative

Harmoni-CA has been supported by the DG-RTD as a large scale *concerted action*[9] with the objective, amongst others, to create a forum for communication, information exchange and harmonisation of Information Communication & Technology (ICT)-tools for integrated river basin management and the implementation of the WFD. This initiative is described in detail in Chapter 2.4. The idea of linking WFD requirements and RTD outputs through an operational 'interface' came directly from this initiative, in the first instance through the building of a web-portal, originating from this project (see below).

5.1.3.2 WISE-RTD Web Portal

Resulting from the Harmoni-CA initiative, the WISE-RTD (http://www.wise-rtd.info/) has been conceived as a platform for accessing scientific information of potential use to water policy implementation.[10] This portal is described in Chapter 4.2. It aims to give access of policy relevant scientific information of interest to policy officers, RTD managers and scientific stakeholders. This web-portal is to be seen as an operational tool supporting the interface.

5.1.4 General Principles of the Science–Policy Interface

In the sketch below (Figure 5.1.1), the water policy process (centre of the figure) is closely linked to a larger circle concerning *research* and technological developments [FP7 for EU research, ERA-NET for transnational research (Chapter 1.4) and national research], *demonstration* (LIFE programme, possible INTERREG support), *interface* through a 'single web portal' (WISE-RTD and possible other linked websites) and *research policy* development (reviews, integration, research needs).[6]

This diagram, however, shows an ideal situation that is far from being operational. In other words, there are few or no coordination(s) between

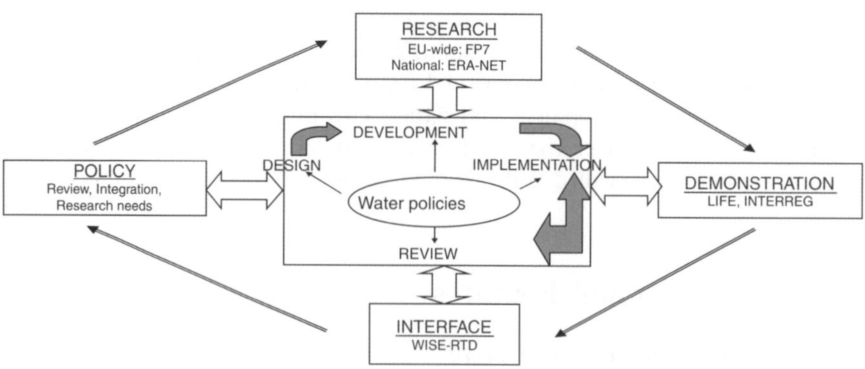

Figure 5.1.1 Links between water policies and research & demonstration (after ref. 6 with permission).

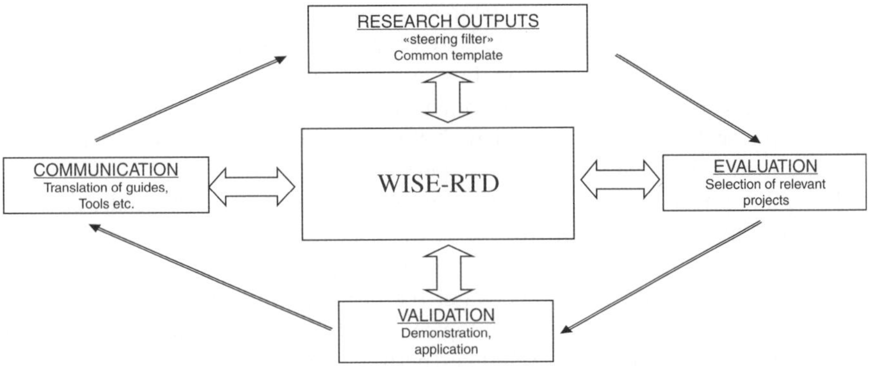

Figure 5.1.2 WISE-RTD and science–policy interfacing needs (after ref. 6 with permission).

different funding mechanisms. To make it workable, a strengthened coordination has to be developed among the responsible services, as shown in Figure 5.1.2.

The goal would be to operate a science–policy interface using 'WISE-RTD' as a tool (providing access to RTD information) as displayed in Figure 5.1.2. In this configuration, a first phase would act as a 'steering filter' for RDT and demonstration projects (analysis of relevance to water) – Phase 1 'Research Outputs' – followed by an Evaluation (selection of the most relevant projects), a Validation (demonstration of successful RTD) and Communication in the form of technical guides translated into the EU languages (addressed to policy managers).[2,6] It is most likely that only a limited number of projects would be selected until the last step. In other words, there are many projects that may be of high scientific value without necessarily fulfilling concrete policy needs. The evaluation and demonstration steps would then be of critical importance to judge those projects that deserve to be transferred at wide-scale to water managers in the EU and beyond. The overall system depicted above could only be developed on the basis of operational national and/or regional relay platforms functioning in networks. Examples of national and regional approaches are described in the literature[6] and in Section 3 of this book (Links to Water National or Regional Research, Policies and Management).

5.1.5 Pilot Initiative: the CIS-SPI Activity

Acquiring a capacity to develop a global and strategic perspective in support of water policies, developing the WISE-RTD web portal as a tool, ensuring operational links between the scientific community and decision makers (policy, civil society and other stakeholders) are complex and challenging tasks that require dedicated and coordinated efforts. The development of an operational interface would only be possible through interactions and guidance of a dedicated group mixing policy people, scientists and stakeholders. In this respect, an attempt has been made to propose a Science–Policy Interfacing group (SPI) within the Common Implementation Strategy (CIS) to the EU Water Directors, to constitute an anchorage point for future developments. A pilot phase has been launched in 2009 with voluntary countries, EC DG Research, stakeholders and invited experts.

The pilot phase aims to provide some insight into research needs, competences and gaps, notably on the basis of discussions with the thematic CIS working group in their field of activities (*e.g.* ecological status, chemical monitoring, groundwater, floods, water scarcity and drought).[4] A focus would be put on transversal and methodological developments, for the development of a framework to improve exchanges between science and end-users for the transfer of research results and for the identification of research needs linked to the policy cycle, based on a few successful examples.

This pilot, if successful, would pave the way for a kind of 'institutionalisation' of the group through the CIS, to channel discussions and decide about concrete developments. The issue is essentially linked to a more efficient transfer system, helping Member States to get timely access to scientific information supporting water policy implementation (in particular the Water Framework Directive and its daughter directives, as well as other relevant directives such as, for example, the Floods Directive) and to identify short-, medium- and long-term research needs. To be effective, the entity would call on national and regional relays, so-called 'mirror SPI groups', as a promising way forward in ensuring good coordination. The medium-term perspective for the functioning of this institutionalized European science–policy platform is described in Chapter 5.3, highlighting how science and policy can be (re)connected at several steps of the R&D life cycle, consisting of reviewing RTD projects with regard to their policy relevance, demonstrating their applicability, communicating their outputs, compiling feedback from users, and identifying further research needs linked to policy development, implementation and review.

References

1. Environmental Science & Policy, 2005, Proceedings of the workshop on Research and Technology Integration in support of the Water Framework Directive, *Environ. Sci. Policy*, 2005, **8**(3).
2. Ph. Quevauviller, P. Balabanis, C. Fragakis, M. Weydert, M. Oliver, A. Kaschl, G. Arnold, A. Kroll, L. Galbiati, J. M. Zaldivar and G. Bidoglio, *Environ. Sci. Policy*, 2005, **8**, 203.
3. Ph. Quevauviller, *Environ. Sci. Pollut. Res.*, 2007, **14**, 297.
4. European Commission, 2001, *Common Implementation Strategy for the Water Framework Directive, European Communities*, ISBN 92-894-2040-5, 2003. Final CIS document available at http://europa.eu.int/comm/environment/water/water-framework/implementation.html.

5. European Commission, 2000, Directive 2000/60/EC of the European Parliament and of the Council of 23 October 2000 establishing a framework for Community action in the field of water policy, *Official J. Eur. Communities*, L 327, 22.12.2000, p. 1.
6. Ph. Quevauviller, B. Harris and P. Vervier, in *The Water Framework Directive – Ecological and Chemical Status Monitoring*, ed. Ph. Quevauviller, U. Borchers, K. C. Thompson and T. Simonart, John Wiley & Sons Ltd., Chichester, 2008, pp. 411–444.
7. J. D'Eugenio, P. Haastrup, S. Jensen, A. Wirthmann and Ph. Quevauviller, General Introduction to WISE, 7th International Conference Hydroinformatics, Nice, September 2006.
8. W. de Lange, G. Arnold, P. Willems, F. Provost, F. Hatterman, J. Plyson, M. Mestdagh, P. Swartenbroeckx, P. Balabanis and Ph. Quevauviller, WISE-RTD web portal – A gate to scientific information for WFD implementers and water managers, poster presented at the International Conference on Monitoring under the WFD, Lille, March 2007.
9. HarmoniCA, 2001, Harmonised Modelling Tools for Integrated River Basin Management, Harmoni-CA, EU funded concerted action, Contract EVKI-2001-00192, www.Harmoni-CA.info.

CHAPTER 5.2

*Role of Translators in Science–Policy Interfacing**

ANTONY CHAPMAN,[a] PHILIPPE QUEVAUVILLER,[b, c] WILLEM J. DE LANGE[d] AND PHILIPPE VERVIER[e]

[a] WCA Environment Ltd., Brunel House, Volunteer Way, Faringdon, Oxfordshire SN7 7YR, United Kingdom; [b] European Commission, DG Research, rue de la Loi 200, 1049 Brussels, Belgium; [c] Vrije Universiteit Brussel (VUB), IUWPARE, Building T, Pleinlaan 2, 1180 Brussels, Belgium; [d] Deltares, PO Box 85467, 3508 AL Utrecht, The Netherlands; [e] ECOBAG, 15 rue Michel Labrousse, BP 42353, 31023 Toulouse Cedex, France

5.2.1 Introduction

Making good use of relevant science has always been and always will be a requirement of successful policy making. However, while interaction between scientists and policy makers takes place at various levels in various ways to a greater or lesser extent, there is very rarely an explicit mechanism by which scientific information is provided to policy makers.[1] The implications of this are that links between policy and science in both directions could be refined and improved, that more use could be made of available scientific information and that presenting and interpreting scientific information effectively to a non-expert audience is difficult for both parties. However, while most actors in both communities view better quality integration and interaction between science

*The views expressed in this chapter are purely those of the authors and may not in any circumstances be regarded as stating a formal position of the European Commission.

Water System Science and Policy Interfacing
Edited by Philippe Quevauviller

Published by the Royal Society of Chemistry, www.rsc.org

and policy as a good thing, what is less clear is how these issues can be tackled or who should be responsible for doing so.

In recent years projects such as NeWater (Chapter 2.6), Harmoni-CoP, Harmoni-CA (Chapter 2.4), SPI-Water (Chapter 1.4) and the EUPOL sub-project of AquaTerra (Chapter 2.2) have dealt with a range of issues related to science policy interaction and participation in river basin management and have discussed the processes and mechanisms of interaction between science and policy (Chapter 2.2).[2,3] One recurring theme (also highlighted in Chapter 1.3) that has developed is the notion of the creation of a role of 'translator' to act as a mediator, communicating information and improving cooperation between science, policy makers and also practitioners in the field. The key aim of the translator's activities would be to improve the use and understanding of science by the policy community, and by implication improve the quality of policy. It is also the case that policy makers may request advice and information on a specific topic from scientists and/or scientific groups, *e.g.* the recent 'Veerman Commission' in the Netherlands, which advised the Government on adaptation measures to protect Dutch coastlines against the effect of climate change on sea levels.[4,5] This sort of policy–science interaction is something that could be co-ordinated by translators. To date the general view in the discussions that have been held on the subject is that such translators would be beneficial, although it is difficult to quantify exactly what this benefit would be. However, there has been little discussion about exactly how the role of the translator would operate. The aim of this chapter is to move on from general discussions and begin an in-depth debate on the potential role of the translator, the range of responsibilities they may have, the opportunities that may be created by the role, the type of interactions they might stimulate, tools that can facilitate their role and, most importantly, their value to both the science and policy communities.

5.2.2 Rationale behind Translators

Science is a fundamental part of modern society, affecting all of us to some extent or other and probably more than many of us are aware. Science influences our lives and our surroundings in two principal ways. Firstly, it is responsible for technical advances that improve quality of life, but which may also need control and regulation. Secondly, it provides evidence to explain the functionality of people, flora, fauna and environments that can be used to guide Governmental policy in various fields. If this scientific evidence is used well, it helps to produce informed policy that protects people and the environment and results in the optimal use of resources.[6] However, if policies are developed on the basis of poor quality science, a field of science that has yet to be fully understood, misinterpreted scientific information or are not based on any scientific evidence at all, then the result could possibly be very negative. In our current society, which is often sceptical or wary of scientific advances, it is not unusual for the blame for misguided or misinformed policies to fall on the science (or lack of science) that supported them, whether this is justified or not.[6] This is detrimental to the scientific community, as it could affect the public image of a particular area of science, and possibly the funding that supports it, restricting further opportunities for development and innovation.

These concerns raise issues about how science is communicated beyond its immediate sphere of experts to those for whom it is of interest and who may wish to make use of it, but who may not have the time and/or the capacity to fully appreciate the detail of the work. Yet at the same time, these people must use the information available to them to make important decisions. The fact that the concept of a translator is being talked about within the science and policy communities implies that the current level and means of communication between the two groups is not completely satisfactory.

Thus, a translator should be a person, or group of people, whose knowledge and experience straddles both the scientific and policy communities, who should be comfortable in dealing with both groups and who has the time and the capacity to be able to interpret scientific data for the benefit of policy makers and to make clear the needs of policy to the scientific community. A key question in relation to such a role is what it offers above and beyond what already takes place: what systems are already in place to communicate science to policy and vice versa and is 'translation' not a job that could be taken on by other people rather than being a specific role? In describing the requirements of the role of a translator and of the type of person who would fulfil such a role in the following sections of this chapter we hope to demonstrate that the role of a translator is both specialised and would add value to any existing mechanisms for communicating science to policy.

5.2.3 Role of the Translator

The role of a translator would be likely to encompass a wide range of activities for it to be effective. Table 5.2.1 describes the most important roles that would be required. Clearly, it would be impossible for any one individual to have sufficient detailed knowledge of all science and policy areas, so it is assumed that translators would have a specific area of science–policy expertise and probably a good understanding of broader science issues. This implies that more than one person in any one organisation would fulfil the role of a translator, either full or part-time. Thus a translator could be an individual, a group of people or possibly an organisation; the nature of the particular issue and the time available would dictate the need for one or more people to be involved.

Besides these very specific roles that should be part of the translator's responsibility, there are several broader, more philosophical issues that should be considered in relation to the role that a translator, or group of translators, should be expected to play. A key issue is the group to whom a translator should ultimately be responsible: should translators be responsible to policy, science or both? The implications of this have to be considered. According to the detailed description of the role of the translator in Table 5.2.1, translators would be providing their services largely (but not exclusively) to policy makers, which would suggest that they should be responsible to this group. However, this could arguably generate a conflict of interest. As the translator has to present scientific information to policy, it is their responsibility to present this information in a fair, concise, balanced, accurate manner that gives all the relevant information, together with any areas of uncertainty, to the policy maker. Thus the translator is, in this context, the voice of science and as such should be responsible to the scientific community. In a similar fashion, translators may also play a role as the 'voice of policy', presenting the goals, obligations and requirements of policy for good quality, well-balanced scientific knowledge. There are many in the science community who would value such information and a clearer route for dialogue with the policy community. However, in reality translators would be responsible to the group that provides their funding and this is most likely to be the policy group.

A related issue for translators to deal with would be neutrality. It is important that the scientific information provided to policy makers should be the best available and it should not be overly influenced by a particular viewpoint without clear justification or qualification. Policy makers must make an informed decision using the information that is provided to them; if they are presented with information that does not truly represent the spectrum of scientific knowledge and opinion, it is possible that any decisions made using this information could later be called into question and the translator would be at fault, although it is more likely that public opinion would question the science itself.[6] It is possible to argue that translators could never be truly neutral, as a truly neutral response would require an absolute knowledge of all available information; in reality this would be all but impossible, but it emphasises the need for the translator to have a very extensive knowledge of their subject area(s). In this respect a group of translators working together would operate more effectively than an individual.

An important feature of the translator's role would be that they are politically neutral as well as scientifically neutral. The translator should not intentionally or unintentionally act as an unofficial lobbyist for a particular scientific concept or viewpoint. Nor should the translator be placed in a position where they are under political pressure to adopt a particular view of scientific information to support certain political standpoints. For this reason it is important that translators do not have any direct connection with either policy making decisions or scientific funding so as to avoid any political lobbying and to maintain a purely technical role. The responsibility of the translator is purely to provide information that is useful for the context required.

To some extent neutrality of translators and the translation process is ensured by the fact that elements from both science and policy are involved. This creates a broad base of experience from which to translate information. With all stakeholders having at least some involvement and interest in the process, translation is self-regulating as neither science nor policy stakeholders are in a position to influence the process to an extent that is unreasonable.

Table 5.2.1 Description of the potential tasks involved in the role of the translator.

Activity	*Description*
Information provision	The key role of the translator is to provide scientific information to policy makers in a form that is clear, accurate, unbiased and, above all, helpful for the policy maker
Research	A translator would be required to undertake a large amount of personal research into both science and policy issues to be well informed about relevant areas of science. This would involve a large amount of reading and investigation of web-based scientific information resources. It is assumed that the translator would begin with a good level of scientific knowledge and/or policy experience and that this knowledge should be developed further in order to communicate effectively with both groups
Shadowing	Shadowing key people in the relevant science and policy spheres could be an important means of improving understanding about how such people operate and what their personal and professional goals and objectives are. This would be especially important in the early stages of a translator's career. It is assumed that while a translator would be expected to be very knowledgeable, they would not be experts in any particular field; however, they should understand how the experts operate and why
Meetings	Direct contact with the scientific community at all levels would be a vital aspect of the translator's role as this would give a grass roots knowledge of emerging issues and concerns within the scientific and policy communities. Attendance at a range of meetings, workshops and conferences to speak directly to scientists and policy makers in both a formal and informal manner would be a regular task
Committees	As a more formal means of establishing communication, science and policy committees could be set up to discuss how best to develop and maintain links between the groups. The management of these groups would be the responsibility of the relevant translator

Communication of ideas	The translator should be a high-profile figure within the relevant communities. Both science and policy groups should be made aware of who the translator is and what their role should be. Regular presentations to a range of audiences to raise awareness of the existence of translators and to explain their role would therefore be a key part of the translator's role, particularly in the early period when this role is being established. If science and policy groups are aware of the translator and their role, it would also encourage them to make use of the translator's acquired knowledge
Two-way communication	As well as ensuring that policy makers are fully briefed on relevant science and that good use is being made of science in the policy environment, translators would have a role in communicating policy needs more effectively to the scientific community. They would be in a position to communicate both the requirements and interests of policy more directly to science and any scientific concerns can be raised with policy at an early stage; hopefully, where appropriate, this may speed the uptake of science into policy
Generation of enthusiasm	While many people view the idea of a translator as a good one in principle, the role needs to be seen positively across the science and policy communities for it to gain the type of response that will make the position worthwhile. Both the science and policy communities should be aware of the translator, the role that they play and the benefit they can gain from making use of them. Scientists need to be predisposed towards the translator to ensure that information is forthcoming and policy makers need to want to make use of the translator's knowledge and experience. It is an important role of the translator, particularly in the early stages of building working relationships, to ensure that interest and enthusiasm is generated in the translator and their role, and that this enthusiasm is turned into useful information

A further issue is whether translators already effectively exist, but are not known by this name. If there are examples of individuals, groups or activities that already perform the role described above, at least in part, then there could be much to be learned and applied in other fields. While linking science and policy is something that is discussed and undertaken in various ways, the notion of a translator is not necessarily made explicit in this work.

In the environmental field the connection between science and policy was stimulated during the sixth EU Framework Programme[7] and has been reinforced within the seventh Framework Programme. There are now several examples within Europe of an increasingly integrated approach to the management, presentation and dissemination of scientific information in what might be described as its 'post-production' phase. For example, in the Netherlands, the Water Service of both Rijkswaterstaat and Deltares undertake activities that can be described as translation, assuming that such organisations can be considered as translators as well as individuals or groups. For example, they provided much support and advice in the production of the 'Veerman Commission' report (referred to earlier in this chapter). Activities of groups such as the Common Implementation Strategy (CIS) of the Water Framework Directive (WFD) are intended to help harmonise the implementation of the EU water policies across Europe.[8] The COMMON FORUM is a network of European contaminated land managers and advisors who provide expert advice on the development and implementation of policies such as the Soil Framework Directive. Research and Technology Development (RTD) consortia such as the Water Information System for Europe (WISE) present, combine and connect to information from various sources, while the Shared Environmental Information System (SEIS) currently being developed is another European-wide initiative to refine environmental reporting throughout the EU and make such information more widely accessible for policy making.[9] At a national level, parliamentary committees such as the Associated Parliamentary Sustainable Resources Group (APSRG) in the UK, which specialises in waste management issues, provide a forum for politicians to meet and be informed by a range of industry experts and academics on contemporary issues in the field.[10] In such cases the committee manager would be an example of a prototype translator, or at least this is an example of the type of role a translator could be expected to fulfil. In terms of policy informing science, publications outlining research requirements and mid- to long-term strategies may be produced by national or European Government departments.[11] In some cases research projects themselves may make efforts to disseminate their work to a wider audience of policy makers and other stakeholders. Most large scale, multinational projects now produce a website as part of their work, containing introductory information about the project, the staff and possibly some publications and/or newsletters. Some, such as the AquaTerra Integrated FP6 Project (GOCE 505428), go further, including sub-projects that actively aim to link science, policy and stakeholders.[4,12,13]

Although the examples above are not exhaustive, they illustrate the groups, initiatives and the types of activities that aim to provide some sort of link between science and policy, or aim to make scientific data more accessible to the wider world. However, although very valuable they do not necessarily perform the role of a translator as described in Table 5.2.1 in its entirety. In some cases they merely present information in a different form and may not necessarily be policy focussed; in other cases they offer expert advice to policy based on personal knowledge and experience, but not necessarily specific investigation into the available science. In many cases these means of communication have not been established long and it may not be possible to judge the extent of their success. Nonetheless, these groups do communicate science to the wider community and the role of the translator should learn from, be involved with and build on such knowledge and experience. Advice from science to policy and vice versa should be transferred and discussed by means of active participation of stakeholders from all sides and regular communication

5.2.4 Integration of Knowledge

To fulfil the requirement to present information in a usable form for the intended recipients, a core role of the translator would be the co-ordination and integration of knowledge and information from a range of scientific sources so that the overall message coming from the science community, validated by the science community, is clear, concise and comprehensible. As policy increasingly requires an integrated response on a particular issue (*e.g.* the 'Veerman Commission' discussed previously) the translator (or group of translators) would have to work with a broad range of scientists and be able to understand them effectively and give a balanced judgement on the available information.

One means by which this may be achieved, or which may provide assistance to the translator and/or end users, would be the range of Internet resources currently being developed that assist the translation process. WISE-RTD is a web-based tool that acts as a means of linking information related to water management to the milestones and tasks of the Water Framework Directive (Chapter 4.2). This provides an effective link between science, technology, water management and policy. The Shared Environmental Information System (SEIS)[9] provides a means of standardising and accessing water monitoring data across Europe. EUGRIS has developed and expanded into a significant resource for soil, water and contaminated land information across Europe (Chapter 4.3). These and other Internet-based resources are a vital tool in the integration of scientific information and the linkage of science and policy. Recent discussions led by AquaTerra and WISE-RTD aim to improve and enhance the quality of the service provided by these individual resources by increasing the flexibility with which it is possible to search for information across different resources. If successful, this will provide a highly significant next step on the route to integrating information and translating scientific and policy-based information.

5.2.5 Value of Translators

While there is clearly a potential role for translators to facilitate linking science and policy, a more difficult issue is identifying and quantifying the value they give to both the science and policy communities. The cost of operating a 'translators department' at national or European level would potentially be very high, due to the need to travel extensively for meetings and the costs of operating committees, as well as the cost of funding extra staff. The key question is whether the costs of funding translators would be worth it? While the benefits of the translator role are quite clear, could these benefits not be gained in some other way without the significant expense? It could be argued that many communication and dissemination activities already take place and that many individuals work in roles that are at the fringes of the science and policy communities and that the two communities are not so clearly separated in practice as it is easy to separate them in theory. The use of a translator implies a straightforward linear relationship in the transfer of scientific information to policy, whereas in reality the relationships between science and policy are many and varied and do not necessarily function in this way. The question remains whether these lines of communication are sufficiently effective in their own right or whether a translator would make such communication explicit. The fact that the notion of a translator has been raised implies that communication could be improved.

If the translator role genuinely adds value to the science–policy interface, how can this value be quantified? It would probably be difficult to answer this question as much of the benefit is not likely to be directly quantifiable. How, for example, could the improved uptake of science be quantified? The uptake of science is a highly subjective issue, being dependent upon the individual policy makers' need, desire or interest in making use of it. Both scientists and policy makers say that they are interested in the other field, but do not have time to study it in depth to find out as much as they would like rather than lack the will to do so. If a translator existed, it may make science more readily accessible to policy makers and vice versa. However, it would be difficult to separate out the direct effect of the translator in most cases.

One means by which the value of a translator could be quantified might be in the effectiveness of funding. Hopefully, a translator would highlight the information that is already available from contemporary scientific research; this includes scientific information that is not available in English, which represents a significant body of work that is often underexploited. One aspect of this could be to clarify where funding for research would be best directed to address key issues and produce information that is more readily adapted to the needs of policy. This is not to say that the translator role would mean the end of funding for 'blue sky' projects or pure research work, which is almost certainly one of the major criticisms of the translator notion from the scientific perspective. The aim of the translator would be to co-ordinate scientific research to the greater benefit of policy. This might avoid duplication of research efforts, improving the efficiency of science funding, which in turn may release more funds for science through more effective funding of work; however, as with much of the case for translators, this is speculative. While such activities may result in the more efficient use of available funds it is probable that any such saving, if it occurred, would not be directly attributable to the efforts of the translator.

Not being able to quantify exactly how a translator adds value to the science–policy interface makes it difficult to justify the expense involved. While the instinctive reaction to the notion of translators is that they are a good idea, it is more difficult to actually demonstrate that they are good value.

5.2.6 Requirements of a Translator

The success of the role of a translator would rest almost exclusively on the individuals who would be appointed. It is unlikely that the ideal translator could be found ready made: to fulfil the many and varied requirements of the role and thereby justify their value, translators would have to have a wide range of qualities and experiences. Table 5.2.2 summarises the core requirements for a suitable candidate. While this list is not exhaustive, it shows the breadth and depth required for the role.

While it is unlikely that the ideal candidate for the role of a translator would be found anywhere, there may be some groups of people who may be better qualified for such a job. It is probable that translators would come with either a strong science or policy background and be expected to learn the other side of the role. People with broad rather than in-depth experience may be better suited to the role as they would be less inclined to be partial to one area of science or a particular viewpoint. EU Scientific Officers may be well suited to the role, as would political researchers. University researchers of various descriptions may also be well qualified, provided they have the capacity to remove themselves from their particular field of expertise and provide an objective view.

5.2.7 Need for Appropriate Relays

One of the identified difficulties regarding 'translation' of scientific results is linked to the lack of appropriate 'relays' at international, national and regional levels,[14] which often hamper operational links being established among different types of instruments/projects, actors and stakeholders. When considering possible 'relays', it should be kept in mind that the relationship between science and policy may be influenced by contextual factors such as interest at ministerial level, political pressures and the science culture in a country, which may be significant.[15] For example, systematic differences were detected between big, small, Nordic and associated countries, considering that smaller countries may enjoy shorter lines of communication and Scandinavian countries are 'good at consultation with citizens and industry', however the size of a country, its location and culture may not be as important as participation in EU

Table 5.2.2 Personal characteristics and experience required for an individual in the role of a translator.

Qualities/ experience	*Description*
Knowledge	Knowledge of both science and policy in at least one area of expertise is essential. The translator must be capable of being taken seriously by both parties, even though they would not necessarily be expected to have an equal understanding. The translator should have a broad knowledge base rather than a highly specialised one, but they should be expected to have a reasonable depth of knowledge as well. Awareness of a wide range of available resources, including library reference material and online information websites (such as EUGRIS, WISE-RTD, SEIS and other portals) would also be important, as the translator will be required to do an extensive amount of research
Neutrality	While it is almost certain that the individual would have a background in a particular field, they must not have a vested interest in particular science or policy, but have a balanced view of their field so as not to favour any one point of view
Objectivity	The translator must be capable of making clear judgements about the relevance and quality of work and ideas, rather than be persuaded by presentation or personality
Diplomacy	As the job will involve working intermittently with a large number of people from a range of backgrounds, a translator must have the ability to deal with potentially differing points of view, while maintaining a positive working relationship with all parties. A breakdown in relations could mean a reduction in the knowledge transferred, reducing the effectiveness of the translator
Pragmatism	It may not always be possible to do everything; an understanding of what can be achieved and the best way of managing a large amount of information is an important skill
Communication skills	Good communication skills would be essential, both in terms of presenting information and listening to different points of view. The ability to speak more than one language would be valuable. While many people in the scientific and policy community across Europe use English as a *lingua franca* it should not be assumed that this will always be possible. It is quite possible that good, relevant science may have been published in a language other than English and the translator should have the capability to cope with this

Framework Programmes.[15] In this sense, the strength and efficiency of the relays may vary widely from one country to another and even from one sector to another.

Very few countries use standard national guidelines or procedures for the science-into-policy process. The UK was one of the very few to report on a 'central government guide' for science into policy practices.[15] However, these guidelines are rarely used on 'a day-to-day basis' and tend to be considered at a higher level rather than by staff in the system. Moreover, the guidelines address 'proprietary issues' rather than the full range of processes followed in using science to inform policy. Finally, they do not address how to work with other scientists in other organisations. This suggests that where organisations such as government departments and agencies have different organisational cultures, structures and processes in place, there are difficulties in communication (through proper 'translation'), networking and collaboration. Consequently, there is considerable room for improvement and, in the specific water monitoring field, a possible opening for a concrete science–policy interface development that should be backed-up by guidelines on the use of scientific outputs (at the most appropriate level, *i.e.* from regional to international) and demonstration activities.

The efficiency of appropriate relays and translation could be judged on the basis of the time needed from the production of scientific results to their recognition and actual use by policy-makers and stakeholders. An example of success in this respect is the BRIDGE project (funded under FP6), which developed the scientific foundation for a methodology for the establishment of groundwater threshold values (linked to the implementation of the Groundwater Directive 2006/118/EC). By linking scientific and policy-making networks and using a core group of 'translators', the scientific results could be made available in a usable form to the policy-making community in less than two years after the end of the project in the form of a policy CIS guidance document (CIS Guidance Document N°18). If we compare this with the awareness about tributyltin contamination and its recognition at EU level in the form of an EQS, we may consider that the translation work had been much less effective since the first scientific evidence of TBT impact on marine ecosystems came at the beginning of the 1980s and the EU Directive on Priority Substances (2008/105/EC) including a TBT-EQS was adopted in December 2008!

5.2.8 Conclusions

Modern policy making has to reflect the modern world; science is an increasingly important part of this world and good quality science should be used as a source of information for policy makers to support their work in adapting human society to future environmental change. Likewise, the requirements of policy should be clearly presented to the science community so that the needs of both groups are understood. To provide good quality science in a useful form to the policy community and local stakeholders, communication between the relevant groups is essential, something that has been promoted across Europe through the Sixth and Seventh EU Framework Programmes, which have led to various initiatives to improve the connectivity of scientific research across the EU and make it more accessible to a wider audience. In particular the development of a range of Internet-based resources has improved the presentation of science to the wider community.

Communication between science and policy is not straightforward; it takes place in various ways at various levels and some branches of science appear to communicate to policy and application more effectively than others.

There is probably no single mechanism that would be appropriate to all forms of science–policy communication at the national or supra-national level. The translator, whose role is specifically to make science accessible and useful to policy makers and other groups for whom it may be of interest, could be a very effective means of science–policy interfacing.

In the recent past the notion of a specialist translator role to make the link between science and policy explicit has gained credence. If there were such a role, it would be easier to create a route through which scientific information could reach the policy community and the policy community could transmit its requirements to the scientific community. Arguably, at least some of the tasks of the translator, as outlined in this chapter, are already undertaken to some extent or other by individuals already working in the science and policy communities, such as EU science project officers, scientific researchers and political researchers. However, these roles do not make the translation explicit. The fact that the role of translator as described would require a very broad range of skills and would encompass a wide range of responsibilities implies that the translator is a specialist role and it could be undertaken more effectively by a specific individual rather than through *ad hoc* communications between individual scientists or policy makers on top of their other responsibilities. This is not to say that this sort of communication should be discouraged or would be superseded by the translator. Nor should the range of Internet-based resources currently available be discounted as a valuable resource to help the communication of scientific information.

While instinctively it is possible to see that the role of a translator would be beneficial, it is far more difficult to quantify what this benefit would be. It would also be difficult to separate out the exact effect of the translator. As a result it will be difficult to provide a justification for what could potentially be an expensive role, despite the fact that it might well be an effective use of resources. This chapter has set out specific ideas and discussed the issues surrounding the role of the translator, stating the case for making the role explicit rather than implicit and has defined the nature of the work a translator would undertake.

We recommend that work identifying how different organisations address the issue of translation and quantifying their success should be undertaken. Raising awareness of the need for translation, improving opportunities for training and education that would increase the knowledge range of potential translators and the setting up of a translators network (focusing initially on major international issues) are additional activities that should be addressed in the immediate future.

References

1. Ph. Quevauviller, P. Balabanis, C. Fragakis, M. Weydert, M. Oliver, A. Kaschl, G. Arnold, A. Kroll, L. Galbiati, J. -M. Zalvidar and G. Bidoglio, *Environ. Sci. Policy*, 2005, **8**(3), 203–211.

2. T. Raadgever and E. Mostert, Public participation in information management, NeWater Project, Delft University of Technology, NL http://www.usf.uni-osnabrueck.de/projects/newater/downloads/newater_rs10.pdf, 2005.
3. HarmoniCOP, *Learning Together to Manage Together: Improving Participation in Water Management*, ISBN 3-00-016970-9, 2005; available at http://www.harmonicop.uos.de/HarmoniCOPHandbook.pdf.
4. Anon, press release: Delta Commission: working with water, Delta Commission, Netherlands, September 3 2008 http://www.deltacommissie.com/doc/pressrelease_080903.pdf.
5. Anon, Advice Summary, Delta Commission, Netherlands, September 3 2008, http://www.deltacommissie.com/doc/summary.pdf.
6. B. Ballentine. Enhancing the role of science in the decision-making of the European Union, European Policy Centre Working Paper 17, March 2005, EPC, Brussels.
7. Ph. Quevauviller, *J. Soils Sediments*, 2006, **6**(4), 259–261.
8. Common Implementation Strategy (CIS), Common Implementation Strategy for the Water Framework Directive (2000/60/EEC) Strategic Document, 2001. http://ec.europa.eu/environment/water/water-framework/objectives/implementation-en.htm.
9. European Commission, Towards a Shared Environmental Information System (SEIS), Communication COM(2008)46 Final, European Commission, Brussels, 2008. http://www.gmes.info/pages-principales/library/reference-documents.
10. The Associate Parliamentary Sustainable Resource Group (APSRG) home page http://www.policyconnect.org.uk/content/environment/apsrg.
11. The Department for the Environment, Food and Rural Affairs, Defra. *Evidence and Innovation: Defra's Needs from Science over the Next 10 Years*, Defra, London, 2004.
12. C. Merly, A. S. Chapman and J. Brils, *Report on Potential Collaboration with other FP6 Funded Projects/Integration within projects on same Scientific Iss*ues, Deliverable INTEGRATOR I3.5 EC Framework 6 Project AquaTerra, Project no. 505428 (GOCE). October 2007, www.eu-aquaterra.de.
13. A. S. Chapman, R. P. Bardos and C. Merly, Finalised Website Report, AquaTerra Online Information System, Atois, Deliverable INTEGRATOR I3.13, April 2009, EC Framework 6 Project AquaTerra, Project no. 505428 (GOCE), October 2007, www.eu-aquaterra.de.
14. Ph. Quevauviller, R. Harris and P. Vervier, in *The Water Framework Directive – Ecological and Chemical Status Monitoring*, ed. Ph. Quevauviller, U. Borchers, K. C. Thompson and T. Simonart, John Wiley & Sons Ltd., Chichester, 2008, Chapter 10.1.
15. A. Scott, J. Holmes, G. Steyn, S. Wickham and J. Murlis, Report on the 'Science-meets-policy' conference, London, http://www.theknowledgebridge.com, 2005.

CHAPTER 5.3

Lessons Learnt and the Way Forward*

PHILIPPE QUEVAUVILLER,[a, b] PATRICK SWARTENBROECKX,[c] KEES J. M. KRAMER,[d] MICHIEL W. BLIND[e] AND MARIE-PERRINE DUROT[f]

[a] European Commission, DG Research, rue de la Loi 200, 1049 Brussels, Belgium; [b] Vrije Universiteit Brussel (VUB), IUWPARE, Building T, Pleinlaan 2, 1180 Brussels, Belgium; [c] Hydroscan, Tiensevest 26/4, 3000 Leuven, Belgium; [d] Mermayde, Zakedijkje 60, NL-1862 HC Bergen, The Netherlands ; [e] Deltares, PO Box 85467, 3508 AL Utrecht, The Netherlands; [f] ONEMA, the French National Agency for Water and Aquatic Environments, Hall C, 5 square Felix Nadar, 94300 Vincennes, France

5.3.1 Introduction

Various chapters of this book illustrate the extensive discussions about science–policy interfacing that have taken place over the last five years in the water sector. In particular, the Harmoni-CA project[1] (described in Chapter 2.4) expressed recommendations linked to uptake of catchment modelling[2] and led to the development of the WISE-RTD web portal[3] (Chapter 4.2). This is a good step forward. However, lessons learnt so far show that interfacing science and policy is much more complex to achieve due to the 'human side' of the problem.

* The views expressed in this chapter are purely those of the authors and may not in any circumstances be regarded as stating a formal position of the European Commission.

Water System Science and Policy Interfacing
Edited by Philippe Quevauviller

Published by the Royal Society of Chemistry, www.rsc.org

Various facets and simple facts have been highlighted in different chapters, for example:

- Scientists starting strategic research with policy implications should – before they draft the outline of a research proposal – discuss with relevant advisors in the policy fields concerned to ensure that the final scientific outputs will answer to defined policy needs (this can be technical implementation, review, *etc.*).
- Scientific results that could potentially be used in the policy making and implementing process should be better disseminated (not only in scientific journals but also in a way that is 'readable' by policy makers) and adequate training on the use of research outputs could be envisaged.
- Policy makers and implementers should establish contacts with relevant scientific fields to express their research needs linked to technical policy milestones (possibly with the help of scientific experts), *e.g.* specific technical implementation steps, review of technical annexes, *etc.*
- Exchanges between scientists and policy-makers are, however, not 'naturally occurring' and a (new?) mechanism is needed to enable sustainable dialogues at regional, national and supra-national (EC) levels.
- Dedicated research in support of policies (SSP) may be successful, providing that consortia involve policy-makers as well, *i.e.* networks have to be a mix of representatives from the scientific and policy communities.
- Linked to this, research aiming to support policies should develop a work plan mixing good science and demonstration of its applicability through testing against policy deliverables/milestones.

This chapter describes further lessons recently learnt as well as a way forward to develop an operational interface.

5.3.2 Recommendations Expressed within 'Science-meet-Policy' Events

General recommendations expressed in the framework of 'Science-meets-Policy' conferences[4] are directly relevant to the water sector, as highlighted in Chapter 1.2 and elsewhere:[5]

- Working relationships between scientists and policy makers should be encouraged and supported. Many researchers only publish in the scientific literature (which is rarely read by non-academics), while many policy makers and policy implementers do not specialise in a specific area and often move posts, making the establishment of stable working relationship difficult. Moreover, policy oriented reports are rarely read by scientists. It is therefore necessary to develop/strengthen mechanisms to enable those working on either side to engage. We have seen in separate chapters that this is possible in the framework of 'stable' networks such as the expert groups operated under the WFD Common Implementation Strategy (*e.g.* production of guidance documents that involved policy-makers, stakeholders and scientists). Linked to this, when research is aimed to support a given policy, the participation of researchers in the policy-making processes should be better recognised, *e.g.* while scientist's works are rewarded through scientific publications, the recognition of possible inputs in the policy sector is rarely valued at the same level.

- Science–policy relationships are often thought to be uni-directional, *i.e.* scientists delivering their findings at the end of their (policy supporting) research, trusting that these findings will then directly influence policy. This is an unrealistic representation of the policy-making process. Developments are based on policy agendas that are influenced by many inputs of which research is only one component. An example has been given in the groundwater policy area (Chapter 1.4) where a scientific report was used as a basis for developing a policy guidance document, which took more than 18 months after the delivery of the report (this is actually to be considered as very fast!). There is hence a need for constant interaction and dialogue between science and policy from the start of the policy supporting research process, including identifying together the research questions. Although this may present extra work, there is hardly any other alternative when dealing with policy-relevant research. In other words, through a proper dialogue with stakeholders, researchers may reach a better understanding of the policy issue(s) they are investigating, and this can actually help to enhance the relevance, quality and applicability of their research.
- A dialogue between policy-makers and researchers is, however, not only helpful to clarify expectations and possibly improve research dissemination. It is indeed also relevant to the policy process, in particular when scientific evaluation of policy impacts is required, or in the case of reviews of long-standing policies in the light of new findings. This is typically the case in the water sector with the cyclical review processes of the WFD and its daughter directives.
- Training and education are also key aspects, in particular for researchers to improve science communication and efficient transfer of research output, and for policy-makers to learn how to better use scientific information and results. The need for training to develop specialists that are able to bridge the science–policy gap has also been highlighted, *e.g.* in the form of post-graduate courses that could be stimulated at EU level (*see* Chapter 4.4 as an example).
- Interdisciplinary partnerships (including socio-economics) are essential to tackle the complex and interconnected nature of current environmental issues, which requires the design of multidisciplinary research programmes. Not only support is necessary but also a review about research effectiveness.
- Planning and managing research programmes in support of policies need to be improved, in particular processes for identifying research needs that

should timely respond to well-identified policy milestones (again the WFD provides a clear roadmap in this respect). In other words, where research is specifically designed to support policy, criteria should relate to policy relevance, timeliness and usefulness to be able to evaluate project results for their 'fit-for-purpose' character. This should not lead to sacrificing scientific quality but should rather help fix clear evaluation criteria that are tailor-made to address policy needs. This may be achieved by ensuring the presence of policy-makers and other stakeholders on research steering boards, policy mentors for research teams to enable linkages directly into policy networks, and regular exposure of scientists to the policy-making and implementing process. This has occurred in several water research-related projects, in particular projects designed to support specific water policies (Chapter 1.4).

- The final set of recommendations relates to the accessibility and uptake of research into the policy process. Intermediaries and translators are needed to 'translate' research output into inputs that may be used by policy-makers (Chapter 5.2). In general, most research results are published in academic papers, journals and books, few of which are read by policy-makers. The language is often technical, full of jargon and the findings are not presented in ways that make clear the policy implications to non-specialists. Therefore, without a process of 'translation' (*e.g.* development of guidance documents like those under the CIS process, see above), much useful knowledge is likely to remain locked up in an inaccessible form. These intermediaries may be found in either research or policy-making/implementing organisations, or in the private sector (*e.g.* consultant companies). Science 'storytellers' in the journalist sector are also active in translating information for non-technical audiences.[6] This is discussed in Chapter 5.2.
- In certain scientific fields a continuous policy relevant synthesis is required to achieve adequate overviews of vast amounts of developments. This is similar to the IPCC, which is in actual fact continuously evaluating new scientific insights to improve the IPCC reports.
- In a different context, linguistic translation is also required. In the European setting, the English language is commonly used for science as well as policy communication. However, this may be a barrier for science–policy discussions and uptake of research results. Not all stakeholders master English sufficiently well as a working language.
- Finally, the need for better scientific databases is becoming more acute; they need to be easily accessible and searchable and summaries should be written in non-scientific language. There is a risk of a lack of memory of past research if no system enables us to trace research back. Electronic databases of scientific publications currently go back 10 or 20 years and mostly cover English-language publications. Moreover, today we might believe that all information is easy at hand using the Internet. As has been demonstrated in Chapter 1.6, project websites tend to have a limited shelf-life. The updating and maintenance of (EC) project websites is generally

poor and dissemination products (research findings) are often not uploaded to the site (or visible for 'partners only'). In addition, once the project has been completed (funding has stopped) the interest of scientists vanishes rapidly. As a consequence, much research output tends to become 'invisible'. Tools such as WISE-RTD are (so far) based on decentralised storage of information. They use hyperlinks to (project research) information. When project websites are closed, the links become a dead-end, and the information is lost. Therefore, there should be developed strategies to maintain in a structured way the research output in an accessible form, preferably using the vast opportunities of the World Wide Web.

5.3.3 Successes and Drawbacks

The way research outputs have been integrated into WFD implementation[7] has been discussed in recent publications.[2,5] While research needs were generally well considered in successive Framework Programmes (FPs), projects could often not anticipate the way policy steps (*e.g.* monitoring programmes, programmes of measures) would be designed by Member States. This has sometimes resulted in scientific outputs that, although they had the potential, were not directly fit for policy implementation. In addition, the co-ordination among different actors has not been sufficient to allow a full integration of input from RTD projects into the policy discussions – neither did they make them accessible in time to meet policy needs. This was due to a lack of a clear 'science–policy interface'.

Notably, the continuously developing science process is not necessarily tailor-made to respond to policy needs. Tension may occur when attempting to achieve scientific 'independence', on the one hand, and policy 'relevance' on the other hand.[4] Scientists are often required to provide advice and/or conduct research in response to policy needs. Some worry has, however, been expressed that by allowing policy to define research objectives it may narrow the scope and thereby reduce benefits from 'free inquiry' that may produce unexpected findings or different options for policy that may ultimately be more useful than policy-led research.[4] Managing expectations is therefore an important aspect of research. Clearly distinguishing between fully independent research, end-user involved and end-user driven research right from the inception of research topics (topics open for tender) and/or research proposals may help to avoid disappointment and widening science–policy gaps.

The relationship between science and policy may also be influenced by the nature of the scientific results or the stage that a particular scientific or policy field may have reached. Strong science (*i.e.* a critical mass of clear evidence and consensual results from many scientists) is more readily taken up and utilised by policy-makers, while emerging science (new problems identified, tentative and/or conflicting results) may need to be strengthened and more widely disseminated before being acted upon. This also has been reflected in the framework of research projects providing direct inputs to the water policy process (Chapter 1.4).

Finally, interactions among policy and scientific communities vary according to the phasing of research programming, project development and of course timing of policy milestones. To date, these interactions are not co-ordinated in a systematic way and are rather functioning on an *ad hoc* basis (as links among 'individuals' rather than 'institutional' structured links), which may create an obstacle in the pursuit of effective collaboration.

5.3.4 The Way Forward – An Operational Science–Policy Interface

Several initiatives have taken place in the recent years that might lead to more efficient links between science and policy. We have already mentioned WISE-RTD as a web portal linking WFD (policy) questions to EU research, to national and river basins information. In addition, several ERA-nets (Chapter 1.4) deal with nationally funded research with linkages to water issues, aiming at development platforms between research funding, scientists and end-users. As was already mentioned, science–policy has also been addressed in the framework of the Common Implementation Strategy (CIS).[8,9] Different CIS working groups (*e.g.* on ecological status, groundwater, priority substances, floods, *etc.*) often lack a cross-cutting vision of available research and R&D tools, sometimes not initially WFD-related but still WFD-relevant. In addition, CIS working groups are focused on short-term implementation challenges and tools and do not have the mandate and resources to explore research questions that would be able to support their work to respond to future challenges.

Acquiring capacity to develop a global and strategic perspective in support of water policies, further developing the WISE-RTD web portal as a tool, ensuring operational links between the scientific community and decision makers (policy, civil society and other stakeholders) are complex and challenging tasks which require dedicated and co-ordinated efforts. The development of an operational interface would only be possible through interactions and guidance of a dedicated group integrating policy people, scientists and other stakeholders (including consultants). In this respect, to constitute an anchorage point for future developments, an attempt has been made to propose a Science-Policy Interfacing group (SPI) for approval by the EU Water Directors within the CIS framework. A pilot phase has been launched in 2009 involving some member States on a voluntary basis, EC DG Research, stakeholders and invited experts. This pilot, if successful, would pave the way for a kind of 'institutionalisation' through the CIS, to channel discussions and decide about concrete developments. The issue is essentially linked to a more efficient transfer system, helping Member States to get timely access to scientific information supporting water policy implementation.

The overall objective of a science–policy interface under the CIS umbrella is to ensure that scientific knowledge and technical solutions will be efficiently transferred to policy advisors, regulators and managers. The objective is to feed the implementation timetable and further reviews of water policies (in particular the Water Framework Directive and its daughter directives, as well as other relevant directives such as *e.g.* the Floods Directive) and to ensure that short, medium and long term research needs are identified.

Such an operational interface could dramatically improve the two-way communication between science/technical experts and policy advisors, regulators and managers and improve synergies between various water related tools and programmes. Supporting and strengthening the two-way communication between science and policy may have a tremendous impact. It may facilitate and speed up improved technical implementation of the WFD via problem-solving (through networking and regular (knowledge) exchanges) and creating solutions that might be too time-consuming or costly to develop in individual Member States, hence representing a significant economic impact. It may ensure that short, medium and long-term research needs are identified, classified according to different funding levels (EU, national or regional) and taken over by the scientific community (and other stakeholders, *e.g.* consultancies) so that the iterative river basin management planning process is constantly fed by relevant and tailor-made research outputs.

By gathering representatives from scientific, policy implementers and stakeholder communities (involved in relevant RTD projects, networks or working groups), a SPI group working at the CIS level would have to provide recommendations for the development of a framework in order to improve exchanges between science and end-users for the transfer of research results and for the identification of research needs linked to the policy cycle. The aim is to guide the cycle consisting of reviewing RTD projects with regard to their policy relevance, demonstrating their applicability, communicating their outputs, assuring dissemination and accessibility of their results, compiling feedback from users, and identifying further research needs linked to policy development, implementation and review.

Also, water policy is quite often at the crossroads of several issues, such as CAP reform, climate change adaptation and/or mitigation policies, biodiversity preservation or energy policies. Integrating these various aspects and science dealing with these issues is crucial for the relevance of future water regulations. A SPI group with the capacity to deliver a proper state of the art of existing scientific knowledge, potentially of interest to water policy implementers and stakeholders, and able to identify further research needs linked to key policy milestones on these aspects, would be of real added-value. This is the aim of the current initiative for which a mandate is being developed and which will be proposed to the EU Water Directors for the next CIS mandate (2010–2012).

A functional science–policy interface on the European level will be even more successful if information and communication reaches out all the way to local water authorities and stakeholders, occasionally even below the level of competent authorities of the WFD river basins. In some countries 'translator' institutes exist which may play a pivotal role. Additionally, much can be gained by coordinating (to some extend) research funding on various levels, in particular allowing quick response to emerging policy research needs. In particular at the European level, funding mechanisms which allow the initiation of research to answer policy needs (say within months) are currently lacking.

5.3.5 Conclusions

This chapter highlights some lessons learnt from various initiatives described in this book, and in particular the needs for better integrating scientific findings into water policy implementation. Difficulties experienced to date stem from the fact that there is no sufficient streamlining of information from the scientific community to policy decision-makers and vice versa due to the lack of proper transfer and of appropriate 'relays' at international, national and regional levels.[5] In this respect, this chapter describes efforts that are ongoing in the framework of the Common Implementation Strategy of the WFD[8] to examine how an efficient and operational 'science–policy interface' could be developed in support of the implementation of the Water Framework Directive and other water policies. The long-term objective is to develop an operational science–policy interface in such a way that it could meet the demand of different levels of users (*e.g.* policy-makers, industry) and stakeholders (*e.g.* the scientific community, NGOs, consultancies), ensuring an efficient dissemination and use of research results. This interfacing goal is ambitious and involves many different actors – hence its complexity. A possible framework, linking water policies in a broad sense to the R&D life cycle, should include research development (links to FP7 and national research), demonstration (testing of R&D outputs at the most appropriate level, *i.e.* regional to international, and under the LIFE Programme), communication (through the WISE-RTD web portal and other exchange platforms) and policy review (taking policy-related research needs into account when establishing research priorities). The challenge over the next few years will be to establish operational links among the different pieces of a puzzle composed of different types of instruments, actors and stakeholders.

Acknowledgement

Considerations expressed in this chapter are the fruit of many discussions with colleagues from the European Commission' Research Directorate-General (among others Panagiotis Balabanis, Christos Fragakis, Elisabeth Lipiatou, Marta Moren-Abat, Cornelia Nauen, *etc.*) and representatives of the scientific community (among others Wim de Lange, Geo Arnold, Jos Brils, Adriaan Slob, Simon Gardner, Peter Allen-Williams, Stephan Midgley, Natacha Amorsi and many others). They are all gratefully acknowledged for sharing views and ideas. Many considerations are also issued from conclusions of 'science-meet-policy' events and in particular the London 2005 report, which are reflected in Chapter 1.2.

References

1. Harmoni-CA, Harmonised Modelling Tools for Integrated River Basin Management, Harmoni-CA, EU funded concerted action, Contract EVKI-2001-00192, www.Harmoni-CA.info (2001).
2. Ph. Quevauviller, P. Balabanis, C. Fragakis, M. Weydert, M. Oliver, A. Kaschl, G. Arnold, A. Kroll, L. Galbiati, J. M. Zaldivar and G. Bidoglio, *Environ. Sci. Policy*, 2005, **8**, 203.
3. W. de Lange, G. Arnold, P. Willems, F. Provost, F. Hattermann, J. Plyson, M. Mestdagh, P. Swartenbroekx, P. Balabanis and Ph. Quevauviller, WISE-RTD webportal – A gate to scientific information for WFD implementers and water managers, poster presented at the International Conference on Monitoring under the WFD, Lille, March 2007.
4. A. Scott, J. Holmes, G. Steyn, S. Wickham, J. Murlis, Report of the 'Science-Meets-Policy' Conference, London, November 2005, www.theknowledgebridge.com.
5. Ph. Quevauviller, B. Harris and P. Vervier, Needs for an operational science-policy mechanism in support of WFD monitoring – National and regional examples, in *The Water Framework Directive. Ecological and Chemical Status Monitoring*, ed. Ph. Quevauviller, U. Borchers, K. C. Thompson and T. Simonart, 2008, John Wiley & Sons, Ltd, Chichester, pp. 411–444.
6. Science journalism – the science storytellers, *research*eu*, European Commission, Brussels, 2008, 56.
7. European Commission, 2000. Directive 2000/60/EC of the European Parliament and of the Council of 23 October 2000 establishing a framework for Community action in the field of water policy, *Official J. Eur. Communities, L* 327, 22.12.2000, 2000, p. 1.
8. European Commission, Common Implementation Strategy for the Water Framework Directive, European Communities, ISBN 92-894-2040-5, 2001, Final CIS document available at http://europa.eu.int/comm/environment/water/water-framework/implementation.html.
9. European Commission, Common Implementation Strategy for the Water Framework Directive, Key issues and research needs under the water Framework Directive, Final document comprising phase 1 and phase 2, 2005.

Subject Index